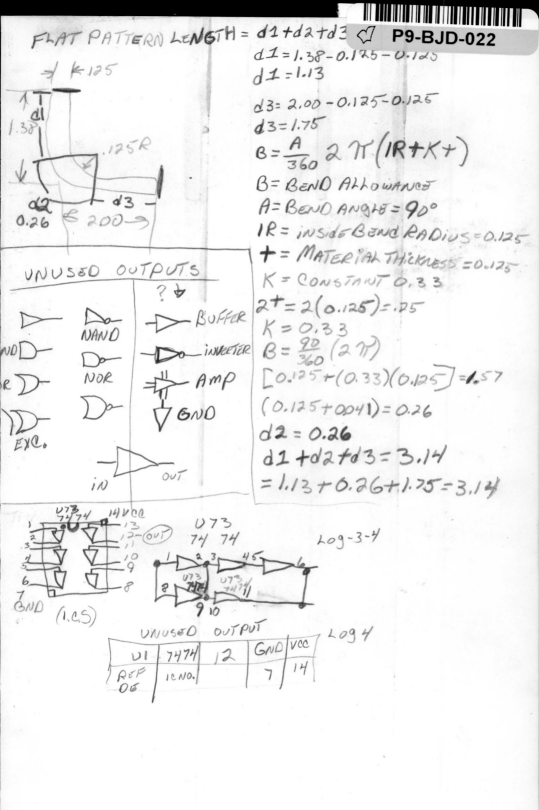

FLAT PATTERN LENGTH = d1 + d2 + d3   ◁ P9-BJD-022

$d1 = 1.38 - 0.125 - 0.125$
$d1 = 1.13$

$d3 = 2.00 - 0.125 - 0.125$
$d3 = 1.75$

→ k .125

↑
d1
1.38

.125R

d2   d3
0.26 ← 200 →

$B = \dfrac{A}{360} 2\pi (IR + K+)$

B = BEND ALLOWANCE
A = BEND ANGLE = 90°
IR = INSIDE BEND RADIUS = 0.125
+ = MATERIAL THICKNESS = 0.125
K = CONSTANT 0.33
$2+ = 2(0.125) = .25$
$K = 0.33$
$B = \dfrac{90}{360}(2\pi)$
$[0.125 + (0.33)(0.125] = 1.57$
$(0.125 + 0.041) = 0.26$
$d2 = 0.26$
$d1 + d2 + d3 = 3.14$
$= 1.13 + 0.26 + 1.75 = 3.14$

UNUSED OUTPUTS

? ⊽

NAND   ▷— BUFFER
              ▷o— INVERTER
NOR   AMP

NO   GND

R

EXC.

in   OUT

U73
7474   14 VCC
1   13
2   12—(OUT)   U73
3   11   74 74
4   10
5   9
6   8   Log-3-4
7
GND (I.C.S)

1 2 3   4 5   6
   U73   U73
   7474 7474
8   9 10

UNUSED OUTPUT   Log 4

U1	7474	12	GND	VCC
REF OS	IC NO.		7	14

ELECTRONIC DRAFTING AND DESIGN

Fourth Edition

NICHOLAS M. RASKHODOFF

Formerly with
Naval Research Laboratory
Washington, D.C.

PRENTICE-HALL, INC., *Englewood Cliffs, New Jersey 07632*

Library of Congress Cataloging in Publication Data

RASKHODOFF, NICHOLAS M.
    Electronic drafting and design.

    Bibliography: p.
    Includes index.
    1. Electronic drafting. 2. Electronic apparatus and
appliances—Design and construction. I. Title.
TK7866.R28 1982   621.381'022'1   81–18264
ISBN 0–13–250621–1   AACR2

To The Grandchildren

Editorial/production supervision and
    interior design by *Virginia Huebner*
Manufacturing buyer: *Gordon Osbourne*
Cover design by *Jorge Hernandez*
Cover art: Courtesy of Kollmorgen Corporation,
    PCK Technology Division

Printed in the United States of America

10 9 8 7 6 5 4 3 2

ISBN 0-13-250621-1

PRENTICE-HALL INTERNATIONAL, INC., *London*
PRENTICE-HALL OF AUSTRALIA PTY. LIMITED, *Sydney*
PRENTICE-HALL OF CANADA, LTD., *Toronto*
PRENTICE-HALL OF INDIA PRIVATE LIMITED, *New Delhi*
PRENTICE-HALL OF JAPAN, INC., *Tokyo*
PRENTICE-HALL OF SOUTHEAST ASIA PTE. LTD., *Singapore*
WHITEHALL BOOKS LIMITED, *Wellington, New Zealand*

CONTENTS

2

DRAWING PRACTICES

42

3

DRAFTING ROOM ROUTINE

64

4

COMPONENTS IN ELECTRONICS

79

5

FASTENERS, MATERIALS, AND FINISHES

132

6

GOVERNMENTAL REQUIREMENTS

156

7

ENGINEERING STANDARDS AND SPECIFICATIONS

167

8

GENERAL ELECTRONIC DESIGN

174

9

GRAPHIC SYMBOLS

231

10

REFERENCE DESIGNATIONS

274

11

SCHEMATIC DIAGRAMS

291

12

CONNECTION OR WIRING DIAGRAMS

348

13

PRINTED CIRCUITRY

395

14

INDUSTRIAL ELECTRONIC DIAGRAMS

444

15

WIRING HARNESSES

466

PREFACE

This book is intended for the student, the electronic draftsman, or the technician who has a basic knowledge of engineering drawing and who is interested in applying that knowledge to the specialized field of electronic drafting. It will introduce the design and drafting techniques involved in the production of electronic equipment for consumer, commercial, and military applications. The material presented should also be helpful to persons employed in the electronics industry who wish either to further their careers or to use the book for reference purposes.

The text presents information on the specialized electronic drafting practices, components, printed circuitry, semiconductor details, and materials peculiar to this branch of engineering drawing. The various specialized details of mechanical drawing, design, circuitry, graphics, military requirements, and national industry standards will provide the student with a broad perspective of the entire field of electronic drafting. The sectionalized approach offers the material in a form that is easily adapted to individual classroom needs. The summaries, review questions, and exercises at the ends of the chapters further emphasize the important phases of the subjects covered.

This book also will serve as a reference book since it treats many military and industry standards in detail and devotes considerable space to their listing and availability. No effort has been spared to update such information just prior to the book's publication.

The rapid technological advances made since the original publication of this book have also made it necessary to expand the coverage of such subjects as semiconductors, printed circuitry, modules, and specialized drafting. Semiconductors have become more complex and have developed into configurations such as flat packs and dual-in-line packages. The description of these devices, together with their circuit use, should prove helpful to the student who may not be familiar with this branch of electronics. The applications of printed circuitry have become more diversified and complex, and devices such as multilayer boards and flat cables are now more reliable and economical to produce.

Considerable space has been devoted to the military aspect of electronic drafting. Since military electronics represent a large share of expenditures for electronic equipment, it is advisable that the reader become familiar with the various military specifications, standards, and design requirements. In many electronic manufacturing facilities the draftsman is expected to know these requirements and to follow them.

Details of engineering standards have been presented at some length. A detailed treatment of the graphic symbols has been given because they are such an important part of circuitry diagrams. This diagrammatic presentation plays a significant part in electronic drafting, and requires a logical approach so that it will be understood easily by shop and service personnel and other persons actively engaged in the construction or maintenance of equipment. A practical approach to the problems and the techniques used to resolve them will help the newcomer in drafting work in industry.

The detailed presentation was chosen so that the book can be used either as a self-teaching aid or as a text for students enrolled in trade schools or technical institutes. Supplementary and reference texts have been included in the Appendix to aid the student further.

In addition to most of the material covered in the third edition, this book contains updated information that conforms with the latest changes in drafting, graphic, and military standards. The format of chapters, followed by summaries, review questions, and exercises has been maintained, although some of the material has been rearranged to improve the sequence of presentation.

One major subject added to this edition is the metric system of measurement as it applies to drafting. This system has become important as industry and the Department of Defense convert to its use. Multiwire wiring, in some instances the only practical method of dense interwiring is another emerging topic. A more comprehensive coverage of semiconductor materials and integration should also prove helpful to the electronic draftsman in his design efforts.

NICHOLAS M. RASKHODOFF

ACKNOWLEDGMENTS

The author wishes to express his gratitude to the various company and corporation officials and engineers who provided the necessary photographs, drawings, and data for use in this book. Lack of space has prevented presentation of some of these drawings in full.

Accuride, Div. Standard Precision, Inc.

Alpha Wire Corporation

AMP, Inc.

Amerace Corporation, Control Products Div.

AMPHENOL NORTH AMERICA DIV., Bunker Ramo Corporation

Analogic Corporation

Barry Controls Div., Barry Wright Corporation

Beckman Instruments, Inc.

Berol USA/RapiDesign, Div. Berol Corporation

Bivar, Inc.

Chomerics, Inc.

Consul & Mutoh, Ltd.

Control Technology Corporation

Cooper Electronics Div., Cooper Industries, Inc.

Dialight

Douglas Electronics

Dow Corning Corporation

Dzus Fastener Co., Inc.

Eldre Components, Inc.

Electronic Arrays, Inc.

Faber-Castel Corporation

Falstrom Co.

The Gerber Scientific Instrument Co.

Grayhill, Inc.

Hewlett-Packard, Optoelectronics Div.

Hughes Aircraft Co., Electro-Optical & Data Systems Group, Strategic Systems Div.

IBM, General Systems Div., International Business Machines

Kepco, Inc.

Keuffel & Esser

Koh-I-Noor Rapidograph, Inc.

Kollmorgen Corporation, PCK Technology Div.

Langley Corporation

Panduit Corporation

Parlex Corporation, Methuen, Mass.

RCA Solid State Div.

RCA Corporation

Raychem Corporation

Raytheon Co., Submarine Signal Div.

Rexnord, Inc., Specialty Fastener Div.

Robinson-Nugent, Inc.

Rockwell International, Collins Divisions
Rockwell International, Electronic Devices Div.
T & B/Ansley Corporation
TRW/Inductive Products
Teledyne Electro-Mechanisms
Texas Instruments, Inc.
Texas Instruments, Inc., Military Products Dept.
ULANO CORPORATION
Union Carbide Corporation
Unitrack Div., Calabro Plastics
Useco Div. Litton Industries
Vero Electronics, Inc.
Weltronic Company
Westinghouse Electric Corporation, Marine Div.
Zenith Radio Corporation

The following governmental agencies and industry associations also furnished useful data:

Department of Defense
Aeronautical Radio, Inc.
Aluminum Association
American National Standards Institute
American Society of Mechanical Engineers
Electronic Industries Association
Institute of Electrical and Electronics Engineers, Inc.
National Cable Television Association
National Electrical Manufacturers Association
National Machine Tool Builders' Association
Tubular Rivet and Machine Institute
Canadian Cable Television Association

My special thanks are extended to Norma J. Turrill, who contributed advice and assistance in the preparation of this book, to Charles M. Thomson, former Dean of Instruction, Wentworth Institute and Wentworth College of Technology, Boston, Mass., for his review and criticism of the original manuscript, and to my editors at Prentice-Hall, David Boelio and Virginia Huebner, who contributed advice and helpful information.

NICHOLAS M. RASKHODOFF

Largo, Fla.

TABLE LIST

1

Drafting Procedures

The broad field of electronics is continually finding new applications in home entertainment equipment, medical, commercial and industrial equipment, and military electronic equipment for use on land, sea, or air. A student entering this dynamic vocation will find that a knowledge of drafting practices, military requirements, graphic symbols, and diagram composition will be helpful whether he or she becomes a full-fledged draftsman, a designer, or a technician in government service or industry.

Intricate electronic devices and assemblies require the cooperative efforts of engineers, technicians, designers, and draftsmen to produce a finished product. During the preliminary work, experimental models are constructed and information obtained from catalogs, handbooks, drafting standards, and other sources. Circuit diagrams, sketches, and graphs are made to record the design features of the equipment. Finally, a set of finished drawings is prepared for use in the manufacture, installation, and servicing of the equipment.

To become an experienced electronic draftsman or designer and to perform work of this nature, the student must have a knowledge of elementary mathematics, a familiarity with components, and a working knowledge of electronic theory. He is expected to be neat, accurate, and patient, and to be able to visualize the proposed construction or layout. He must have the perseverance to solve problems and the ability to organize data as they accumulate during the course of work. He should maintain an open attitude toward new developments and changes on a project and carry tasks to completion.

On any given project, the electronic draftsman is likely to be assigned to an engineering team that may consist of electronic and mechanical engineers, designers, and other draftsmen. They present their ideas in a rough form, such as breadboard layouts, and he translates them into designs that are rugged enough to withstand the rigid requirements of military and industrial service.

In preparing the various types of drawings, he must constantly consider accuracy, ease of fabrication, and production costs. Many equipment failures can be traced to mechanical rather than electronic problems. Thus the importance of good mechanical design cannot be overstressed.

Since the draftsman converts engineering ideas into workable equipment through the preparation of sketches, drawings, and parts lists, he is expected to provide information pertaining to the purchase of raw materials, components, and hardware, and their eventual conversion by the manufacturing facilities into a finished product. He also acts as a link between the engineers and technicians and the purchasing and production personnel, through his clear and understandable rendering on paper of the information they require. His knowledge of components, materials, engineering terminology, and shop manufacturing procedures comes into play as he translates this information into various standardized drawing types, utilizing graphic symbols, technical abbreviations, as well as conventional drafting processes.

The draftsman must also consider the existing shop facilities and possibilities of lowering costs in design and drawing details, and still meet the specification provisions. He is also concerned with the product's maintenance and servicing by its eventual user.

The constantly expanding nature of electronic drafting, in which one is continually called upon to solve new problems, makes the field an extremely interesting and profitable one. Through the acquisition of techniques, experience, and knowledge, the adept draftsman is likely to develop into a well-trained, capable designer.

Initially, however, the student must begin by securing information about drawings and their composition as applied to electronic drafting. A number of different types of drawings are used to delineate electronic equipment and its subdivisions, circuitry, test results, and other information. These drawings are needed to provide complete manufacturing details for the construction of equipment or of its prototype. The draftsman must, of course, be able to prepare these various drawings in accordance with the prevalent government and industry standards.

Finally, decision has to be made whether the drawings are to use inch or metric dimensioning. If metric dimensioning is to be used, it should follow the practices outlined in ASTM Standard E380-76 to comply with standard DOD-STD-100 when government drawings are involved.

Further details are given in this chapter in regard to metric system application.

1.1 DRAWING TYPES

A complete set of engineering drawings is needed to manufacture, test, install, and maintain in service any given piece of equipment. Such drawings vary, depending upon their basic purpose. They are known as *detail* drawings if they illustrate the details of a single part or, depending on their magnitude and complexity, they may be *subassembly* or *assembly* drawings that show several such parts joined together mechanically (by screws or other hardware) or electrically (by wiring).

Other drawings, basically electrical in nature, may diagram the connections between various mechanical and electronic components, which, when combined with the mechanical parts and assemblies, form an integrated electronic assembly or equipment.

Basic Electronic Types

Drawings may be classified into several categories, and may be further subdivided within these categories:

Mechanical *Inch or Metric* *Dimensioning*	*Electrical*	*Other*	*Use*
Layout	Schematic or elementary	Graphic	Commercial
Monodetail	Connection or wiring	Block	Computer
Multidetail	Wiring harness		Consumer
Tabulated	Cable assembly		Industrial
Assembly	Running wire list		Military
Detail assembly	Interconnection		
Photographic assembly	Single-line		
Inseparable assembly	Logic		
Arrangement	Printed wiring		
Specification control			
Source control			
Installation			
Mechanical schematic			
Book-form			
Numerical control			
Outline			
Standard			

Engineering Sketches

Many freehand *sketches* may be made to develop some of the details, to present ideas for discussion, or to arrive at some basic dimensions before a project is started and layouts are made. These sketches are useful in developing various ideas and discarding others, as limitations are discovered in discussions of the project.

Such freehand sketches also help to develop circuit diagram arrangement and space allocation, as discussed in Chapters 11 and 12.

Mechanical Drawings

The complexity of an electronic equipment determines the number and types of mechanical drawings that are needed to illustrate the details of various parts and components and the assemblies of these parts and components.

Regardless of whether the equipment is highly complex or relatively simple, the design always begins with a layout drawing—or, more simply a *layout*. This layout establishes placement of the major components and the overall dimensions of the unit, or if the equipment consists of several major units, shows their positioning. If the basic unit is quite simple, it will not require any extensive layout work.

After the basic layout is established, it is necessary to prepare *detail* drawings of each part, using the layout drawing as a guide. These drawings can take several forms as indicated in DOD-STD-100C, Engineering Drawing Practices.

First are detail drawings, which must give complete details of the part, including any appropriate manufacturing notes. These fall into three categories. A *monodetail* drawing illustrates the configuration and all dimensions, tolerances, materials, surface finishes, symbols, etc., for a single part. A *multi-detail* drawing delineates two or more identified parts on the same drawing with each part giving the same information required on a monodetail drawing. A *tabulated detail* drawing depicts similar items with constant and variable characteristics. The latter are presented in tabular form with dimensions coded by the use of letters. It also provides the basic information required under the monodetail drawing.

When two or more parts are joined together, they form a unit known as a *subassembly*. The drawing representing such a unit is called a *subassembly drawing*. Although there is no sharp line of demarcation between subassembly and assembly drawings, the latter generally show more parts or include one or more subassemblies that form a larger unit. An example of an assembly drawing is shown in Fig. 8-21. Subassemblies and parts are called out on the drawing by their part numbers or by item numbers that are cross-referenced to the part numbers in a table or parts list. When pertinent, assembly drawings should contain references to installation drawings, schematic and connection diagrams, etc.

A *detail assembly drawing* depicts an assembly on which one or more parts are detailed in the assembly view. A *tabulated assembly drawing* shows more than one assembly by tabulating variations in the same manner as the tabulated detail drawing. A *photographic assembly drawing* is made by photographing the assembled item and delineating the items with numerical call outs. An *inseparable assembly drawing* shows details of items that are fabricated separately but permanently joined together by welding, riveting, etc.

In addition, there are specialized drawings. An *arrangement drawing* shows the relationship of major units of the item, with or without controlling dimensions but with major units identified. A *specification control drawing* shows a commercially developed item, listing its sources of supply, detailed dimensions, and such other information that establishes its suitability for a given application. A *source control drawing* depicts a specific commercial item that provides the exclusive performance and other characteristics required for a critical application. The drawing should include a note such as the following: The item described on the drawing, when procured from the vendors specified, is suitable for the applications specified. The words *Source Control Drawing* must also be added next to the title block. An *installation drawing* gives sufficient details of the item to install it relative to its supporting structure. Enough dimensions should be furnished to install all parts. A *mechanical schematic diagram* illustrates the operational arrangement of a mechanical device. A *book-form drawing* is the assembly of a number of drawings, preferably of *A*-size format, bound into book form. It shows the engineering requirements of one or more items or a system by technical tabulation, pictorial views, or both.

A *numerical control* (*N/C*) *drawing* illustrates the complete physical and functional requirements of an item for production by tape or computer control method. It uses the English decimal inch or metric linear measurements and rectangular coordinate system to locate machined surfaces, holes, etc.; see Fig. 1-1.

An *outline drawing* is a simplified version of a detail or subassembly drawing showing only the bare essentials and the overall dimensions. Such drawings are frequently used in trade catalogs and technical publications to illustrate the shape, dimensions, and other characteristics of the part.

A *standard drawing* illustrates and lists the details of items commonly used in electronic equipment applications, such as hardware (screws, nuts, washers, and bolts) and components (capacitors, resistors, and inductors). In military applications, these drawings are known as MS or Military Specifications.

Electrical Drawings

These drawings show the electrical interconnections between the various electronic components in an assembly or equipment or between the equipment units. To help to identify the exact connection points and to clarify the opera-

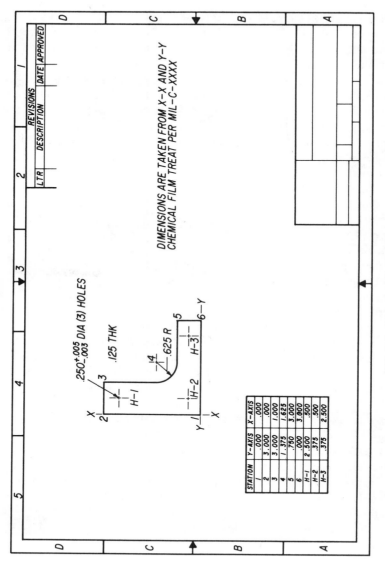

FIG. 1-1. Numerical Control Drawing.

6

tion of the components, these drawings or diagrams may also include mechanical features of the connected components.

A *schematic diagram* (also known as elementary diagram) pictures the various electrical, electronic, and occasionally, mechanical components, in the form of universally adopted graphic symbols. Definite symbols are also assigned to the interconnection wiring. The schematic diagram shows the functions of the circuit and its component parts in a simplified, readily recognizable symbolic form. Consequently, the symbols are not arranged on the diagram in the same physical relationship as the components are in the equipment, nor do they show any of the physical details. An example is shown in Fig. 1-2.

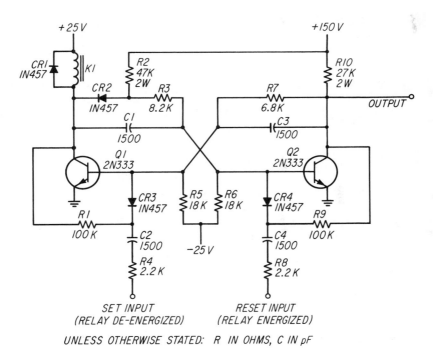

FIG. 1-2. Typical Schematic Diagram.

Some knowledge of electronics will help the student understand the underlying principles in schematic diagram composition, circuit operation, and component functions. Circuit symbols, which are described in Chapter 9, are graphic representations of the various components that enter into the construction of an electronic assembly. Reference designations are used to identify similar graphic symbols and also to distinguish between them. These designations are discussed in Chapter 10. A more complete description of schematic diagrams is given in Chapter 11.

A *connection* or *wiring diagram* shows the point-to-point wiring within the equipment, the module, or other electronic assembly. Unlike schematic diagrams, the electronic or mechanical components are represented by their physical shapes rather than by graphic symbols and are shown in their relative positions on the control panel, equipment assembly, etc. Connections are shown so that they represent the actual wire-by-wire wiring of the equipment. Further details are described in Chapter 12.

A *wiring harness drawing* shows the conductor grouping, the harness pattern, and if necessary, the dimensions of the conductors required for intercomponent wiring of an electronic assembly. The conductors are preformed, cut to required lengths, and held together by lacing or other methods. Further details are given in Chapter 15.

A *cable assembly drawing* shows the power, signal, and radio-frequency or audio-frequency cables used between equipments, units, racks, etc. Cable terminations are plugs, sockets, connectors, etc. Further details appear in Chapter 12.

An *interconnection diagram* is similar to a connection diagram, except that it shows the external connections between units, sets, groups, or systems.

A *running wire list* is a book-form drawing that consists of tabular data and instructions to establish wiring connections within or between units of equipment, sets, or assemblies of a system.

A *single-line diagram*, or *one-line diagram*, shows the equivalent of a simplified schematic diagram by single lines and graphic symbols.

A *logic diagram* depicts the sequence and function of logic circuitry by means of graphic symbols. (See Chapter 11.)

A *printed-circuit drawing* is usually the artwork required to lay out the circuit on printed-circuit boards which serve the dual function of mounting and interconnecting the electronic components. The details for preparing this artwork and other relevant drawings are given in Chapter 13.

Other Drawings

These include graphic and block drawings, as well as component location drawings.

A *graphic drawing* presents mathematical or other technical data in pictorial form, such as a chart or graph. Such drawings are used to display design calculations, to compare values visually, and to present test results. Chapter 17 is devoted to graphic drawings.

A *block diagram drawing* represents components, groups of components, or units of equipment in a series of blocks or rectangles, as discussed in Chapter 18. It reduces the complex circuits in each stage of equipment to a simple block form.

A *component location drawing* may be required as a prototype for a label showing the location of components within the enclosure of a radio receiver (see Fig. 1-3). Similar labels are required in military electronic equipment when space limitations prevent direct identification with reference designations.

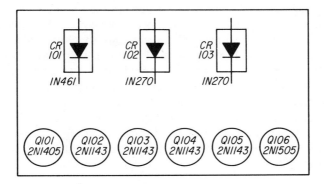

FIG. 1-3. Semiconductor Location Diagram.

Drawings such as these are also made to show component location and other details in technical manuals.

A complete set of working drawings may include nearly all of the mechanical, electrical, and other drawing categories discussed above. The mechanical-type drawings of details and assemblies are, of course, the most important ones in such a set.

Drawing Use

Electronic drawings are also classified according to their end use.

Commercial electronic drawings for example may depict electronic aircraft equipment that has its own specific requirements. These might include specifications governing weight, shock mounting, overall size, connector terminations, and other factors that affect overall design.

Consumer electronic drawings have to meet the special requirements and standards fixed by the various consumer product agencies—such as the Underwriters Laboratories, to name but one.

Industrial electronic drawings use a different approach in circuit diagrams and require more specialized graphic symbols than those used in commercial and military electronic drawings (see Chapter 14).

Military electronic drawings must follow certain prescribed guidelines and adhere to security restrictions. In addition, the equipment itself must be constructed to meet definite military specifications and standards (see Chapter 6). These considerations naturally affect the design and drawing practices.

1.2 TECHNICAL SKETCHING

When a new project is being started, an engineering meeting is usually held to resolve the various problems that might arise. One of the best means of presenting new ideas and approaches at such a meeting is through the use of *technical sketches*.

The preliminary sketching of details to determine the feasibility of some of the ideas can save many valuable hours of drawing later. Such sketches help to establish tentative component placement and to check calculations and clearances between parts. They are useful both as rough layouts for the schematic or connection diagrams and the graphs or charts, and as a means of familiarizing persons handling specific details with the project as a whole.

These sketches may be drawn in orthographic, isometric, or perspective projections, either in rough form or carefully sketched to a definite scale.

Although sketches are used only temporarily until regular drawings can be prepared, they should include such identification as title, date, draftsman's initials or name, job order number, and similar information.

Sketching Materials

The only materials required for sketching are paper, medium-soft pencils, and an eraser. Although a sketch pad or plain paper can be used, graph or cross-section paper is helpful in drawing straight lines and approximating cirles, arcs, and angles. A clip pad is also convenient to retain the sketch sheets and to provide a working surface.

Transparent paper can be used to make prints of the sketches, so that the original may be kept for reference and filing. Graph paper with nonreproducing lines (see page 13) is useful when it is desirable to have copies without a grid pattern. A six-inch scale, together with inside and outside calipers, and a micrometer will help to obtain measurements.

Sketching Procedure

It is important to obtain complete information for sketching purposes. For example, if a manufactured item is being examined, the nameplate data, such as manufacturer's name and address, the catalog and type numbers, the terminal markings, and the color coding of removable connections should all be recorded. If the item is part of an equipment, however, such details as overall dimensions, special tolerances, orientation, and clearances should be noted.

Considerable time and effort can be saved by obtaining all available data before starting a sketch. For example, the details of commercial components can be quickly found in catalogs, thus eliminating the need to take actual measurements if complete identification of such components is provided. Thus, the sketch may be simplified or even found unnecessary.

The application of the sketch will determine the amount of detail required. If it conveys change information, dimensional notations can be reduced to a minimum. Sometimes, a multiview sketch is advisable to bring out all the details and dimensions. In other cases, auxiliary views may be required to illustrate positioning or dimensioning or a pictorial view to supplement the sketch.

An example of a rough sketch drawn to assist in making a series of calcu-

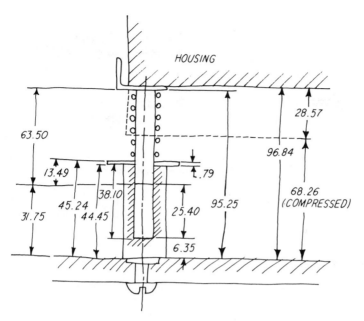

FIG. 1-4. Sketch in mm Used to Assist in Making Calculations.

lations in mm is shown in Fig. 1-4. Here, a housing is displaced a definite distance within its supporting structure during the installation.

1.3 DRAFTING TOOLS

The same basic drawing instruments and materials are required in electronic drafting work as for mechanical drawing, as well as some specialized instruments and accessories.

Special Drawing Pencils

Among improvements of interest to the electronic draftsman are the mechanical drawing pencils which use pencil leads .012 or .019 inch in diameter, Fig. 1-5(a). Such small-diameter leads eliminate sharpening the lead to a point for drawing or layout work. The leads are available in various hardnesses from HB to 4H, and are fed automatically by pressing on a plunger located on top of the pencil.

Special Ink Instruments

When extensive ink work is necessary, such as inking schematics or other drawings, a *technical drawing pen*, such as the Rapidometric®,* Fig. 1-5(b), will

*® Registered trademark of Koh-I-Noor, Inc.

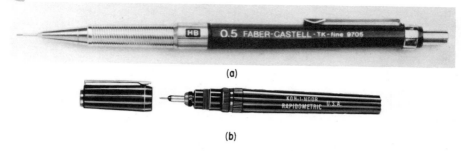

(a)

(b)

FIG. 1-5. (a) Special Drawing Pencil. (b) Rapidometric® Pen.

save time and effort. It is available in nine drawing line widths and can be used with plastic curves or templates to draw many long lines as well as for conventional straight-line work without refilling.

Drawing Papers

There are several varieties of drawing papers available. The selection of the paper depends upon the relative importance and the ultimate use of the drawing. Using a poor-quality material for a frequently used complex assembly drawing or diagram will only result in ultimately having to redraw it.

Engineering drawings should be made on transparent paper so copies may be easily reproduced. Drawing paper should also possess stability; variations in temperature and humidity should cause little or no dimensional change. It should be suitable for writing by pencil and pen, as well as for typing and microfilming. Among the four materials available for drawings are: bond paper, vellum paper, cloth, and polyester film.

Natural bond *tracing paper* is the least expensive, but it lacks durability, good transparency, and wearability. The all-rag variety is the only type that is suitable for pencil, ink, or typewriter work.

Vellum is a rag paper that has been treated to give it greater transparency. The grade generally used comes in reams of 500 sheets, 17 by 22 inches, that weigh about 18 pounds.

Transparent *cloth* is a coated linen. It resists aging better than paper and is very good for microfilming purposes. Some grades are made for either pencil or ink work, and are also treated to be moisture-resistant.

Polyester film is another drafting material. Essentially, it is a Mylar®* film that has a matted surface applied on one or both sides by the film process companies. It is also manufactured by the duPont Company in a precoated form sold under the trade name of Cronaflex®.* The film is extremely durable and dimensionally stable. It is suitable for both pencil and ink and has excellent

---

*® duPont registered trademark.

transparency and resistance to aging. It is available in thicknesses from .002 to .004 inch for general drafting work and from .005 to .008 inch for printed-circuit layouts where extreme dimensional stability is required. Clean, sharp lines can be obtained with either ordinary or special pencils, adding to its desirability for complex drawings and circuit diagrams that require permanence.

Among the special films used for making integrated circuit masks and printed circuit artwork are Rubylith®* (red) and Amberlith®* (amber) brand masking films. These films consist of a polyester backing sheet .005 to .007 inch thick (500- to 700-gage), to which is attached a strippable masking film which can be hand-cut or cut on machine driven coordinatographs, and peeled to form open windows, drop outs, masks, and overlays.

Cross-Section Papers

These papers, available in transparent or opaque types, are used for a variety of purposes, including the drawing of graphs, charts, outlines, test data, and sketches. They can be obtained with blue, red, orange, or green lines in various rulings (Fig. 1-6). Cross-section papers are also suitable for drawing rough circuit diagrams. Background lines printed in blue ink will not print on reproduced copies.

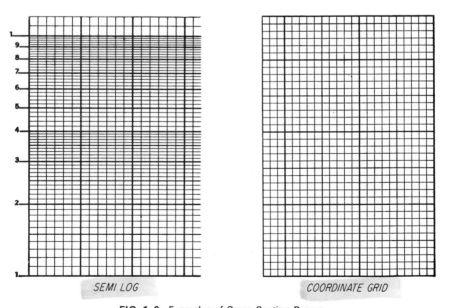

SEMI LOG   COORDINATE GRID

FIG. 1-6. Examples of Cross-Section Papers.

*® Registered trademarks of ULANO Corporation.

Drafting Machines

These machines have replaced the parallel rule and T-square—for many years symbols of the draftsman and the designer—and are available in a variety of styles, sizes, and price ranges. A machine of the type shown in Fig. 1-7(a), mounted on a tilt-top drafting table, permits drawing over practically the entire board area. It uses metal or plastic horizontal and vertical scales, thus eliminating the need for a separate triangle to draw the vertical or inclined lines. The control head, which is calibrated in degrees and includes a vernier, has automatic index stops disengageable by a button every 15 degrees. A locking lever on the control head makes it possible to set intermediate angular positions. To reduce fatigue, the tiltable drafting table can be locked in any position from horizontal to vertical and raised to any desired height for drawing in a seated or standing position.

It is good practice to lift the head of the machine whenever the position of the head is changed on the drawing. This helps to keep the drawing surface clean in pencil work and to avoid smearing in ink work. Another helpful idea is to tape a piece of drawing paper over the drawing area that is not being worked on to keep it clean. This practice is especially desirable on large assemblies or circuit diagrams that involve considerable time and effort.

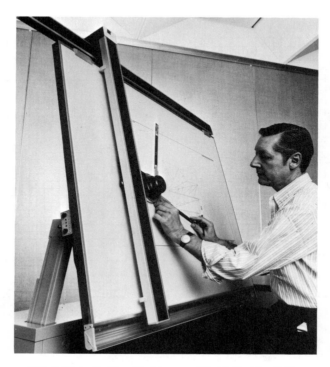

FIG. 1-7(a). Drafting Machine on a Tiltable Drafting Table.

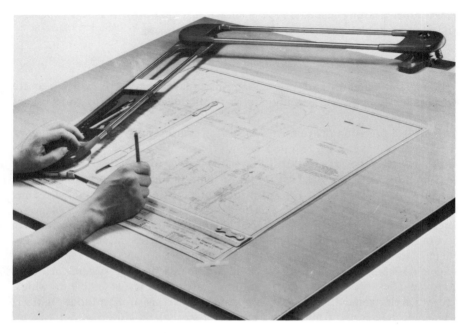

FIG. 1-7(b). Drafting Machine for Horizontal Drafting.

FIG. 1-7(c). MUTOH Micro/Plotter Coordinatograph.

Another drafting machine type is shown in Fig. 1-7(b). It is less expensive and more suitable for horizontal drafting work.

Coordinatographs

Integrated circuits and printed and multilayer circuitry, which require extreme accuracy in the execution of artwork have resulted in wide use of *coordinatographs*.

This is a precision instrument, Fig. 1-7(c), that uses a system of rectangular coordinates to lay out integrated circuits or circuitry 10:1, 20:1, 50:1, or more, with accuracies of $\pm.001$ inch.

Pencil or ink can be used for layout on conventional drafting materials, and special knife blades are used to cut through the coating of the stable-base films to form the desired pattern. The machine locates a given point in a plane referenced to a zero point by rectangular coordinates. For even greater accuracy, the X and Y coordinates can be read directly on an electronic display which is an optional part of the instrument.

Figure 1-8(a) shows an example of an automated drafting system for drawing schematics, printed-circuit artwork, integrated circuitry, assembly drawings, and other electronic drawings. Such a system may consist of an input digitizer, a video display, a minicomputer to process or store the digitizer output, and an automatic flat-top plotter that responds to the digitizer output, directly or from

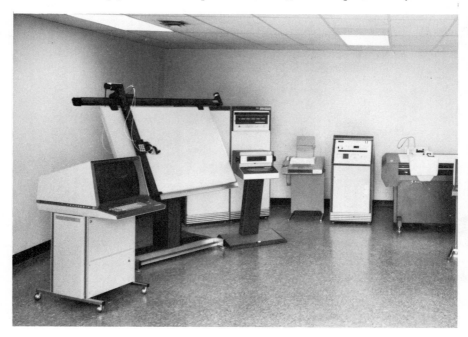

FIG. 1-8(a). An Automated Drafting System.

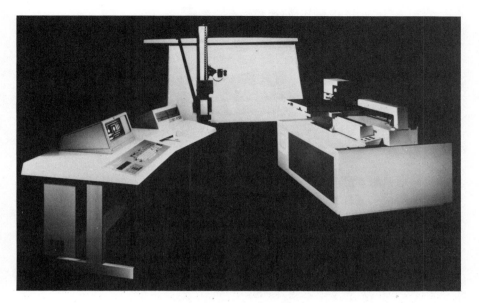

FIG. 1-8(b). Printed Circuit Artwork Drafting System.

the computer memory. The automatic plotter may use ball-point or wet-ink pens on linen, vellum, or Mylar®* films. Very high accuracy is secured in such a process because of high built-in plotting accuracy and scale reduction to the final size of the circuitry being plotted.

Another type of simplified automated drafting system is shown in Fig. 1-8(b). Primarily designed as a low-cost system for developing printed-circuit artwork from a designer's layout, it eliminates the need for manually taped masters. It consists of a coordinate digitizer, a control station with a graphic display terminal, and a photoplotter. The following are available from this system: PC artwork masters, solder masks, assembly drawings, and silk screen masters.

Schematic Guide Plates

Circuitry work in electronic drafting involves repetitive delineation of symbols and other shapes. Considerable time and labor can be saved, however, by using clear plastic *templates*, from .030 to .060 inch thick, that contain variously shaped cutouts. The cutout dimensions include pencil allowances, and the edges are bevelled on one side for ink work.

Templates are made for various specific purposes—electronic, electrical, mechanical, and architectural, to name several. Some representative electronic templates are shown in Fig. 1-9(a). In purchasing or using templates, care should be taken to make sure that they correspond to the latest military or ANSI

*® duPont registered trademark.

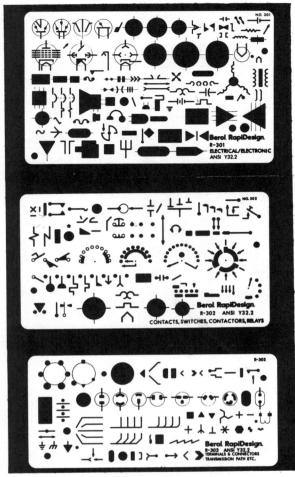

FIG. 1-9. (a) Electronic Symbol Templates. (b) Printed Circuit Component Template.

(a)

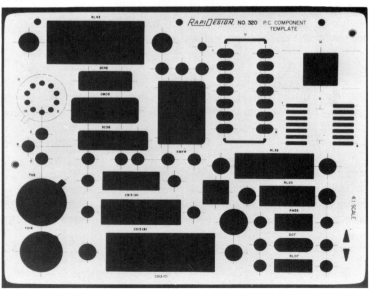

(b)

standards and that their individual symbols are identified by molded-in or engraved markings.

Another template, Fig. 1-9(b), is of particular use in making printed-circuit board component layouts (*see* Chapter 13). It is available in 4:1 and 2:1 enlarged scale to correspond with the enlarged scale layouts of the printed circuit board conductor artwork. Because templates eliminate the details of construction, they save time and effort and achieve a certain uniformity. A well-selected set of templates will thus prove to be a worthwhile investment.

When combining individual parts of symbols that are separated on the template, the draftsman must be careful to draw the symbol correctly. Also, fixed resistor symbols should be limited to three peaks on each side of the axis even though the template cutout may have five or more peaks.

Adhesive Drafting Aids

Another time- and labor-saving device used in electronic drafting is the printed drafting aid, which has a transparent adhesive on the back. These aids are made in solid color shapes and tapes (Fig. 1-10) and are frequently used to produce the printed-circuit artwork described in Chapter 13. They provide uniformity and sharp definition of the required patterns at a reduced overall cost. Conductor strips are available in widths of from $\frac{1}{64}$ to $1\frac{1}{2}$ inches.

The symbols printed on matte acetate film in tape form (Fig. 1-11) represent another type of adhesive aid. An active adhesive on the back acts as a

FIG. 1-10. Precut Adhesive Shapes and Conductor Strips. (Courtesy of Bishop Graphics, Inc.)

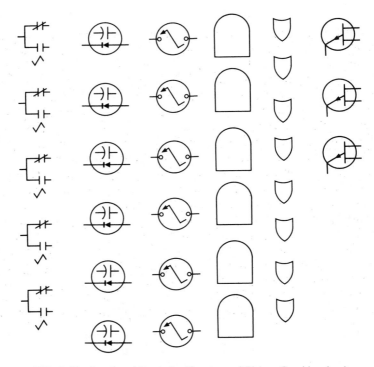

FIG. 1-11. Preprinted Symbols. (Courtesy of Bishop Graphics, Inc.)

pressure-sensitive coating. After removal from the release liner the symbol is pressed firmly into place.

1.4 DRAWING DETAILS

Military specification DOD-D-1000B, *Drawings, Engineering, and Associated Lists*, outlines the specific military specifications and standards to be followed by contractors supplying equipment, related drawings, and other documentation to the Department of Defense.

It classifies engineering drawings into Levels 1, 2, and 3. Level 1 discloses engineering information sufficient to evaluate the engineering concept to enable fabrication of developmentable hardware, with drawings made to military standards; with government or contractor code identification and document numbers assigned—depending upon the contract.

Level 2 (production prototype) has engineering drawings and associated lists of a design approach suitable for manufacturing a production prototype and limited production models—designs that employ standard parts and models made in final form.

Level 3 (production) are engineering drawings and associated lists which provide engineering data for quantity production and permit competitive procurement.

The details that follow have been selected from both commercial and government requirements.

Standard Drawing Sizes

Drawing paper must be in standard size sheets identified by a designation letter A through F (see Table 1-1) or rolls. These sizes and designations, established by the American National Standards Institute (ANSI), their standard ANSI Y14.1–1980, are based on a standard letter-size sheet, $8\frac{1}{2}$ by 11 inches (A size), so that they can be folded to fit standard letter files.

The A-size drawings may be obtained in both horizontal and vertical formats (Fig. 1-12).

Table 1-1 Standard Drawing Sheet Sizes (in inches).

Letter Designation	Size	Margin
A (horiz.)	$8\frac{1}{2} \times 11$	*
A (vert.)	$11 \times 8\frac{1}{2}$	*
B	11×17	3/8
C	17×22	1/2
D	22×34	1/2
E	34×44	1/2
F	28×40	1/2

*Horizontal margins—$\frac{3}{8}''$, vertical margins—$\frac{1}{4}''$.

Table 1-2 Standard Drawing Roll Sizes (in inches).

Letter Designation	Width (X)	Length (min.) (Y)	Length (max.) (Y)	Margin (Z)
G	11	42	144	3/8
H	28	48	144	1/2
J	34	48	144	1/2
K	40	48	144	1/2

Roll-size drawings are standardized in sizes listed in Table 1-2. The width (X), length (Y), and margins (W) and (Z) are illustrated in Fig. 1-13.

Standard metric drawing sheet sizes are listed in Table 1-3, page 39.

Drawing Format

The format for military drawing sheets, such as B size, is shown in Fig. 1-14. Most commercial drawing sheets follow a similar layout using preprinted borders, title blocks, company name and address, scale, and other details, as shown in Fig. 1-15.

Arrowheads are usually used on drawing sheets as indicators for microfilming purposes. Figure 1-16 shows the arrangement for small drawings, and Fig. 1-17 shows the arrangement for roll-size drawings that require multiframe microfilming.

It illustrates *area zoning*, which subdivides larger drawings into smaller

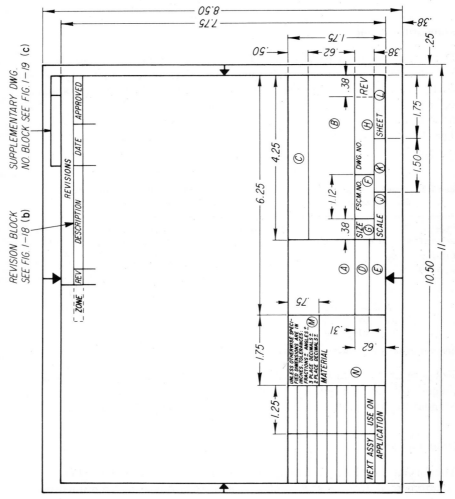

FIG. 1-12. Horizontal and Vertical Formats for A-Size Drawings.

22

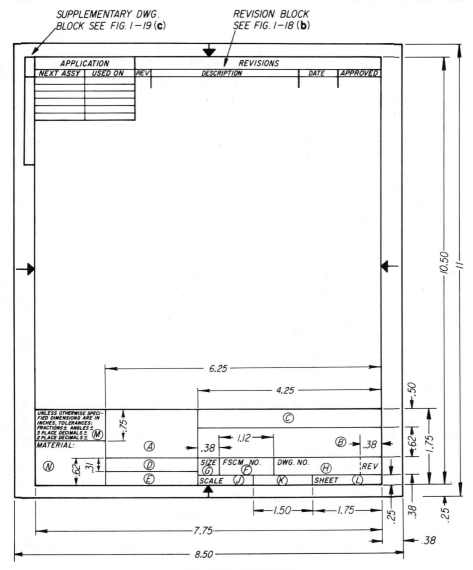

FIG. 1-12. (Continued.)

areas to facilitate the location of revisions, note references, and so forth. Vertical zones are indicated by capital letters, starting at the title block. Horizontal subdivisions are indicated numerically, starting at the right side.

It has become accepted practice on most commercial drawings, and a requirement on all military drawings, to use uppercase gothic lettering. It may be either slanted (oblique) or vertical, and drawn either freehand or by a template or lettering guide. Typewritten characters may be either upper- or lowercase.

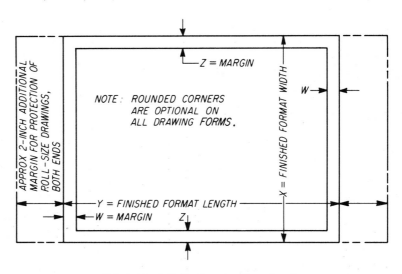

FIG. 1-13. Roll-Size Drawing Sheet Details.

Title Block Details

On military drawings, the title block must be located in the lower right-hand corner of each drawing, as in Fig. 1-17, and arranged according to Fig. 1-18. Each block section has a definite purpose.

The preparation, checking, and approval of the drawing are recorded in Section *A*, along with their respective dates. The drawing title is located in *B*, and the name and address of the design activity (government agency or contractor) is listed in *C*. The activity approval is in *D*, while any other approval is given in *E*.

The code identification number of the design activity (*FSCM*) is located in Section *F*. This is a five-digit number, assigned individually to every company and government agency that controls the design or produces drawings, specifications, or standards that in turn control the design. These numbers are listed in Handbook H4-1, *Federal Supply Code for Manufacturers, Name to Code*.

A few representative code numbers are shown in the accompanying table.

Activity	*Code Identification Number*
Department of the Navy	99993
Naval Sea Systems Command	53711
Naval Research Laboratory	81995
Winfred M. Berg, Inc.	29440
General Electric Co. Ordnance Dept.	24583
Raytheon Co.	49956

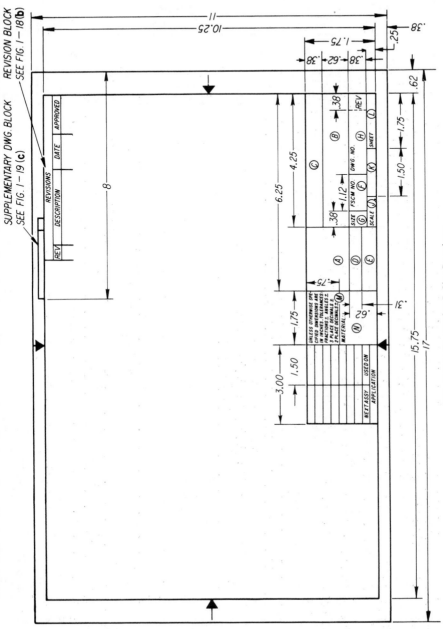

FIG. 1-14. Format for *B*-Size Drawings.

25

TOLERANCES	DR. BY	COMPANY NAME		
	CHECKED BY:	ADDRESS		ZIP CODE
	APPD. BY:	TITLE		
NO. REQ'D.	MECH.			
J.O. NO.	ELECT.			
NEXT ASSY.	STDS.	SCALE	DRAWING NO.	REV. LETTER

FIG. 1-15. Commercial Title Block.

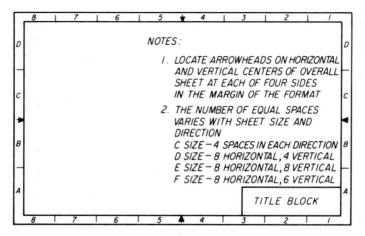

FIG. 1-16. Arrowheads for Microfilming of Alignment of *C*-through *F*-size Drawings and Zoning Subdivisions.

The letter designation of the drawing size, as listed on Tables 1-1 and 1-2, is shown in Section *G*, while the number of the drawing itself appears in Section *H*. This drawing number may also appear in the right-hand margin or on the ends of roll-size drawings and on drawings for microfilming (Fig. 1-17).

The scale of the drawing is listed in *J*, and such additional details as weight, reference to a specification, and so on are given in Section *K*.

The numbers for single- or multiple-sheet drawings are listed in Section *L*. The first sheet is identified as "sheet 1 of 4," while the remaining sheets are numbered consecutively "sheet 2," "sheet 3," and "sheet 4" without specifying the total number of sheets, Fig. 1-19(a).

Commercial drawing sheets follow a similar pattern in title block information, except that the code identification number of the company may be omitted. Small companies are likely to use even less information in the title block.

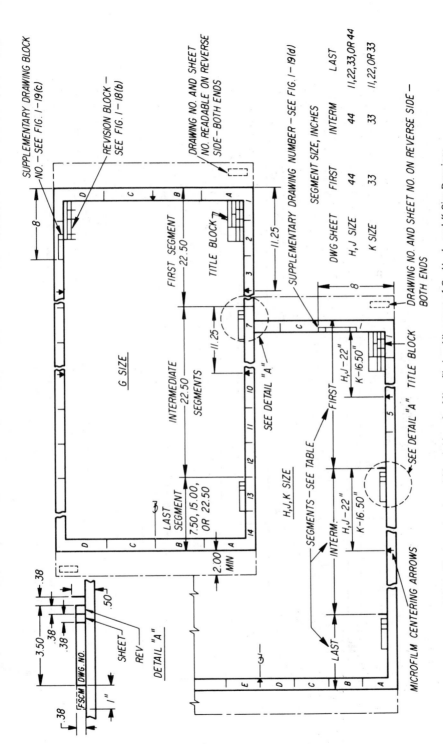

FIG. 1-17. Arrowheads and Match Lines for Microfilming Alignment of G-, H-, J-, and K-Size Drawings.

27

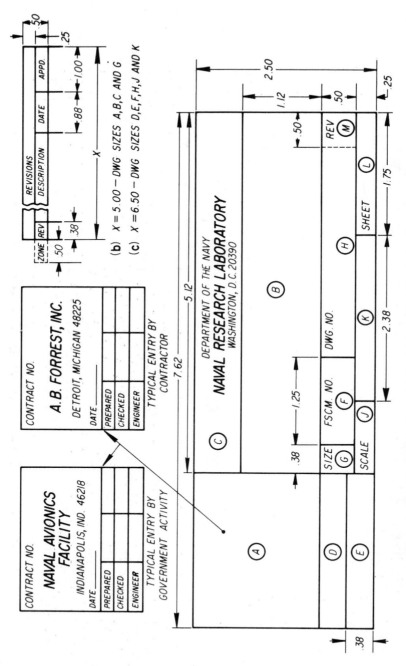

FIG. 1-18. Title and Revision Block Details. (a) Title Block Details for Sheet Sizes *D, E, F, H, J,* and *K.* (b) Revision Block Details for *A, B, C,* and *G_* Sizes. (c) for *D, E, F, H, J,* and *K_* Sizes.

(b) X = 5.00 — DWG SIZES A,B,C AND G

(c) X = 6.50 — DWG SIZES D,E,F,H,J AND K

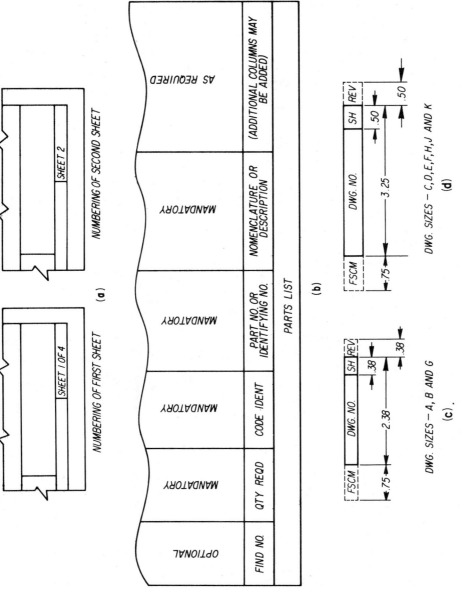

FIG. 1-19. Drawing Sheet Details. (a) Numbering. (b) Parts List Format. (c) Supplementary Drawing Number Block for A, B, and G sizes. (d) Supplementary Drawing Number Block for C, D, E, H, J, and K Sizes.

(a)

NUMBERING OF FIRST SHEET

SHEET 1 OF 4

NUMBERING OF SECOND SHEET

SHEET 2

(b)

FIND NO.	QTY REQD	CODE IDENT	PART NO. OR IDENTIFYING NO.	NOMENCLATURE OR DESCRIPTION	(ADDITIONAL COLUMNS MAY BE ADDED)
OPTIONAL	MANDATORY	MANDATORY	MANDATORY	MANDATORY	AS REQUIRED

PARTS LIST

(c)

FSCM	DWG. NO.	SH	REV

.75 2.38 .38 .38

DWG. SIZES — A, B AND G

(d)

FSCM	DWG. NO.	SH	REV

.75 3.25 .50 .50

DWG. SIZES — C, D, E, F, H, J AND K

Drawing Numbers

Drawings are numbered for reference and quick identification. Sketches are generally identified by a prefix, *SK* or *X*, followed by the number. The drawing number generally consists of several digits, and the designation letter of the drawing sheet appears as a prefix (Fig. 1-20). The first part of the number indicates the code designation for a division of the company, followed by coded product designation—as, for example, *D37-67985*.

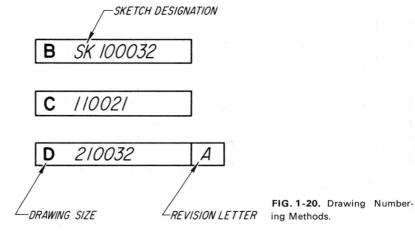

FIG. 1-20. Drawing Numbering Methods.

Military standard DOD-STD-100C, *Engineering Drawing Practices*, specifies that the drawing number may not exceed 15 characters—including numbers, letters, and dashes—and that the drawing number may not contain blank spaces. The letters "*I*," "*O*," "*Q*," and "*X*" may not be used. Letters must be capitals, and only whole arabic numerals are allowed. Dashes should not be used between letters or between letters and numerals.

Drawing Titles

The drawing title, in uppercase letters, consists of the basic name of the part, followed by a modifier to differentiate it from other similar items in the same major assembly. Basic names should be selected from *Handbook* H6, *Federal Item Identification Guides* if drawings are to be supplied to the government.

The basic name is to be a noun or noun phrase, but not the material or method of fabrication. It should be singular form unless the only noun form is plural or the nature of the item is plural, e.g., "GLOVES."

Assembly titles, listed in the above handbook, should be followed by a modifier, e.g., CABLE ASSEMBLY, SPECIAL PURPOSE.

On certain items that imply containers or materials, a noun phrase should be used as the basic name:

	Acceptable		*Unacceptable*
	JUNCTION BOX		BOX, JUNCTION
	CABLE DRUM		DRUM, CABLE
	SOLDERING IRON		IRON, SOLDERING

The following words should be used as the last word of a noun phrase, not alone:

Apparatus	Element	Machine	Tackle
Assembly	Equipment	Mechanism	Tool
Assortment	Group	Outfit	Unit
Attachment	Installation	Plant	Vehicle
Compound	Instrument	Ship	
Device	Kit	Subassembly	

The modifier may be either a single word or a qualifying phrase and is separated from the noun or noun phrase by a comma. Additional modifiers, which indicate the shape of the item, its form, or its function rather than its application or location, may also be used.

Parts List Block

Drawings of mechanical parts (details, subassemblies, and assemblies), as well as electronic components, generally have a table which lists purchased parts, components, and quantities required. This table usually abuts the title block, Fig. 1-19(b), and extends upward as required. The column arrangement is optional and additional columns may be added as required.

The "*Find No.*" column identifies the items on the field of the drawing by numbers assigned to such items.

The "*Qty Reqd*" column shows the number of parts required to make one item or the bulk quantity in terms of weight, length, volume, or as required (*AR*).

The "*Code Ident*" column (code identification number) is for the code number of the manufacturer.

"*Part No.*" or "*Identifying No.*" lists the drawing and dash number (if any) of parts delineated on the drawings. Parts identified by MS or AN numbers and their dash numbers are listed in this column. Bulk material is also entered here and identified by type, class, grade, etc.

The "*Nomenclature*" or "*Description*" column describes each item by a noun, noun phrase, or a government model number if one is assigned.

In addition to the listed mandatory items, columns such as: "*Specification, Material, Unit Weight,*" or "*Zone*" may be included on the drawings.

The "*Specification*" column lists the specification number of the materials used for each part, the specification for parts bearing type designations, and the specifications for bulk material.

The name or description of the material for each part is listed in the *"Material"* column, while weight data are listed in the *"Unit Weight"* column. Items located on the drawings by zoning notations are given in the *"Zone"* column.

Security Marking

Engineering drawings, charts, and artwork that require security classification should be marked in accordance with DOD 5220.22-M, *Industrial Security Manual for Safeguarding Classified Information*. Drawings should have the classification marked under the legend, title block, or scale, and at the top and bottom. For example:

SECRET
Legend Unclassified

In addition, the following notation should appear on classified drawings:

> This material contains information affecting the National Defense of the United States, within the meaning of the espionage laws, Title 18, U.S.C., Sections 793 and 794, the transmission or revelation of which in any manner to an unauthorized person is prohibited by law.

Other Notices

In addition to security notices, the drawing may require a patent notice or a manufacturer's notice that states something similar to the following:

> Except as otherwise provided by contract, these drawings are the property of _____, are issued in strict confidence, and shall not be reproduced, or copied, or used as the basis for the manufacture or sale of apparatus without permission.

When required, drawing prints may be hand stamped to give such notations as "Experimental," "Not for Production," "Checking," etc.

General Notes

Usually placed in the lower left-hand corner of the drawing, these notes read down and include such items as *"Heat treat per Specification No. 000,"* *"Finish all over,"* *"Plating to be in accordance with Spec. No. 000."* These notes should also indicate whether they apply to the drawing as a whole or to only one part of a subassembly shown on the drawing.

Notes on circuit diagrams give such reference information as line input voltage, instruments to be used for measurements, or component values. Notes should be brief, in the present tense, and all in capital letters, with no individual words underlined. Abbreviations should be kept to a minimum and must conform to the latest standards.

Scale

In general, items are drawn *full size*, as long as they are not exceedingly small or large. This helps to keep the true proportions in mind when working on a full size detail or an assembly drawing.

Small objects are drawn on an *enlarged scale*, either 2 : 1 or twice size, 4 : 1 or four times size, or, for very small objects, 10 : 1 or ten times size. An enlarged scale drawing is useful in checking closely related mechanical motions or clearances. It is general practice, however, to draw the object in its true size in one corner of the drawing.

Such objects as a large size electronic unit are drawn on a *reduced scale*, 1 : 2 or one-half size, 1 : 4 or one-quarter size, or even smaller if much of the detail can be omitted.

Frequently, objects that are drawn full size will have some sections enlarged so that the details will be better illustrated. In all cases, however, the drawing scale of the enlarged sections should be given.

The triangular scales, known as mechanical engineer's scales, have a variety of scales that can be used to measure or lay out reduced or enlarged views without calculations.

Revisions

After the drawing has been issued and prints distributed, any drawing revisions made must follow a definite format. They are made by erasure, deletion, or the addition of new or revised information. Where practical, dimensional changes are corrected by drawing the changed part to scale. When this cannot be done, dimensions that are not to scale are underlined by a wavy line.

Changes must be listed in a *revision block*, Fig. 1-18(b)(c), in the upper right-hand corner of the drawing sheet. A *suffix letter* is added to the drawing number (Fig. 1-20), and the change is recorded in the revision block, listing the suffix letter, a description of the changes, the date, and the approval signature. The revision letters are placed close to the location of the change and are circled or enclosed in other-shaped figures to simplify finding the change.

If the drawing revisions are on zoned drawings or on tabular drawings, where the changes can be easily located, it is not necessary to indicate them by revision symbols. However, it may be helpful to record the changes on large or complex drawings by entering each change in the zone column on the same line as the description of the change (Fig. 1-21). On other drawings, these symbols should be located near the changed dimensions, views, or notes.

Revision symbols are a combination of capital letters followed by a numerical suffix and are enclosed in a circle about $\frac{3}{8}$ inch in diameter. The

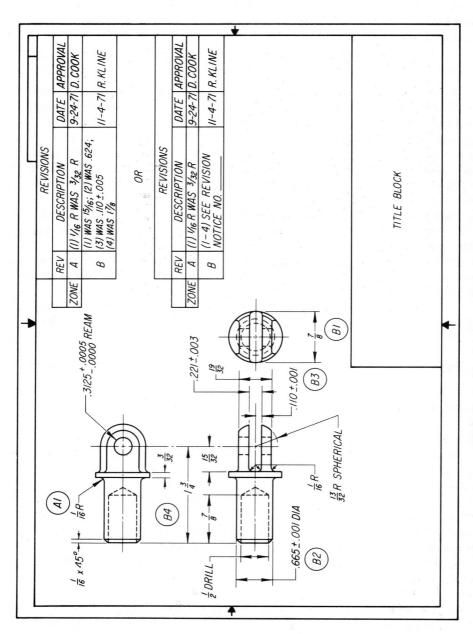

FIG. 1-21. Revision Practices on Production Drawings.

34

letters follow in alphabetical sequence, omitting "*I*," "*O*," "*Q*," and "*X*." The revision following "*Z*" is identified by "*AA*" and followed by "*AB*," "*AC*," etc.

The same revision letter is used to identify one or more changes made at one time, and these changes are numbered consecutively, beginning with suffix number 1 (Fig. 1-21).

The revision letter that appears in the revision block should be added as a suffix letter to the drawing number in the title block.

Sheet Layout

To have a neat, well-balanced drawing, it is necessary to decide which drawing views or parts of a circuit diagram will be shown. However, because most mechanical drawings are drawn full size, they are considerably easier to plan than are circuit diagrams.

After the drawing views have been selected, they should be sketched on a piece of paper. The dimensions of the drawing sheet within its borders and the title block should also be included.

A trial layout is shown in Fig. 1-22. For appearance, greater space is left at the bottom (*A*) than at the top (*B*). Spaces (*C*) and (*D*) should be left sufficient for all dimensioning pertinent to both top and front views. Space (*E*) should

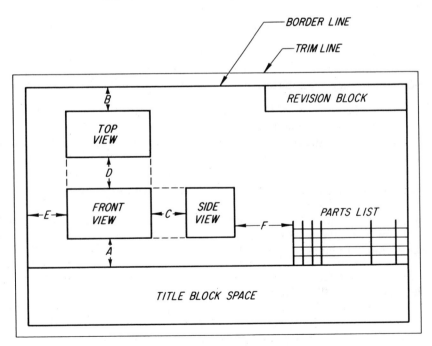

FIG. 1-22. Sheet Layout and View Spacing.

be large enough for dimensions, general notes, drill sizes, etc. Space (*F*) is kept open as a clearance for the parts list.

Once the overall dimensions of the required views and the drawing sheet size are known, the size of these spaces can be roughly calculated and marked on the trial layout. Thus, it is possible to establish adequate space for dimensioning, notes, and other relevant matter before drawing is started. If necessary a larger drawing sheet may be considered, or the drawing may be reduced to half scale.

1.5 METRIC PRACTICES

Various industry agencies have issued their own standards for metric practice. Among these are:

American National Standard Z210.1-1976, *Standard for Metric Practice*

American Society for Testing and Materials (ASTM), Standard E 380-76, *Standard for Metric Practice*

Institute of Electrical and Electronics Engineers (IEEE), IEEE Standard 268-1976, *Metric Practice*

The Department of Defense has approved their use by DOD agencies, and lists metric practice in its DOD Index of Specifications and Standards.

The metric system of measurement, known as the *International System of Units* (*SI*), is being adopted worldwide. Words such as *metre* and *litre* have become commonplace, replacing *meter* and *liter*.

Some of the special applications of the metric system in electronic drafting are:

1. Surface textures should be expressed in micrometres.

2. In mechanical drawing, the millimeter (mm) is used for linear dimensions even when the values are beyond the .01 to 1000 mm range.

3. Unit symbols should have vertical or roman-type lettering regardless of the style in the surrounding text, i.e., mm and not *mm*, without periods except at the end of a sentence.

4. A space should be left between numerical values and the unit symbol, i.e., 20 mm, not 20mm. However, a hyphen is added when it is used as an adjective, i.e., 35-mm film.

The reader is referred to one of the standards listed at the beginning of this section for more of the specified practices.

| METRIC |

The Department of Defense, in its military standard DOD-STD-1476, *Metric System, Application in New Design*, lists the applicable preferred metric units as contrasted with the present inch-pound design.

Contracts for new items, equipment, or systems require a DOD decision on the use of metric unit design to avoid hybrid systems.

When the metric system is specified, the values on engineering drawings are to be expressed in metric units.

In cases of dual dimensioning, the metric value is shown first, followed by the inch-pound value in parentheses.

If a tabular format is used, such a format may be included on the drawing or document and should convert all required values from one unit system to the other.

The identifying word "METRIC," enclosed in a rectangle, should be letters of the same size as the drawing number and located near the title block.

The following metric items are included in the Appendix:

Metric conversion
Small drills, metric
Screw threads, metric

Drawing Conversion to Metric

When existing drawings are changed to the metric system, it is necessary to do the following:

1. Add the word METRIC, in vertical lettering about $\frac{5}{16}$ inch high, next to the title block.
2. Delete the existing tolerance block and add one in millimeters.
3. Add metric dimensions in brackets next to existing inch dimensions.
4. Indicate that the drawing shows third-angle projection, a U.S. standard.

Figure 1-23 shows the standard method to be used on drawings to indicate first- or third-angle projection, ISO Recommendation R128.

Method E, known as the European or first-angle method, is shown in Fig. 1-23(a). With the front view as reference, the other views are arranged as follows:

The view from above is placed underneath.
The view from below is placed above.
The view from the left is placed on the right.
The view from the right is placed on the left.
The view from the rear may be placed on the left or on the right.

Method A, known as the American or third-angle method, is shown in Fig. 1-23(b). With the front view as reference, the other views are arranged as follows:

The view from above is placed above.
The view from below is placed underneath.

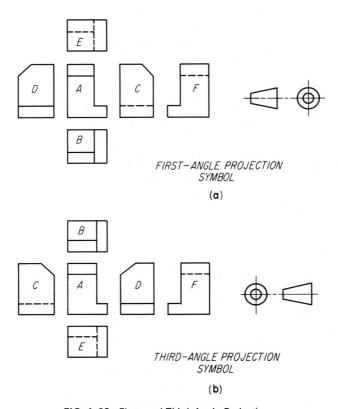

FIRST−ANGLE PROJECTION
SYMBOL

(a)

THIRD−ANGLE PROJECTION
SYMBOL

(b)

FIG. 1-23. First- and Third-Angle Projections.

The view from the left is placed on the left.

The view from the right is placed on the right.

The view from the rear may be placed on the left or on the right.

The method used is indicated on the drawing by its distinctive symbol, see (a) and (b), just above the title block on the right-hand side.

5. Convert tolerance dimensions to their metric equivalents, and present them in brackets. Generally, tolerance dimensions are: English XXX, metric XX; English XX, metric X.

Metric Sheet Sizes

Figure 1-24 illustrates the sizes of standard metric drawing sheets, and Table 1-3 gives their actual dimensions in millimeters (mm).

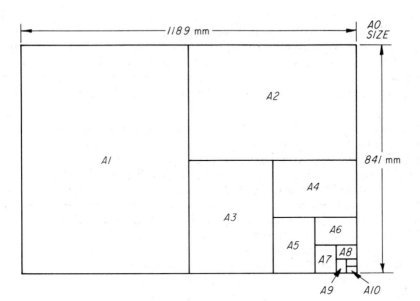

FIG. 1-24. Metric Sheet Sizes

Table 1-3   Standard Metric Sheet Sizes (mm)

Sheet Designation	Length	Width
A0	1189	841
A1	841	594
A2	594	420
A3	420	297
A4	297	210
A5	210	148
A6	148	105
A7	105	74
A8	74	52
A9	52	37
A10	37	26

SUMMARY

Electronic drafting is really an extension of engineering drawing practices, and the techniques involved in becoming an electronic draftsman can be acquired by most persons. Drawing sizes have been standardized for both civilian and military applications. The draftsman should be familiar with the many details involved in their preparation. Various drafting aids are available to simplify these tasks and help execute them quickly.

Metric drafting, although still limited in scope, is gradually entering the drafting field.

QUESTIONS

1.1 List some of the desirable characteristics for an electronic draftsman.

1.2 Describe the differences between detail, subassembly, and assembly drawings.

1.3 What drawing category (mechanical, electrical, or other) are the following drawing types associated with: (a) schematic; (b) connection; (c) outline; (d) graphic; (e) standard; (f) harness; (g) layout; (h) block; (j) printed-circuit; (k) detail; (m) assembly; (n) installation?

1.4 List the uses of engineering sketches.

1.5 What is a layout drawing and how is it made?

1.6 Describe: (a) a schematic diagram; (b) a connection or wiring diagram.

1.7 What are the reasons for making technical sketches?

1.8 What information should be secured before starting a sketch of a manufactured item?

1.9 What are the important characteristics of: (a) vellum drawing paper; (b) polyester film?

1.10 Describe a template and name its applications. What precautions should be taken when templates are used for symbols?

1.11 What are the advantages of adhesive aid devices?

1.12 List the dimensions, in inches, of sizes *B, C, D, E,* and *F* drawing sheets.

1.13 When are arrowheads used on drawing sheets? Why?

1.14 What are the various title block sections on military drawings?

1.15 What procedures are involved in making a drawing revision?

1.16 What are some of the general notes that may appear on: (a) mechanical drawings; (b) circuit diagrams?

1.17 When would a drawing be made: (a) full scale; (b) 4:1 scale; (c) 1:4 scale?

1.18 What determines a good drawing sheet layout?

1.19 Describe "*Code Ident No.*" designation.

1.20 What is the system of measurement in metric drafting?

1.21 List some of the special applications of the metric system.

1.22 List the dimensions of *A*0, *A*4, *A*6, and *A*8 drawing sheets.

1.23 Name some of the details for conversion of a drawing to the metric system.

EXERCISES

1.1 Prepare sheet layouts for: (a) *A*-size vertical; (b) *A*-size horizontal; (c) *B*-size; using Figs. 1-12, 1-14, and 1-19 for reference.

1.2 Make a two-view, half-scale drawing of a chassis 4 by 6 by 1½ inches deep. Use an *A*-size vertical drawing format, and locate the views with equal spacing around and between the views.

1.3 Prepare an *A*-size horizontal sheet layout, and draw, horizontally, two double-size views of Fig. 4-22(c), scaling the illustration for missing dimensions.

1.4 Prepare an *A*-size vertical sheet layout and sketch, vertically, the details in Fig. 4-25, Flat Package, six times size, adding the necessary dimensions.

1.5 Draw a title block, $2\frac{7}{8}$ inches high by 8 inches long for Fig. 1-18, and subdivide in accordance with the figure. Fill in all the block subdivisions for a hypothetical drawing.

1.6 Using Fig. 1-19(b) for reference, prepare a hypothetical parts list containing fifteen parts.

1.7 Select a fixed-paper capacitor from a component catalog and draw an outline drawing of the capacitor in pencil on an *A*-size sheet. Include a tabulation for ten sizes and such notes as required.

1.8 Convert the metric dimensions in Fig. 1-4 to inches and draw a similar freehand sketch on an *A*-size sheet.

1.9 On an *A*-size sheet make a sketch of the various sizes of drawing sheets in the metric system.

1.10 On a *B*-size sheet draw the gear box assembly as in Fig. 1-25, double size, using the metric dimensions shown in the figure. Add tolerance to the shaft diameters and specify the clearance drill diameter for the mounting screws.

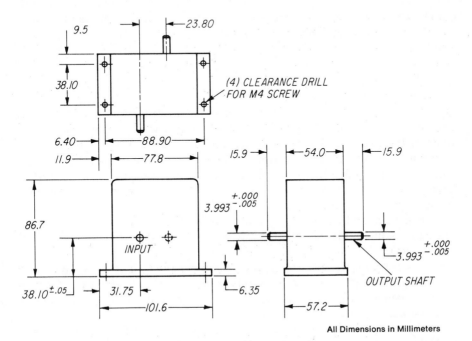

All Dimensions in Millimeters

FIG. 1-25. Servo Gear Box.

2

Drawing Practices

The electronic draftsman is expected to produce neat, legible, correct, and complete engineering drawings, in pencil and occasionally in ink. Consequently, the knowledge of drawing practices, acquired in the field of mechanical drawing, along with the details outlined here, will help the student achieve the desired results, with a minimum of effort.

2.1 PENCIL WORK

This can be subdivided into layout and regular drawing work. The techniques are similar, except that layouts are generally drawn with a finely pointed hard grade of lead, such as 4H, to produce thin layout lines. Because many alterations may be required in layout work before the final solution is achieved, these lines should be drawn lightly to prevent indentations on the drawing surface. Layout lines should extend beyond the view being constructed so that the aligning adjacent views can be drawn.

When a finished detail or assembly drawing is laid out, the center, outline, or extension lines are first drawn lightly, using the hard grade of lead. The individual sections of a circuit diagram can be outlined in a similar manner. If these lines are drawn very lightly, they will not have to be erased later to prevent them from appearing on the prints. Such lines should also extend beyond the view being drawn to align other drawing views and to act as a guide for other derails appearing in these views.

After the student completes the general view layout, he is ready to proceed with the actual drawing. A 2H or 3H lead is satisfactory for this purpose, depending upon personal preference and occasionally upon the manufacturer. The lines should be opaque enough to produce good prints without smudging the drawing excessively. Slowly rotating the pencil as the lines are drawn reduces the need for frequent repointing and produces a uniform line width. This is especially useful in circuitry work, where long lines are the rule on large schematic or connection diagrams. The new pencil, Fig. 1-5(a), eliminates the need for repointing.

Corner junctions should be drawn sharp, without overhangs or gaps. Such terminations or nontangent connections should also be avoided on curves or arc junctions. The ends of dash lines should be well defined, with uniform line lengths and spaces.

Generally, lines are drawn by starting with the thinnest—the center, dimension, extension, and break lines—and followed by outline and dash lines, although many draftsmen have their own personal preferences. The circles or arcs should be drawn next, before they are joined to straight lines. Such similar lines as dimension lines should be all drawn at the same time and followed by arrowheads. Extension and hole center lines are drawn in the same systematic manner.

Similar procedures should be followed on diagrams. Like symbols are drawn at the same time, followed by vertical and horizontal connection lines. Each section of the diagram should be completed at one time and the interconnections filled in as the drawing develops.

Work on larger drawings generally starts at the upper left-hand corner and progresses across and downward. This procedure helps to keep the drawing clean.

The dimensions are filled in after the main part of the drawing has been completed, and then general notes, title block information, and other details are added. Such information as component values, reference designations, and general notes are added to circuit diagrams after the diagram layout has been finished.

2.2 LETTERING TECHNIQUES

Because most engineering drawings use freehand lettering, mechanical lettering sets, such as Wrico or Leroy, are used only on drawings for technical manuals and other formal presentations.

To present a professional appearance, lettering should be uniform in height, line weight, inclination, and letter separation. A 2H or H lead is preferable for lettering because it will not leave indentations in the drawing paper if corrections are made later in the notes or dimensions.

Lettering guides, such as Ames, can be used to draw horizontal lettering guide lines rapidly and to maintain uniformity in lettering height and line spacing. These guide lines are drawn lightly. Inclined lines should be added at about 67 degrees from the horizontal to act as a guide for inclined lettering (Fig. 2-1), and vertical lines added for vertical lettering.

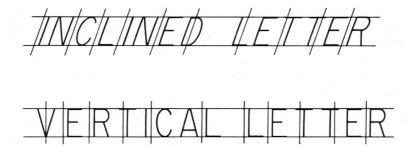

FIG. 2-1. Guidelines Used to Produce Uniform Lettering.

Lettering Style

Single-stroke, commercial gothic-style, inclined lettering has two basic advantages. After it is mastered, it is easier to do than vertical lettering, and, in addition, variations in the inclined lettering are less pronounced than in the vertical, both in declination and in letter width.

Inclined uppercase (Fig. 2-2) and/or vertical uppercase lettering is specified for all military drawings with a minimum height of $\frac{5}{32}$ of an inch when reduced. Uppercase lettering has become almost standard in all commercial drawings. The same type of lettering should be used consistently on any given drawing, and it should be opaque for full-scale or reduced-size copy.

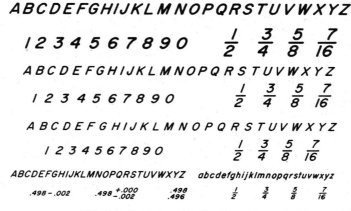

FIG. 2-2. Examples of Inclined Lettering.

Lettering Placement

All horizontal dimensions and dimension notes should be readable from the bottom of the drawing sheet. Angular and vertical dimensions should be readable from the bottom or right-hand side of the drawing.

Lettering Size

Such factors as the size of the drawing sheet, the amount of reduction expected, and the extent of detail shown all govern the size of lettering, which should not be less than $\frac{3}{64}$ of an inch after reduction.

The spacing between letters should be from one-half to three-quarters of the lettering height, Fig. 2-3(a), to give a well-balanced appearance. The recommended lettering height for A-, B-, and C-size sheets, Fig. 2-3(b), is as follows:

Drawing and part number, "*A-A*," "*B*"	$\frac{5}{16}$ inch
Title	$\frac{1}{4}$ inch
Subtitle, "*Section*," "*Detail*"	$\frac{1}{4}$ inch
Figures and notes	$\frac{3}{16}$ inch
Fractions and tolerances	$\frac{5}{32}$ inch

Larger characters may be used to give larger drawings a well-balanced appearance.

*THIS LINE SPACING IS
ONE-HALF LETTERING*

*THIS LINE SPACING IS
THREE-QUARTERS LETTER*

(a)

1 4 2 7 3 6 5 8 $\frac{5}{16}$

SUBASSEMBLY $\frac{1}{4}$

*THIS PART MUST BE
HEAT TREATED* $\frac{3}{16}$

(b)

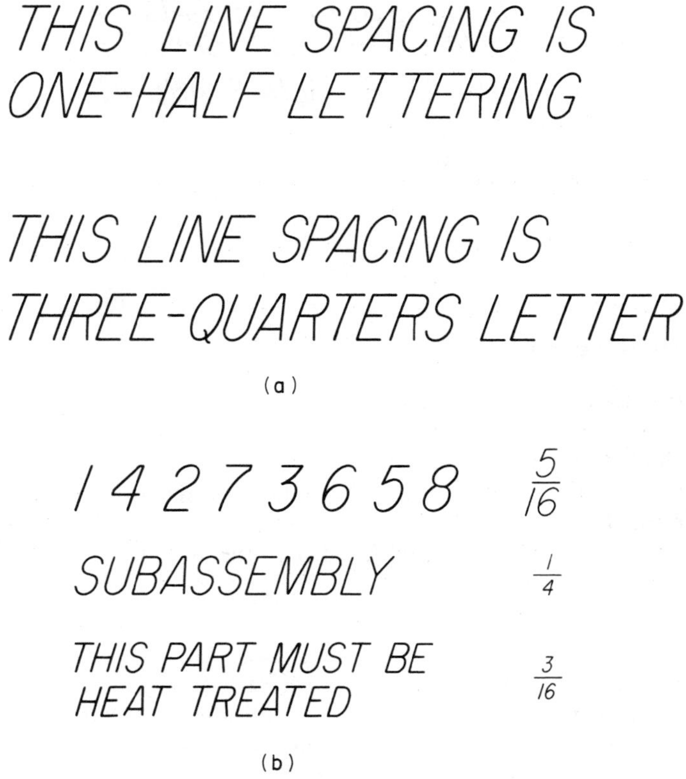

FIG. 2-3. (a) Line Spacing. (b) Lettering Size Applications.

Character Proportions

The width of most inclined uppercase letters (Fig. 2-2) is equal to about two-thirds their height—a proportion that results in a pleasing appearance and an easily read drawing. The student should try to establish a style that is suitable for the drawings he prepares and that he can execute rapidly.

Lettering Layout

One or more lines may be needed on the body of the drawing for notes and similar information. This material should first be written in longhand and then checked for grammar, spelling, meaning, and completeness, and subdivided into shorter notes, if necessary. The lettering layout for a general note is shown in Fig. 2-4. The individual notes should be aligned vertically, using a vertical guide line.

NOTES:

1. ALL RESISTORS ARE HALF-WATT UNLESS OTHERWISE SPECIFIED.

2. INDUCTANCE VALUES ARE IN HENRYS.

3. CAPACITANCE VALUES ARE IN pF.

FIG. 2-4. Typical Note Layout.

Symmetrical Lettering

In such lettering applications as titles, the title word or words must be centered within the title block. One of the easiest ways to do this is to make a trial layout of the lettering on a separate piece of paper, using guide lines and lettering of desired height and measuring the centers of the title lettering and the block. The trial layout can be placed beneath the title block on the drawing with these centers matched and the title traced over the layout.

Typing

If an engineering drawing requires extensive lettering—such as a parts list, long notes and instructions, or other data—considerable drafting time may be saved by typing this information. Lettering on *A*- and *B*-size drawing sheets can be done on a long-carriage typewriter.

2.3 DIMENSIONING METHODS

Dimensions may be specified in a number of ways. They may be given from a surface to a center, to another surface, or from center to center. Hole centers are generally used for location, except when the holes are slotted, square, or rectangular, or when other irregularly shaped openings have to be located. A dimension *tolerance* is the maximum amount by which a given dimension may vary.

Details of *Dimensioning and Tolerancing* are given in American National Standards Institute Standard Y14.5-1973. The Department of Defense has made use of this standard mandatory.

Datum or Base-Line Dimensioning

In this dimensioning method, used for close-tolerance work, datum points, lines, or surfaces are assumed to be exact references from which locations of other points, lines, or surfaces may be established. Each dimension is given from a *datum reference surface* or *datum line*, Fig. 2-5(a), and because each dimen-

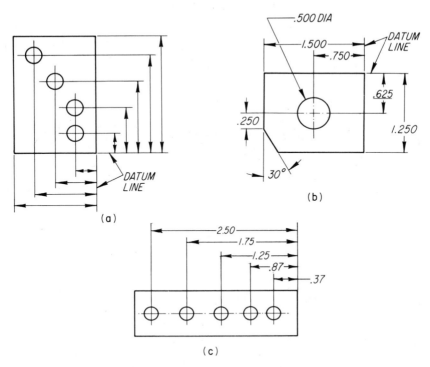

FIG. 2-5. Datum or Base-Line Dimensioning.

sion starts at the same line, the datum line, the dimension has no added cumulative error. Mating parts should use corresponding points, lines, or surfaces as datums to ensure correct assembly and to facilitate tool and fixture design.

The preferred or unidirectional method of dimensioning where all the dimensions are read from the top of the drawing is shown in Fig. 2-5(b). Note that the .625 dimension is underlined to indicate that it is not drawn to scale. Dimensioning in one straight line from a datum line (vertical line, extreme right) to all of the hole centers is shown in Fig. 2-5(c) with single arrowheads used to indicate the distance from the datum line. If possible, the longest dimension line should be separate and complete.

Center-Line Dimensioning

The locations of small holes are shown in Fig. 2-6 in reference to the large hole, which is located from datum lines. These small holes are maintained in a definite relationship to the large hole. This method is useful when it is necessary to match several holes in two separate pieces, as for example the mounting holes

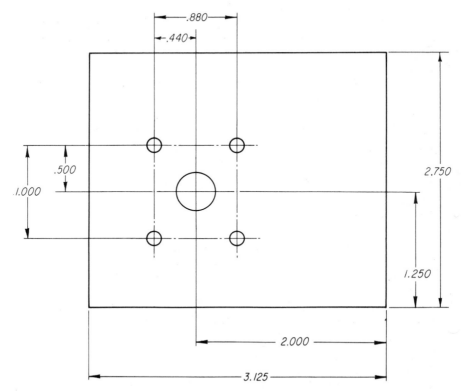

FIG. 2-6. Center-Line Dimensioning.

in a printed-circuit board that will be mounted on a chassis having a similar set of holes. Note, too, that a dimension extends across each pair of holes, thus eliminating tolerance accumulation.

Continuous Dimensioning

A continuous string of dimensions, Fig. 2-7(a), may be used to locate holes or surfaces if such considerations as close accuracy and tolerance accumulation

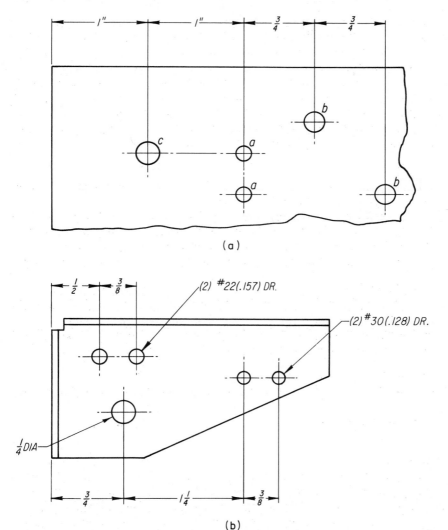

(a)

(b)

FIG. 2-7. Continuous Dimensioning.

are not critical. This method is generally used in combination with center-line dimensioning for chassis drawings. Multiple chassis holes are identified by lower-case letters as in Fig. 2-7(a), with a letter assigned to each hole size. A hole legend, listing the number of holes of each size and their diameters, is located to one side of the chassis detail.

The hole identification method shown in Fig. 2-7(b) is used when the holes are few and sufficiently different in size to prevent error.

Another method of locating a series of equally spaced, identical holes is shown in Fig. 2-8. By placing a close tolerance on the overall dimension, the intermediate spaces must also maintain a close tolerance instead of accumulating a series of tolerances.

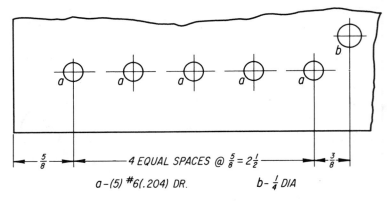

$a-(5)$ #6(.204) DR. $b-\frac{1}{4}$ DIA

FIG. 2-8. Repetitive Hole Dimensioning.

The dimensioning given in all of these methods is known as the *system* of *rectangular coordinates*. This term signifies that the linear dimensions are measured parallel to reference lines or planes that are perpendicular to each other.

2.4 DIMENSIONING OF HOLES

Round holes are specified by diameter and are understood to be through holes unless the depth is specified on the drawing. The following information is also furnished, as required: size, diameter tolerance, type (drilled, reamed, tapped, punched, counterbored, etc.), depth unless through, and number (if holes are identical).

Round holes are dimensioned in the various ways shown in Fig. 2-9. The dimension leaders should be in line with the center of the hole. Counterbored holes, Fig. 2-9(b), should indicate the diameter and depth of the counterbore. Countersunk holes, Fig. 2-9(c), should have the diameter and angle of countersink indicated. Spot-faced holes should have the diameter of the faced area indicated, but not the depth.

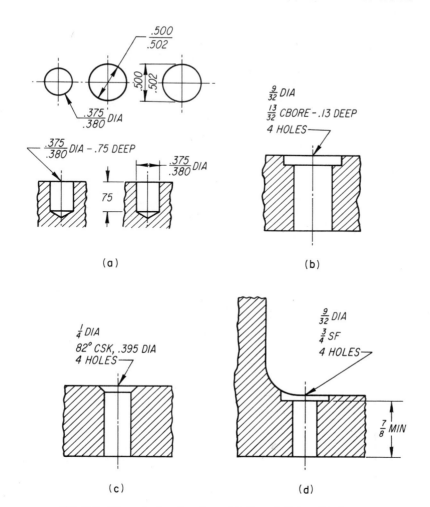

FIG. 2-9. Dimensioning Practices. (a) Round Holes. (b) Counterbored Hole. (c) Countersunk Hole. (d) Spot-face.

Note that, when leaders are used, the abbreviation *DIA*, signifying diameter, is added to the dimension, Fig. 2-9(a), which also gives hole diameter limits. Holes may have their diameters indicated directly on the hole.

Symmetrical Holes

Holes located symmetrically around a common center are shown in Fig. 2-10(a). The note identifies the size of the holes and their equal spacing on a $1\frac{3}{4}$-inch-diameter bolt circle (*BC*). The orientation is defined by the center lines on which one or more holes appear.

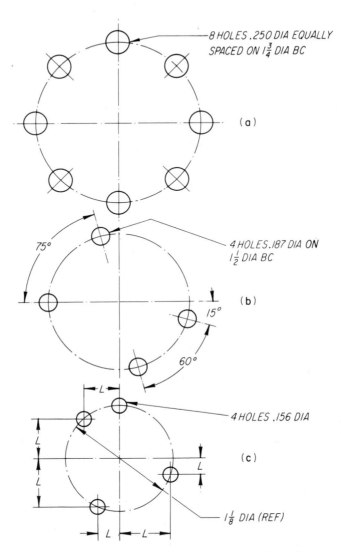

FIG. 2-10. Holes Located on a Common Center. (a) Equally Spaced.
(b) Unequally Spaced. (c) Coordinately Spaced.

The unequally spaced holes, Fig. 2-10(b), have one hole located on a center line to serve as the reference point. The other holes are located angularly from this point. Again, the note indicates the hole diameters and their locations, in this case on a 1½-inch bolt circle. If none of the holes appear on the center line, a reference angle should be given to one of the ordinates.

Unequally spaced holes on a common center, Fig. 2-10(c), are dimensioned by the system of rectangular coordinates. The bolt circle in this case serves as a reference dimension. The manufacturing department may prefer the coordinate layout in view of greater accuracy in locating the holes.

Hole Data

It is a common practice on detail drawings that have several sizes of holes to identify each of the individual hole sizes by letters, generally lowercase. This hole information, called the *hole legend* or *hole data*, is placed to one side of the drawing outline and lists the holes in ascending diameters when practicable. The number of holes of each size is listed (Fig. 2-11) after each identifying letter, along with the drill size, and corresponding decimal equivalent or fraction of an inch. The latter two may be replaced by the type of hole, such as *"ream .375 dia"* or *"¼-20 UNC-2B THD."*

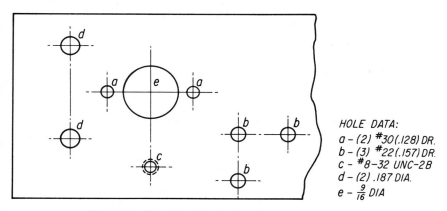

HOLE DATA:
a – (2) #30(.128) DR.
b – (3) #22(.157) DR.
c – #8–32 UNC-2B
d – (2) .187 DIA.
e – $\frac{9}{16}$ DIA

FIG. 2-11. Holes Identified by Letter Symbols.

This legend is helpful in production and checking because the total number of holes can be noted quickly, and similar hole sizes clarified. It also eliminates the need for numerous leader lines and notes on the drawing and is very adaptable to programming for numerical control.

Hole Location by Tabulation

Some electronic drawings, such as those for printed-circuit and component boards, chassis layouts, and other similar drawings, have a high hole density and a complex hole pattern. This makes it necessary to locate holes by a different method. Otherwise, the numerous dimensional lines required would make the drawing diffcult to read and increase the possibility of error.

The hole location and identification method shown in Fig. 2-12 is a tabulation of each hole size and its location from a definite reference point. The reference axes are formed by one vertical and one horizontal edge of a surface or by

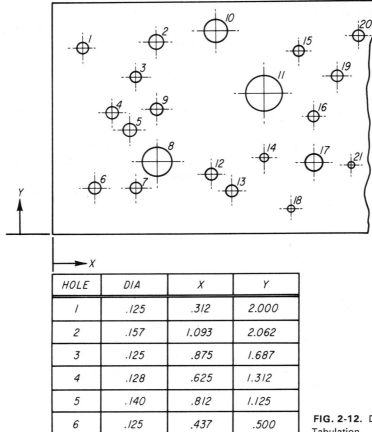

HOLE	DIA	X	Y
1	.125	.312	2.000
2	.157	1.093	2.062
3	.125	.875	1.687
4	.128	.625	1.312
5	.140	.812	1.125
6	.125	.437	.500

FIG. 2-12. Dimensioning by Tabulation.

a vertical and a horizontal reference line located on the layout. Each hole is listed on the table by number, with its diameter or drill size, its location along each reference axis from the starting or reference point, and any other required information. The holes are drawn to scale in their positions on the drawing with their centers shown. They are identified sequentially by numbers, proceeding from left to right and from top to bottom. The direction of measurement from each reference axis is indicated by arrows.

2.5 TOLERANCES AND LIMITS

Tolerances for decimal dimensions are given in decimals to the same number of places that are included in the dimensions.

Decimal dimensions are used to signify dimensions that must be held to a close *tolerance*—to within two or three decimal places, and rarely to four.

Title Block Tolerance

The dimensional tolerance information is given in the title block on the drawing sheet. It may state "Tolerances, unless otherwise specified, $\pm.03$ for two decimals, and $\pm.005$ for three decimal places." Or it may be shown as in Fig. 2-13(a). Notice that the tolerances change as the basic dimension increases.

UNLESS OTHERWISE SPECIFIED		
DIMENSIONS ARE IN INCHES AND INCLUDE CHEMICALLY APPLIED OR PLATED FINISHES		
TOLERANCES		
BASIC DIMENSION	DECIMALS	
	2 PLACE	3 PLACE
UNDER 6	$\pm.02$	$\pm.005$
6 – 24	$\pm.03$	$\pm.010$
OVER 24	$\pm.06$	$\pm.015$
ANGLES $\pm\frac{1}{2}^{\circ}$		
COMMERCIAL TOLERANCES APPLY TO STOCK SIZES		

FIG. 2-13(a). Typical Tolerance Block.

It is generally an accepted practice to specify tolerance to two or three places in inch dimensioning. On metric drawings, see Fig. 2-13(b), the equivalent tolerances are specified to one or two decimal places. A three-figure tolerance is

UNLESS OTHERWISE SPECIFIED		
DIMENSIONS ARE IN MILLIMETERS AND INCLUDE CHEMICALLY APPLIED OR PLATED FINISHES		
TOLERANCES		
BASIC DIMENSION	DECIMALS	
	1 PLACE	2 PLACE
UNDER 15	$\pm.1$	$\pm.01$
15 – 60	$\pm.2$	$\pm.02$
OVER 60	$\pm.3$	$\pm.04$
ANGLES $\pm\frac{1}{2}^{\circ}$		
COMMERCIAL TOLERANCES APPLY TO STOCK SIZES		

FIG. 2-13(b). Metric Tolerance Block.

UNLESS OTHERWISE SPECIFIED		
DIMENSIONS ARE IN MILLIMETERS AND INCLUDE CHEMICALLY APPLIED OR PLATED FINISHES		
TOLERANCES		
BASIC DIMENSION	DECIMALS	
	2 PLACE	3 PLACE
UNDER 150	±.50	±.125
150–600	±.75	±.250
OVER 600	±1.50	±.375
ANGLES ± $\frac{1}{2}$°		
COMMERCIAL TOLERANCES APPLY TO STOCK SIZES		

FIG. 2-13(b). (Continued.)

rounded out to two places when the third figure is 5 or less, i.e., a ±.275 becomes ±.27.

Toleranced Dimensioning

In a continuous chain of dimensions with individual tolerances, Fig. 2-14(a), the overall variations that may occur in hole locations are equal to the sum of the tolerances on the intermediate distances, as between points X and Y. In datum dimensioning, Fig. 2-14(b), the same tolerance accumulation is possible between points X and Y on two dimensions from the datum because of larger tolerances, while in direct dimensioning, Fig. 2-14(c), the same points are directly dimensioned with a tolerance to avoid both accumulation and the use of extremely small tolerances.

Unilateral tolerances, which allow variations in only one direction from the basic dimensions, Fig. 2-15(a), are used in specifying closely fitting holes and shafts. *Bilateral tolerances* allow variation in both directions from the basic size, Fig. 2-15(b). When the plus variation is equal to the minus variation, the combined plus-and-minus sign is followed by a single tolerance figure, Fig. 2-15(c).

Limits

Only the largest and the smallest allowable dimensions are stated in the limit dimensioning method, Fig. 2-15(d), the tolerance being the difference between the limits. The high limit is placed above the low limit unless the dimensions are given in the form of a note.

Limit and tolerance figures should be aligned with or below the dimension figure on the right or with the dimension line between the dimension number and tolerance as shown in the preceding illustrations.

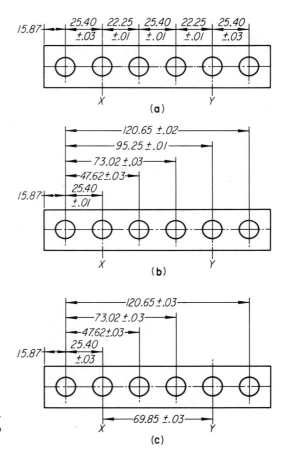

FIG. 2-14. Toleranced Dimensioning in mm. (a) Point to Point. (b) Datum. (c) Direct.

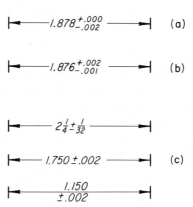

FIG. 2-15. Tolerances. (a) Unilateral. (b) Bilateral. (c) Combined Plus and Minus Sign.

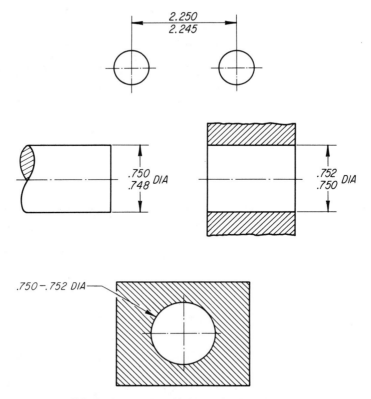

FIG. 2-15. (Continued.) (d) Limit Dimensioning.

2.6 DRAFTING SELF-AIDS

Certain electronic and mechanical components are so frequently used in electronic drafting that they have become rather commonplace. They include fixed capacitors and resistors, inductors, potentiometers, transformers, rotary and toggle switches, terminal boards and blocks, plug-in relays, jacks, connectors of all types, transistors, all types of knobs, dial lights, and transistor sockets.

To eliminate the laborious effort of drawing these components every time a layout or an assembly drawing is made, paper templates of them may be prepared. These templates, Fig. 2-16, are drawn full size on A-size sheets and show the major dimensions and all the views and details, such as mounting and terminations. The dimensions may be given in tabular form, and a connection diagram may be included to show the terminal connections or variations. The views should be accurately drawn to scale, preferably on polyester film to withstand possible rough usage. Prints of the originals can be slipped beneath the drawing sheet to trace the desired views.

Over a period of time, such templates can be accumulated to form a sizable

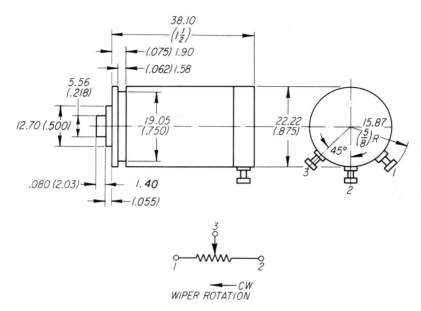

FIG. 2-16. An Example of a Component Template in Inch and Metric
Dimensioning.

collection. The information needed to prepare them can be obtained from various catalogs or by measuring the actual components if catalog data is not available.

Transparent copies can be made of the original drawings and used in layouts of assemblies, chassis, component boards, and panels. These copies can be superimposed upon each other to check clearances between the components or to obtain a better arrangement.

Another useful method is to prepare cardboard templates of the major components in a plan or side view or to use the cutout views of these components made from their prints. The templates, cardboard or paper, can be placed directly on an outline of a chassis, panel, or assembly layout and shifted around until a satisfactory layout arrangement is obtained. Thus, no actual drawing is required until the desired component disposition is established.

Commercially made templates to simulate transistor, integrated circuit, and other component outlines are available; see Chapter 13.

SUMMARY

Persons engaged in electronic drafting should become proficient in freehand lettering. This skill can be acquired through practice, developing a style that can be executed rapidly. Dimensioning and tolerance practices vary according to the drawing requirements and the fabrication methods employed. Dimensioning of complex hole patterns presents a special problem that can be solved by the tabulation method. Metric dimensioning on a drawing requires a special tolerance table as well as its identification as being metric.

QUESTIONS

2.1 Describe the features of good line work and list three common faults.

2.2 List in proper sequence the preferred order of pencilling.

2.3 What are some of the helpful practices to follow in layout work?

2.4 What features determine the appearance of lettering? What style is recommended? What is the minimum height after reduction?

2.5 Discuss the following dimensioning methods and list some of their advantages: (a) base line; (b) center line; (c) continuous.

2.6 What information is included in specifying hole diameter?

2.7 Make a freehand sketch showing the method of dimensioning: (a) symmetrical holes; (b) unequally spaced holes; (c) hole dimensioning by rectangular coordinates.

2.8 What is: (a) a hole legend; (b) the hole location and tabulation method?

2.9 Describe the difference between unilateral tolerance and bilateral tolerance.

2.10 What are component templates and how are they used?

2.11 How does the tolerance block for a metric drawing differ from a conventional tolerance block?

2.12 What is the common metric unit for indicating length on drawings?

EXERCISES

2.1 Prepare an *A*-size horizontal sheet and rule 10 guide lines, using a lettering guide. Space the lines $\frac{1}{4}$ inch apart. Using inclined uppercase letters between alternate lines, letter the following: amplitude, coaxial, decibel, dynamotor, frequency, henry, integrated circuit, junction, kilowatt, momentary, ohmmeter, permanent, potential, potentiometer, push-pull, suppressor, synchronous, transmitter, variable, wirewound. If necessary, use inclined guide lines to maintain the correct lettering slope, and repeat words with any faults.

2.2 On an *A*-size horizontal drawing sheet, draw a printed label, 2 by 4 inches, to represent a transistor location diagram. Show the transistor sockets in Fig. 8-22, and letter the transistor type designation within transistor circles. Add a title to lower third of this label, reading: *TRANSISTOR LOCATION—REFERENCE REGULATOR.*

2.3 On an *A*-size vertical drawing sheet, draw guide lines $\frac{1}{4}$ inch apart for lettering and $\frac{1}{8}$ inch apart for space between lettering lines. Letter 10 items from the following list, using all uppercase letters: (a) 3 PH 208V OUTPUT; (b) #10–32 FIL-LISTER HEAD SCREW; (c) CONNECTION DIAGRAM; (d) SWITCHING POWER SUPPLY; (e) UNIFIED FINE THREAD; (f) TOLERANCE $\pm\frac{1}{64}''$; (g) MILITARY STANDARD; (h) HARNESS CABLE; (j) #10–32 UNC-2A; (k) #22 (.157) DR.; (l) SOLID STATE RELAY; (m) ELEMENTARY DIA-GRAM; (n) PRINTED BOARD ASSEMBLY. Repeat any words with errors.

2.4 Prepare a sheet as in Exercise 2.3, and divide with a vertical center line. Letter any ten items from the following list, centering them on the vertical line: (a) channel; (b) heterodyne; (c) subcarrier; (d) magnetomotive; (e) intercommunication; (f) momentary contact; (g) frequency modulation; (h) phase modulation; (j) direct current volts; (k) beat frequency (l) amplitude modulation; (m) motor generator; (n) large-scale integration. Repeat any items that are off center. Use uppercase letters.

2.5 Draw nameplate rectangles, 2 by 4 inches, and letter items from the following list. Use $\frac{3}{8}$-inch lettering and center each line.

(a) Transistor 2N3866 oscillator   (b) Dynamotor—DY-15/ART-2, 1000 volts   (c) Range finder, fire control—M15   (d) Power supply switching 20 kHz

2.6 Make a cardboard template for each of the following components: (a) a toggle switch; (b) a potentiometer; (c) a transistor outline of TO-5, TO-78, and TO-85; (d) a transistor socket; (e) a single-section rotary switch. Make templates for a plan and a side view.

2.7 Make a half-scale drawing of the panel drawing given in Fig. 2-17(a), using a

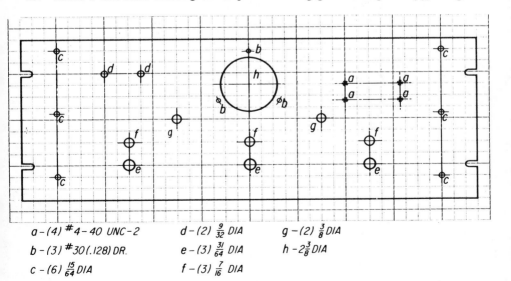

$a-(4) \, {}^{\#}4-40 \, UNC-2$ $d-(2) \, \frac{9}{32} \, DIA$ $g-(2) \, \frac{3}{8} \, DIA$

$b-(3) \, {}^{\#}30(.128) \, DR.$ $e-(3) \, \frac{31}{64} \, DIA$ $h-2\frac{3}{8} \, DIA$

$c-(6) \, \frac{15}{64} \, DIA$ $f-(3) \, \frac{7}{16} \, DIA$

FIG. 2-17. (a): Details of Panel.

B-size sheet and the standard block shown in Fig. 1-14. Do not draw the grid lines. Interpolate the intermediate dimensions. Use base-line dimensioning, except for *a* and *c* holes, which are to be dimensioned by the center-line dimensioning method. Include a hole legend in the lower left-hand corner of the drawing, using the following hole sizes: (a) #4-40; (b) .128; (c) $\frac{15}{64}$; (d) $\frac{9}{32}$; (e) $\frac{31}{64}$; (f) $\frac{7}{16}$; and (g) $\frac{3}{8}$. Specify the drill sizes for up to $\frac{1}{4}$ inch diameter. See Table A-4 in the Appendix for twist-drill sizes. Arrange hole sizes in ascending order. Panel size is 19×7 inches.

2.8 On an *A*-size horizontal sheet, make a complete double-size dimensioned drawing of a nameplate to fit *a*-size mounting holes. Letter the following in 3 lines, with a letter height of $\frac{1}{4}$ inch: POWER SUPPLY UNIT, ABC RADIO CORPORATION, BALTIMORE, MD. 21205. Use vertical lettering; nameplate is 1 by 3 inches; *a* holes—#4-40 clearance (see Table 5-2).

2.9 Make a full-size drawing of the bracket shown in Fig. 2-17(b) on an *A*-size drawing sheet. Do not draw the grid lines. Interpolate the intermediate dimensions. Use fractional base-line dimensioning for the bracket and decimal center-line dimensioning for all holes relative to large hole. Include a hole legend in lower left-hand corner and list the following hole diameters: (a) .098; (b) .204; (c) $\frac{21}{64}$; (d) $\frac{7}{32}$; (e) $\frac{1}{2}''$. Use drill sizes for holes up to $\frac{1}{4}$ inch in diameter. See Table A-4 in the Appendix for twist-drill sizes. Arrange hole sizes in ascending order.

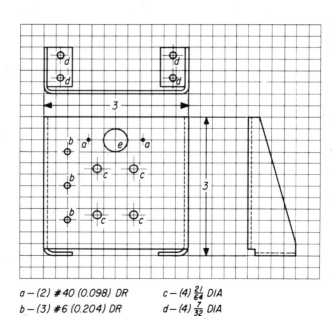

$a - (2)$ #40 (0.098) DR $c - (4) \frac{21}{64}$ DIA
$b - (3)$ #6 (0.204) DR $d - (4) \frac{7}{32}$ DIA

FIG. 2-17. (b) : Details of Bracket.

2.10 Prepare a hole legend with 12 different hole sizes on an *A*-size sheet. List the holes in ascending order, and note the number of holes for each hole size. Holes of $\frac{1}{4}$ inch or less should be given in decimals and by the nearest drill number (see Table A-4, Appendix). Select the hole sizes from the following list: (2) $\frac{1}{4}$; (12) $\frac{5}{64}$; (8) $\frac{1}{32}$; (10) $\frac{3}{64}$; (6) $\frac{11}{64}$; (2) $\frac{1}{16}$; (11) $\frac{9}{64}$; (3) $\frac{3}{8}$; (4) $\frac{5}{16}$; (7) $\frac{1}{8}$; (2) $\frac{9}{16}$; (4) $\frac{1}{2}$; (2) $1\frac{1}{8}$; (7) $\frac{23}{32}$; (3) $\frac{7}{16}$; (11) $\frac{3}{16}$.

2.11 Select 8 sizes of holes from the list in Exercise 2.10, and lay out a panel, 8 by 10 inches, with holes located by the tabulation method. Assign coordinate dimensions to the various holes, and prepare a table as in Fig. 2-12.

2.12 Make a double-size drawing of TO-5 transistor and 14-lead JEDEC-TO-116 package (see the Appendix), using the unilateral tolerance method for dimensioning wherever possible.

2.13 Select six sizes of holes from the list in Exercise 2.10, in metric dimensions nearest to their inch equivalents, and lay out a panel, 7 by 12 inches, with holes located by the tabulation method, both in inch and metric equivalent dimensions. Assign coordinate dimensions to the various holes and prepare a table similar to Fig. 2-12.

2.14 Make a full-size drawing of a bracket shown in Fig. 2-17(b) on an $A4$-size drawing sheet (see Table 1-3). Use a 12 mm border. Do not draw the grid lines. Interpolate the intermediate dimensions. Use metric dimensions throughout. Include a hole legend in the lower left-hand corner and list the following hole diameters, converting to nearest metric drill sizes (see Table A-6 in the Appendix): (a) .090; (b) .200; (c) $\frac{21}{64}$; (d) $\frac{7}{32}$; (e) $\frac{1}{2}$. Arrange hole sizes in ascending order.

2.15 On an $A5$ sheet (see Table 1-3), with mm borders, make a tabulation of the plain and tapped holes listed in Fig. 2-17, converted to the nearest metric equivalents (see the Appendix, Table A-7).

2.16 Convert the tabulation table in Fig. 2-12 to metric dimensions.

2.17 Prepare a legend with twelve different hole sizes on an A-size sheet. List the holes in ascending order, and note the number of holes for each hole size. Select the holes from the nearest drill number (see the Appendix, Table A-6) of twist drills in millimeter sizes. Select the hole sizes from the following list: (2) $\frac{1}{4}$; (12) $\frac{5}{64}$; (8) $\frac{1}{32}$; (10) $\frac{3}{64}$; (6) $\frac{11}{64}$; (2) $\frac{1}{16}$; (11) $\frac{9}{32}$; (3) $\frac{3}{8}$; (4) $\frac{5}{16}$; (7) $\frac{1}{8}$; (2) $\frac{9}{16}$; (3) $\frac{7}{16}$; (11) $\frac{7}{32}$.

2.18 Add metric dimensions on the panel drawing shown in Fig. 2-17(a), using a B-size sheet and the standard block shown in Fig. 1-14. Add the word "METRIC" notation to the drawing and replace the tolerance block with a metric tolerance block; see Fig. 2-13(b). Using the tables in the Appendix, select the nearest metric hole sizes and make a new hole legend in the lower left-hand corner next to the existing legend.

3

Drafting Room Routine

A well-organized drafting room provides certain facilities, follows definite routines, and establishes explicit requirements for the electronic draftsman to follow. These well-defined practices and procedures are needed to make sure the drawings conform to existing commercial and military standards.

Such practices include the numbering and filing of drawings, the assignment of drawing titles, the standardization of abbreviations used on drawings, and the use of such reference materials as drafting room engineering standards, catalogs, reference books, engineering magazines, and other aids. Facilities are also provided to maintain records of drawing revisions and reproduction.

3.1 NUMBERING AND FILING DRAWINGS

The drafting room is an important segment of the engineering department. The efforts of draftsmen to design a new or an improved article result in a set of production drawings that frequently represent a considerable investment to the company. Thus, it is important for these drawings to be properly numbered and filed.

Although drawing systems vary, the vast majority of them use only a single detail or assembly on each separate drawing sheet. The drawing number designation is based upon the size of the drawing sheet. Thus, all $8\frac{1}{2}$ by 11 inch drawings are ordinarily identified by a capital "A" as the first part of the drawing number. The other letter designations for various sheet sizes are given in Tables 1-1, 1-2, and 1-3.

The numerical designation that follows the sheet-size designation letter is usually assigned in numerical order. In some instances, blocks of numbers are assigned to each company subdivision within a large company or corporation. Thus, it is possible to identify the subdivision issuing the drawings and to prevent the possibility of duplicate drawing numbers being issued.

To simplify filing, drawings are generally separated by size and filed in numerical order.

Drawings that have a security classification stamp, such as CONFIDENTIAL or SECRET, have to be filed in special safes or vaults where they will be available only to security-cleared personnel.

3.2 DRAFTING ROOM ENGINEERING STANDARDS

Many of the larger drafting rooms have well-established engineering standards for drawing, materials, shop procedures, quality control, and so forth. The scope of these standards is determined by the personnel available to prepare and maintain them, the extent of standardization practiced within the plant or agency, and the size of the engineering department. These standards may contain detailed information on such subjects as ball bearings, bends in sheet metal, brazing, cable assemblies, casting details, finishes, gears, hardware, machining practices, marking, nameplates, springs, stampings, stock material, symbols, test specifications, and welding.

3.3 ABBREVIATIONS

Abbreviations should not be ambiguous. The basic rule in using them is, "When in doubt, spell it out."

Accepted practices follow such standards as MIL-STD-12D, *Abbreviations for Use on Drawings, Specifications, Standards, and in Technical Documents,* Department of Defense, and ANSI Standard Y1.1-1972, *Abbreviations for Use on Drawings and in Text.*

As a rule, capital letters are used for all abbreviations on drawings and documents produced in uppercase characters. Lowercase abbreviations are used on documents produced in lowercase characters but such abbreviations may be capitalized when required for emphasis.

Subscripts are used for such applications as the letter symbols for quantities in semiconductors or for quantities on charts or graphs. Abbreviations are used without periods, unless they form words such as ACT., AFT., AT., BET., DISC., FIX., where a period is necessary for clarity.

The diagonal bar is also avoided, except in B/M for bill of material; 1/C for single conductor; T/E for tables of equipment; 2/C for two conductor; and W/O for without.

Although most abbreviations are for single terms, there are two-, three-, or four-word combinations in use. For example, AFC for automatic frequency control, B & S for Brown & Sharpe Wire Gage, DOD for Department of Defense, and others.

Abbreviations should not be used for short words. For instance, RD should not be used in place of RED because RD is a standard abbreviation for road. To avoid confusion, all short words should be written out.

Abbreviations are not to be confused with reference designations, e.g., SW is the abbreviation for a switch, and *S* is its reference designation. Abbreviations are used to save lettering space on the drawing, whereas reference designations are used on schematic and connection or wiring diagrams to identify the individual graphic symbols.

Abbreviations for Drafting Terms

Table 3-1 contains mechanical and drafting terms and their abbreviations that are in common use in electronic design and drafting work.

Table 3-1   Abbreviations for Drafting Terms

above baseline	ABL	azimuth	AZ
acrylic	ACRYL	back of board	B OF B
addendum	ADD	back view	BV
advance engineering order	AEO	backlash	BL
advance material request	AMR	ball bearing	BBRG
Air Force-Navy	AN	base diameter	BDIA
Air Force-Navy Design	AND	base line	BL
alignment	ALIGN	bend line	BL
alteration	ALTRN	between	BETW
alternate	ALTN	between centers	BC
aluminum	AL	between perpendiculars	BP
American Wire Gage	AWG	bill of material	B/M
and so forth	ETC	binding head	BDGH
anneal	ANL	bolt circle	BC
anodize	ANDZ	both faces	BF
antifriction bearing	AFB	both sides	BS
appendix	APX	bottom	BOT
approved	APVD, APPD	bracket	BRKT
approximate	APPROX	branch	BR
Army-Navy	A/N	brass	BRS
Army-Navy Design	AND	Brinnel hardness number	BHN
article	ART	Brown & Sharpe Wire Gage	B&S
as drawn	AD	by (used between dimensions)	X
as required	AR	cabling diagram	CAD
as soon as possible	ASAP	case harden	CH
assembly	ASSY	cast iron	CI
attention	ATTN	cast steel	CS
auxiliary	AUX	casting	CSTG
average	AVG	catalog	CAT

center	CTR	drafting room manual	DRM
center distance	CD	drawing	DWG
center line	CL	drawing change request	DCR
center of gravity	CG	drawing list	DL
center section	CS	drawn	DWN
center to center	C TO C	each	EA
chamfer	CHAM	each face	EF
change notice	CN	eccentric	ECC
change order	CO	electroplate	EPL
change request	CR	elevation	EL
check	CHK	encapsulated	ENCAP
checker	CHKR	end to end	E TO E
chemical milling	CM	engineer	ENGR
chemically pure	CP	engineering	ENGRG
chromium plate	CRPL	engineering change order	ECO
circle	CIR	engineering change proposal	ECP
circuit card assembly	CCA	engineering field change	EFC
circular pitch	CP	engineering memorandum	EM
circumference	CRCMF	engineering order	EO
class	CL	engineering work order	EWO
clearance	CL	equally spaced	EQ SP
clockwise	CW	equipment	EQPT
column	COL	except	EXC
commercial	CML	extra fine (threads)	EF
component board	CB	fabricate	FAB
concentric	CNCTRC	far side	FS
confidential	CONF	fastener	FSTNR
connection diagram	CONN DIA	Federal	FED
continue	CONT	Federal Specification	FS
corrosion-resistant steel	CRES	Federal Stock Number	FSN
counterbore	CBORE	federal supply class	FSC
countersink	CSK	Federal Supply Code for	FSCM
countersink other side	CSKO	Manufacturers	
cross section	XSECT	figure	FIG
dated	DTD	fillet	FIL
datum	DAT	fillister head	FILH
decalcomania	DECAL	finish	FNSH
decimal	DEC	finish all over	FAO
dedendum	DED	flange	FLG
department	DEPT	flat head	FLH
depth	DP	for example	EG
design	DSGN	fractional horsepower	FHP
design change notice	DCN	front	FR
detail	DET	front view	FV
deviation	DEVN	full size	FS
diagonal	DIAG	gear	GR
diagram	DIAG	general note	GN
diameter	DIA	government-furnished equipment	GFE
diameter bolt circle	DBC	grind	GRD
diametral pitch	DP	group	GP
dimension	DIM	gusset	GUS
drafting request	DR	half-hard	1/2H

half-round	1/2RD	panel	PNL
heat treat	HT TR	parallel	PRL, PAR
height	HGT	part	PT
hexagonal head	HEX HD	part number	PN
hinge line	HL	parts catalog	PC
horizontal center line	HCL	passivate	PSVT
horizontal reference line	HRL	pattern	PATT
indentured parts list	IPL	per (between words)	/
inside diameter	ID	performance evaluation and	PERT
inside radius	IR	review technique	
instruction book	IB	perpendicular	PERP
internal pipe thread	IPT	phillips head	PHH
job order	JO	phosphor bronze	PH BRZ
left hand	LH	piece	PC
length	LG	pitch	P
length overall	LOA	pitch circle	PC
lightening hole	LTGH	pitch diameter	PD
list of drawings	LD	plain washer	PW
list of material	LM	plus or minus	PORM
list of specifications and standards	LSST	pound	LB
lock washer	LK WASH	preliminary	PRELIM
long	L, LG	prototype	PROTO
machine screw	MSCR	publication	PUB, PUBN
manual	MNL, MAN	purchase order	PO
manual change order	MCO	qualified products list	QPL
master cross-reference list	MCRL	quantity	QTY
material	MATL	radius	RAD, R
maximum	MAX	rear view	RV
mechanical	MECH	reference line	REFL
military standard (book)	MIL-STD	release	RLSE
military standard (sheet)	MS	request for engineering change	REC
minimum	MIN	required	REQD
model	MOD	revision	REV
modification	MOD	revolutions per minute	RPM
national coarse (thread)	NC	Rockwell hardness	RH
national extra-fine (thread)	NEF	root mean square	RMS
national fine (thread)	NF	round	RND, RD
national special (thread)	NS	roundhead	RDH
national taper pipe (thread)	NPT	scale	SC
nomenclature	NOMEN	screw	SCR
nominal	NOM	section	SECT
not to scale	NTS	serial	SER
number	NO	setscrew	SSCR
obsolete	OBS	sheet	SH
opposite	OPP	shop order	SO
original equipment manufacturer	OEM	socket head	SCH
outside diameter	OD	spare part	SP
outside radius	OR	specification	SPEC
oval head	OVH	specification control drawing	SCD
overall	OA	spherical	SPHER
page	P	spot face	SF
pan head	PNH	spot weld	SW

spring	SPR	unified coarse thread	UNC
square	SQ	unified extra fine thread	UNEF
superseded	SUPSD	unified fine thread	UNF
symmetrical	SYMM	used on	U/O
tensile strength	TS	vertical center line	VCL
thick	THK	vertical reference line	VRL
thread	THD	wide	W
threads per inch	TPI	width	WD
title block	T/B	width across flats	WAF
tolerance	TOL	wire assembly	WA
total indicator reading	TIR	wire list	WL
trade name	TRN	zone	Z

Abbreviations for Electrical and Electronic Terms

Table 3-2 lists the commonly used terms for diagrammatic and electronic design work and their abbreviations.

Table 3-2   Abbreviations for Electrical and Electronic Terms

a programming language	APL	antisubmarine warfare	ASW
actuating	ACTG	anti-transmit-receive	ATR
actuator	ACTR	apparatus	APPAR
adapter	ADPTR	armature	ARM
adjust	ADJ	array	ARY
advance defense communications satellite	ADCS	arrestor	ARSR
		attenuation, attenuator	ATTEN
advanced tactical ballistic missile	ATBM	audio	AUD
		audio frequency	AF
air circuit breaker	ACB	automatic brightness control	ABC
air decoy missile	ADM	automatic checkout equipment	ACE
air position indicator	API	automatic data link	ADL
air-to-ground missile	AGM	automatic data processing	ADP
airborne early warning	AEW	automatic direction finder	ADF
airborne surveillance radar	ASR	automatic frequency control	AFC
alarm	ALM	automatic gain control	AGC
alternating current	AC	Automatic Instrument Landing Approach System	AILAS
alternating current volts	VAC		
altitude transmitting equipment	ALTE	automatic noise limiter	ANL
ammeter	AMM	automatic phase control	APC
ampere-hour meter	AHM	automatic phase lock	APL
amplifier	AMPL	automatic sensitivity control	ASC
amplitude modulation	AM	automatic test and check out equipment	ATCE
analog to digital	A/D		
anode	AD	automatic volume control	AVC
anode (electronic device)	A	automatic volume expansion	AVE
antenna	ANT	auxiliary power unit	APU
antiaircraft fire control	AAFC	back-connected	BC
antiballistic missile	ABM	ballistic missile	BM
antisubmarine rocket	ASROC	band-elimination	BD ELIM

bandpass	BP	crystal unit, piezoelectric	CU
base (electron device)	B	current	CUR
battery (electrical)	BAT	current transformer	CT
beat frequency	BF	dash pot (relay)	DP
beat-frequency oscillator	BFO	data link	DL
binary	BIN	data list	DL
binary-coded decimal	BCD	data storage and retrieval	DS & R
bits per second	BPS	data terminal equipment	DTE
blocking oscillator	BO	data transmitting equipment	DXE
blower	BLO	daughter board	DTRBD
breaker	BRKR	delay line	DL
broadcast	BC	delayed automatic volume control	DAVC
bubble memory	BUBMEM	demodulator	DEM
bulk (substrate) (electron device)	BU	detector	DET
bypass	BYP	differential amplifier	DIFA
capacitor, capacity	CAP	digital to analog	D/A
cathode	CATH	digitizer	DGTZR
cathode ray	CR	diode	DIO
cathode-ray oscilloscope, oscillograph	CRO	direct current	DC
		direct-current test volts	VDCT
cathode-ray tube	CRT	direct-current working volts	VDCW
center tap	CT	direction finder	DF
channel	CHAN	disconnect	DISC
circuit	CKT	discriminator	DSCRM
circuit breaker	CB	display and control unit	D & CU
coaxial	COAX	distant early warning	DEW
collector (electron device)	C	distribution amplifier	DAMP
color code	CC	distribution box	DB
common battery	CB	double contact	DC
communication	COMM	double pole	DP
commutator	COMM	double-pole, double-throw	DPDT
complementary metal-oxide semiconductor	CMOS	double-pole, front-connected	DPFC
		double-pole, single-throw	DPST
computer	CMPTR	double sideband	DSB
computer control unit	CPCU	double throw	DT
conductor	CNDCT, COND	doubler	DBLR
connector	CONN	dynamotor	DYNM
contact	CONT	early warning radar	EWR
contactor, starting	COS	electric power supply	EPS
continuous wave	CW	electroluminescent	EL
control	CONT	electromagnetic interference	EMI
control read-only memory	CROM	electromagnetic radiation	EMR
converter	CONV	electromotive force	EMF
counter countermeasures	CCM	electron-coupled oscillator	ECO
counter electromotive force	CEMF	electronic	ELEK
countermeasures	CM	electronic countermeasures	ECM
counter-radar measures	CRM	electronic differential analyzer	EDFA
coupling	CPLG	electronic switching	ES
cross connection	XCONN	electronically alterable read-only memory	EAROM
cryptography	CRYPTO		
crystal	XTAL	electronics	ELEX
crystal oscillator	XTLO	embedded wiring board	EWB

equal	EQL, EQ	junction field-effect transistor	JFET
equipment and spare parts	E & SP	light emitting diode	LED
erasable programmable read-only memory	EPROM	limit switch	LIM S
		limiter	LMTR
exciter	EXCTR	line printer	LPTR
extremely high frequency	EHF	linear integrated circuit	LIC
facsimile	FAX	link	LK
field-effect transistor	FET	local oscillator	LO
filament	FIL	logic unit	LU
filament center tap	FCT	loudspeaker	LS
filter	FLTR	low frequency	LF
Fire Control System	FCS	low-frequency oscillator	LFO
formula translation	FORTRAN	low pass	LP
four conductor	4/C	low voltage	LV
frequency	FREQ	low-voltage protection	LVP
frequency (combined form)	F	lower sideband	LSB
frequency modulation	FM	lowest usable high frequency	LUHF
front-connected	FC	magnetic amplifier	MAGAMP
gate (electron device)	G	magnetic modulator	MAGMOD
generator	GEN	magnetomotive force	MMF
glass reenforced plastic	GRP	magnetron	MAGN
ground	GND	manual gain control	MGC
guided missile	GM	manual volume control	MVC
gyroscope	GYRO	master oscillator	MO
handbook	HDBK	master oscillator power amplifier	MOPA
heat sink	HTSK	master switch	MSW
high frequency	HF	matrix	MAT
high-frequency oscillator	HFO	maximum working voltage	MWV
high pass	HP	mean time between failures	MTBF
high-pass filter	HPFL	medium frequency	MF
high potential	HIPOT	medium high frequency	MHF
high tension	HT	memory	MEM
high voltage	HV	metal-oxide semiconductor	MOS
identification friend or foe	IFF	metal-oxide semiconductor field-effect transistor	MOSFET
immediate access storage	IAS		
indicator	IND	metal-oxide semiconductor transistor	MOST
inductance-capacitance	IC		
inductance-capacitance-resistance	ICR	microcomputer	MICMPTR
inertial guidance system	IGS	microelectronics	MELEC
infrared	IR	microphone	MIC
infrared laser	IRASER	microprocessor	MIRRCS
input/output register	I/OR	missile tracking radar	MTR
instrument	INSTR, INST	mode	M
instrument landing system	ILS	modulated continuous wave	MCW
integrated circuit	IC	modulator	MOD
interlock	INTLK	module	MDL
intermediate frequency	IF	molecular electronics	MOLELEX
intermediate power amplifier	IPA	momentary contact	MC
interrupted continuous wave	ICW	motherboard	MTHBD
isolated-gate field-effect transistor	IGFET	motor	MOT
jack	JK	motor generator	MG
junction box	JB	multivibrator	MV

negative	NEG	radio-frequency choke	RFC
network	NTWK	radio-frequency interference	RFI
neutral	NEUT	random-access indestructive advanced memory	RAIPM
no connection	NC		
no voltage release	NVR	random-access memory	RAM
normally closed	NC	range height indicator	RHI
nuclear weapon	NW	range marks	RM
number of bits	(N)	readout	RDOUT
numerical control	NC	real time	RT
ohmmeter	OHM	receiver	RCVR
oscillator	OSC	receptacle	RCPT
oscilloscope	SCOPE	rectifier	RECT
overload	OVLD	regulator	RGLTR
page printer	PPTR	relay	RLY
parallel	PRL	remote control system	RCS
phase	PH	resistance capacitance	RC
phase modulation	PM	resistor	RES
phonograph	PHONO	resolver	RSLVR
piezoelectric-crystal unit	CU	reverse	RVS
plan position indicator	PPI	rheostat	RHEO
plug	PL	schematic	SCHEM
polarity	PLRT	secondary	SEC
position indicator	PIN	selenium rectifier	SR
positive	POS	semiconductor-controlled rectifier	SCR
potential	POT	series	SER
potentiometer	POT	shift register	SR
power	PWR	short wave	SW
power amplifier	PA	side-looking radar	SLR
power factor	PF	signal	SIG
power oscillator	PO	single contact	SC
power supply (combined form)	PS	single phase	1PH
power unit	PU	single pole	SP
preamplifier	PREAMP	single-pole, double-throw	SPDT
precision approach radar	PAR	single-pole, single-throw	SPST
primary	PRI	single sideband	SSB
printed circuit board	PC	slow release (relay)	SR
printed wiring	PW	socket	SKT
printing wiring assembly	PWA	solenoid	SOL
printout	PTOUT	solid-state switching	SSS
pulse-amplitude modulation	PAM	speaker	SPKR
pulse-code modulation	PCM	standing wave ratio (voltage)	SWR
pulse frequency	PF	supply	SPLY
pulse-repetition frequency	PRF	switch	SW
pulse-repetition rate	PRR	symbol	SYM
pulse-width modulation	PWM	target	TGT
pushbutton	PB	target identification	TI
push-pull	PP	telephone	TEL
radar	RDR	television	TV
radar (combination form)	R	terminal	TERM
radar countermeasures	RCM	terminal board	TB
radio	RAD	terrain following radar	TFR
radio direction finder	RDF	tertiary	TER
radio frequency	RF	test point	TP

test switch	TSW	unijunction transistor	UJT
thermocouple	TC	upper sideband	USB
time delay	TD	variable frequency oscillator	VFO
transceiver	XCVR	very high frequency	VHF
transformer	XMFR	very low frequency	VLF
transistor	XSTR	video	VID
transistor transistor logic	TTL	video display terminal	VDT
transmitter	XMTR	video frequency	VIDF
transmitter-receiver	TR	voice frequency	VF
triple pole	3P	volt	V
tuned radio frequency	TRF	volt, alternating current	VAC
tuning	TUN	volt, direct current	VDC
tunnel diode	TNLDIO	voltage regulator	VR
tunnel diode amplifier	TDA	voltmeter	VM
tunnel diode logic	TDL	volume	VOL
twin sideband	TWSB	wafer	WFR
two-phase	2PH	waveguide	WG
ultra-high frequency	UHF	Weapon System	WS
under voltage	UNDV	wire-wound	WW

Color Abbreviation Table

Table 3-3 contains the letter abbreviations for various colors specified in MIL-STD-12D.

Table 3-3   Abbreviations for Colors

		Alternate				*Alternate*
amber	AMB		orange	ORN	O
black	BLK	BK	red	RED	R
blue	BLU	BL	slate	SLT	S
brown	BRN	BR	violet	VIO	V
gray	GRA	GY	white	WHT	W
green	GRN	G	yellow	YEL	Y

3.4 LETTER SYMBOLS

Many of the physical, electrical, magnetic, and other quantities are represented by letter symbols or Greek alphabet. These symbols are used in mathematical expressions, graphs, circuits, and other applications. Tables 3-4, 3-5, and 3-6

Table 3-4   Physical Quantities

Quantity	*Symbol*	*Unit*
area	A	circular mil, square mil, square inch, square centimeter
frequency	Hz	hertz
length	l	inch, foot, centimeter, meter
temperature	T	degrees Fahrenheit (°F), degrees Celsius (°C)
time	t	second
volume	V	cubic-inch, cubic-foot

Table 3-5   Multiplier Prefixes

Prefix	Symbol	Quantity	Value
tera	T	10^{12}	1,000,000,000,000
giga	G	10^{9}	1,000,000,000
mega	M	10^{6}	1,000,000
kilo	k	10^{3}	1000
hecto	h	10^{2}	100
deka	da	10	10
deci	d	10^{-1}	.1
centi	c	10^{-2}	.01
milli	m	10^{-3}	.001
micro	μ	10^{-6}	.000001
nano	n	10^{-9}	.000000001
pico	p	10^{-12}	.000000000001

Table 3-6   Electrical Quantities

Quantity	Symbol	Unit	Quantity	Symbol	Unit
capacitance	C	farad	inductance	L	henry
conductance	G	mho	power	P	watt
current	I	ampere	reactance	X	ohm
electromotive force	V	volt	resistance	R	ohm
impedance	Z	ohm			

list, respectively, several physical quantities, multiplier prefixes, and electrical quantities.

3.5 INDEX SYSTEMS

As an aid in locating a desired drawing, an index and cross-reference system is frequently set up in the drafting room which lists all drawings by number and title. This may take the form of a card index system that provides such information about a drawing as its first use, the subassemblies listing it, and so on. The ease with which a drawing can be located is determined by the extent to which the system is applied.

3.6 CATALOGS

Manufacturers' catalogs are invaluable for locating the latest information on available components. They give overall or mounting dimensions, ratings, weight ranges, model designations, applicable military specifications, and many other details that are not necessarily found in drafting standards or other reference material.

Large annual publications that previously could be obtained upon request by an engineer or a draftsman as part of a magazine subscription must now be purchased separately in most cases. These publications may be specialized according to classification and subclassification, as for example "switches." These may be subdivided into: coaxial, float, interlock, lever, limit, power, printed circuit, pushbutton, rotary, toggle, and many others. There may also be a separate alphabetical listing of manufacturers' names, addresses, distributors, number of employees, trademark names, branch offices, and other information.

The same charging practice is true of certain technical magazines, which now charge for a yearly subscription unless the requestor is a design engineer or an engineering manager.

Catalogs are generally maintained in a separate file in the drafting room. It is a great time saver to have such a file cross-indexed both by products and by names of manufacturers.

Most companies will supply catalogs on request, so that it might be well to obtain copies of those most frequently used.

One of the recent newcomers in the catalog field is the combination catalog viewer and printer. It consists of a large number of microfilm spool records—containing the latest manufacturers' catalog information—which are divided according to subject, such as fasteners, shock mounts, capacitors, transformers, etc. These spools can be inserted separately into a viewer, which projects each catalog page on the screen. Any desired information can be photographed directly within the viewer cabinet to give the user a print of the desired catalog page. Since the microfilmed information is reissued at frequent intervals, the user is kept abreast of all the changes in the products.

3.7 REFERENCE BOOKS AND MAGAZINES

No one can be expected to rely upon his memory alone for all the information needed for the solution of a particular problem. On occasion, the electronic draftsman will find it necessary to consult reference handbooks on such subjects as mathematics, drawing, electronics, shop practices, and materials. Some of these are available from manufacturers; others may be secured from the library or from a central source at the company or a government agency.

Because of the rapid changes occurring in the electronic field, it is advisable to keep informed of these changes through the drafting, electronic, and product-design magazines. Among them are:

Circuits Manufacturing	*Electronic Packaging and Production*
Electronic Component News	*Electronics*
Electronic Design	*Machine Design*
Electronic News	*Reprographics*

The publishers and addresses are listed in the Selected Bibliography.

3.8 ENGINEERING INFORMATION SOURCES

New information relating to the work of the draftsman can be obtained from the various technical sources that publish lists of standards and other information. These may be obtained from them at a nominal cost. Among these sources are:

Aeronautical Radio, Inc.

American Society of Mechanical Engineers (ASME)

American National Standards Institute (ANSI)

Electronic Industries Association (EIA)

Institute of Electrical and Electronics Engineers (IEEE)

National Electrical Manufacturers Association (NEMA)

National Machine Tool Builders' Association (NMTBA)

3.9 GOVERNMENT STANDARDS, SPECIFICATIONS, AND DRAWINGS

These may be a part of the reference material available in the drafting room for the engineers and draftsmen. A description of such standards, specifications, and drawings is given in Chapter 6. It is very important to become familiar with these specifications because they determine the design facets of electronic equipment, drawing details, and circuitry diagrams.

3.10 DRAWING REPRODUCTION

There are several processes used to reproduce copies from original drawings—by black-and-white prints, microfilm copies, or microfiche copies, for example. Prints can either be produced in the drafting room or purchased through outside sources.

Black-and-White Prints

These prints have replaced the older blueprint process and have several distinct advantages over the former. Among these are better readability and better adaptation for such drafting work as checking of mechanical or wiring harness drawings and circuit diagrams.

Black-and-white prints are made rapidly by a direct-process machine. The drawing is placed over a sensitized paper and carried on a continuously rotating belt over a mercury-vapor lamp that acts as the light source. The lines on the drawing shield an identical area on the sensitized paper, while the remainder of the drawing is exposed to full light. The exposed sensitized paper passes through ammonia fumes that develop the print. By using special types of sensitized paper, this same machine can make prints with red, blue, or green lines on white background. The colors can be used to indicate the department to which the prints are to be forwarded.

Microfilming Drawings

Microfilm is being used on a large scale to reproduce drawings. It not only saves space but preserves the drawings themselves, since a microfilm copy can be placed in a viewer and printer to obtain additional enlarged copies.

In this process the original drawing is photographed and recorded on 35-mm film. Duplicate rolls, made from this film, are then cut up and inserted into individual aperture cards. These cards may be viewed in a microfilm reader or used to make paper prints for distribution. The cards can be stored manually, or key punched for storage and retrieval by a computer.

Drawings to be microfilmed must be clean and have a minimum number of erasures. Lettering height must be at least $\frac{5}{32}$ inch, with lines spaced a minimum of $\frac{3}{32}$ inch. Lettering and drawing lines must be opaque and pencil lines uniform in width and density.

SUMMARY

A modern drafting room has to provide certain facilities such as drafting standards, which in turn provide the draftsman with information regarding the form of accepted abbreviations, drawing numbering, engineering and drawing practices, reference material, and other data relevant to his work. It also has facilities for him to reproduce the necessary prints of his drawings.

QUESTIONS

Sketches should be included with answers when necessary.

3.1 Describe some of the facilities found in an up-to-date drafting room.

3.2 What is included in the drawing title of military-type drawings?

3.3 List some of the subjects likely to be found in drafting-room standards.

3.4 Make a table listing the symbol, quantity, and value for the following prefixes: (a) giga; (b) mega; (c) centi; (d) micro; (e) kilo; (f) milli.

3.5 What electrical quantities do these units represent: (a) ampere; (b) henry; (c) watt; (d) farad; (e) volt; (f) ohm?

3.6 List the abbreviations for the following: (a) alternating current; (b) list of material; (c) normally closed; (d) single contact; (e) both sides; (f) base line; (g) white; (h) assembly; (i) between centers; (j) brown; (k) single pole; (l) switch; (m) phase; (n) negative; (o) integrated circuit; (p) equal; (r) immediate access storage; (s) bits per second; (t) electroplate; (u) wire list; (v) bubble memory; (y) microprocessor; (z) logic unit.

3.7 Describe the combination catalog viewer and printer. How is the information separated?

3.8 What two processes are generally used to make drawing prints?

3.9 What are the lettering requirements for drawings that are to be microfilmed?

EXERCISES

These exercises will be of help in practicing lettering, spelling, and abbreviating. Vellum should be used, and the necessary guide lines should be drawn lightly with a hard-grade finely pointed lead pencil.

3.1 Select six items from the following list and letter them in uppercase letters $\frac{5}{32}$ or $\frac{3}{16}$ inch high. Use one or two lines, and abbreviate where possible: (a) number 30 drill; (b) countersink 82 degrees, $\frac{1}{4}$-inch diameter; (c) tap number 6-32 Unified National Coarse; (d) tolerances plus or minus $\frac{1}{32}$ inch unless otherwise specified; (e) red dot identifies collector terminal; (f) leads four inches maximum length; (g) countersink other side $\frac{3}{8}$-inch diameter by $\frac{1}{32}$ inch deep; (h) adjust for minimum signal; (j) select master oscillator frequency; (k) three inches to center line; (l) maximum diameter four inches; (m) four hundred hertz; (n) six equally spaced holes on two-inch-diameter bolt circle.

3.2 Select seven items from the following list of general notes, and letter them in uppercase letters $\frac{3}{16}$ inch high. Lines should be approximately three inches long, and line spacing equal to letter height. Use abbreviations sparingly: (a) all resistors are one-half watt, ten percent tolerance, except where otherwise specified; (b) capacitor values below unity are microfarads, all others are picofarads; (c) all test voltages are measured to ground; (d) reference designations should be given for all components on the diagram; (e) inductance values are in microhenrys; (f) all surfaces are to be anodized except where shown masked; (g) all dimensions are in reference to the center line; (h) omit all components marked with an asterisk (∗) in the main assembly; (j) make all measurements with a thousand-ohms-per-volt voltmeter; (k) iridite all surfaces; (l) use only noncorrosive flux; (m) finish zinc chromate and black wrinkle on all exterior surfaces.

3.3 Select six items from the following list of drawing titles and letter in $\frac{3}{16}$-inch uppercase inclined characters, using two or three symmetrical lines as indicated: (a) support, mounting|modulator unit|1-kW transmitter; (b) power supply unit|AN/BRC-2X; (c) schematic diagram|intermediate-frequency amplifier|100-mHz receiver; (d) connection diagram|junction box|AN/ALQ-6; (e) chassis assembly|modulator unit; (f) block diagram|oscilloscope assembly; (g) harness assembly|power supply unit|AN/APS-3A; (h) main bearing|parabolic antenna|AN/GRP-7X; (j) chassis layout|transistorized power supply|television receiver.

3.4 Select six items from Exercise 3.3 and arrange them as nameplate data within a rectangle 2 by 3 or 3 by 4 inches. Use decreasing lettering sizes, starting with $\frac{3}{8}$ inch, with symmetrical lettering lines and an even border. Make vertical characters.

3.5 Select five items from Exercises 3.3 and 3.4 and repeat them in pencil using a Leroy lettering aid or a lettering template.

4

COMPONENTS IN ELECTRONICS

Originally developed for applications in the electrical industry, components have gradually evolved into a great variety of types, each tailored to meet a particular application or requirement. Many of these specialized components were developed for use in electronics.

Both electronic and mechanical components are governed by *specifications* that list the dimensions, construction details, materials, and requirements for the device. These specifications may be in two forms: government specifications for specific individual components, or *data sheets* for commercial applications of electronic equipment components used in the home or in industry.

Some components are referred to as *standard*. This designation applies to components that have been widely used and accepted over a period of time or that meet the design and performance requirements of a specification for a particular component. *Nonstandard* components may not comply with the specification for a specific component and its application, or there may merely be no specification governing the component.

4.1 PREFERRED VALUES SYSTEM

A preferred value system has been established to reduce the number of stock variations in component values. In this system each value differs from its predecessor by a definite amount.

There are three series in this numbering system (Table 4-1) with a plus or minus deviation or *tolerance* from the nominal values in each series. Thus, a component may have a greater or lesser value than the nominal and still be within the total tolerance. Standard tolerances are ±5, ±10, and ±20 percent for each series of values. The multiplying factor for each step is 1.10 for the ±5 percent tolerance series; 1.21 for the ±10 percent; and 1.46 for the ±20 percent.

Table 4-1 Preferred Value Numbering System Tolerance Series

±5%	±10%	±20%
10	10	10
11		
12	12	
13		
15	15	15
16		
18	18	
20		
22	22	22
24		
27	27	
30		
33	33	33
36		
39	39	
43		
47	47	47
51		
56	56	
62		
68	68	68
75		
82	82	
91		

Based upon EIA RS-385, *Preferred Values*, by permission of the publisher, Electronic Industries Association. Approved for DOD use.

This system of preferred values is standard for military and commercial electronic applications. It may be used for small capacitor and resistor values, with either the values listed in the table or decimal multiples of 10, 100, 1000, 10,000 and 100,000, e.g., 1100, 68,000, etc. For lower component cost and availability, ±10 percent is the most commonly used tolerance.

4.2 COLOR-CODE SYSTEM

A universally recognized color-code system has been established to identify the electrical values. It is used to identify through colored bands or dots the voltage rating, tolerance, and other special characteristics of such fixed value components as capacitors and resistors. It is also used for the numerical color identification of wiring leads and terminals in chassis wiring.

Definite colors have been assigned from zero through nine, starting with black and extending to lighter colors (Table 4-2). The color may be used to identify either a definite number or a multiplier. Gold and silver are used as multiplier or tolerance indicators and for other purposes.

Table 4-2   Standard Color Code

Color	Number	Decimal Multiplier
Black	0	1
Brown	1	10
Red	2	100
Orange	3	1000
Yellow	4	10,000
Green	5	100,000
Blue	6	1,000,000
Violet	7	10,000,000
Gray	8	100,000,000
White	9	—
Gold	—	.1
Silver	—	.01

Based upon RS-359, *Standard Colors for Color Identification and Coding,* by permission of the publisher, Electronic Industries Association. Approved for DOD use.

4.3   CAPACITORS

In its basic form, a capacitor consists of two metal plates that are separated by an insulating material, commonly known as a dielectric. The type of insulation or dielectric establishes the capacitor designation—air, vacuum, paper, mica, ceramic, plastic film, or electrolytic if the dielectric is a thin chemical film.

Capacitors may be *fixed, variable,* or *adjustable,* and they may be molded or assembled in a variety of containers. Terminal arrangements also vary. The symbols used for capacitors are given in Fig. 9-3.

Capacitance Value

This is based upon three factors: the area of the metallic plates, the spacing between them, and the dielectric value. This last value, known as the *dielectric constant* or *K* of the material, is the ratio between a capacitor with air as a dielectric, which is 1, and a capacitor with a dielectric material that is 1 or greater.

As the dielectric constant of the material increases, the area of the metallic plates required for a given capacitance decreases and thus reduces the physical size of the capacitor. Consequently, capacitors using very thin chemical films as dielectrics—such as electrolytics—have large capacity and yet are small in volume.

Capacitor Ratings

Capacitors are rated on the basis of capacitance value, tolerance, and voltage rating. Other rating factors are low loss and leakage, operating temperature range, size, construction, and terminal arrangement. Breakdown voltage and losses are established by the dielectric.

The capacitor size increases as capacitance values and voltage ratings go up because of the corresponding increase in the metal plate area and the dielectric thickness.

Paper-Type Capacitors

One of the most widely used capacitors is the paper-foil type. It consists of two metallic foil strips separated by two or more layers of very thin paper that have been impregnated with wax, mineral oil, or some other material. The number of paper layers determines the voltage breakdown value of the capacitor. This whole assembly is rolled into a cylinder. The whole length of foil extends beyond the paper to form a terminal projection. This construction eliminates the inductance effect that would reduce the effectiveness of the capacitor. Radial lead wires connect to each foil termination.

Another variation is the metallized paper or plastic capacitor. In this capacitor, aluminum foil is deposited on the insulation material and replaces the metallic plates.

The cylindrical construction of the capacitor makes it possible to enclose the assembly in a cardboard, plastic, or ceramic tube with radial leads projecting through the end seals. This *tubular* capacitor type is marked with a black band on one end signifying the lead connected to the outer foil (Fig. 4-1). This lead is normally connected to the grounded side of the circuit because the foil connected to it acts as a shield.

Paper dielectric, in many instances, has been supplanted by such plastic materials as: polycarbonate, polystyrene, Teflon, and Mylar®.*

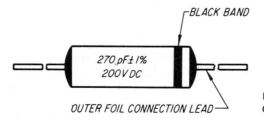

BLACK BAND

270 pF ± 1%
200 V DC

OUTER FOIL CONNECTION LEAD

FIG. 4-1. Paper and Plastic Capacitor Markings.

*duPont registered trademark.

Capacitance is indicated in picofarads (pF); 1 pF is one millionth of a microfarad or μF.

Small rectangular and tubular paper and plastic capacitors are wired into equipment using their leads as sole support.

Mica Capacitors

Color coding of these capacitors is done with dots, with a separate arrow or a similar indicator on the dots to denote the reading sequence. The dots are read from left to right on the top row and right to left on the bottom row. The first dot identifies the capacitor type as mica and indicates its use, black denoting military, white commercial usage. Additional colored dots are included on the opposite side of the capacitor, Fig. 4-2, to complete the mica capacitor identification. As with paper or plastic capacitors, mica capacitors are also designated in picofarads.

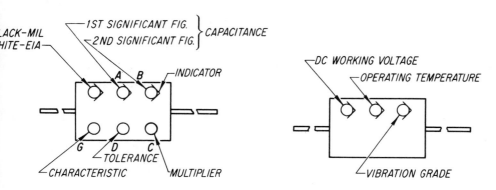

FIG. 4-2. Mica Capacitor Color Code Markings.

The color code in Table 4-3 applies to mica capacitors for commercial applications. The tolerance (D) and the characteristic letter (G) of relative temperature stability are also identified by color dots on the capacitor; i.e., a red dot in the *tolerance* space indicates a tolerance of ± 2 percent. According to the code in Table 4-3 a red dot in the *characteristic* space on the capacitor indicates a temperature coefficient of -200 to $+200$ parts per million per degree C, and a maximum drift from nominal capacitance of $\pm .5$ percent.

Ceramic-Type Capacitors

When ceramic is used as the dielectric material, silver coatings are deposited on the opposite sides of the ceramic in the disc capacitor, Fig. 4-3(a), or on the inner and outer surfaces of the tubing, Fig. 4-3(b).

The tolerance of ceramic capacitors, column D of Table 4-4, is based on

Table 4-3 Mica Capacitor Color Code for Commercial Use

Color	Capacitance (pF) Significant figures (A) (B)	Capacitance (pF) Multiplier (C)	Tolerance ±% (D)	Characteristic (G) Temp. Coeff. PPM per °C	Characteristic (G) Max. drift from nom. cap.
Black	0	—	20	±1000	5% + 1 pF
Brown	1	10	1	±500	3% + 1 pF
Red	2	100	2	±200	.5% + .5 pF
Orange	3	1000	3	±100	.3% + .1 pF
Yellow	4	—	—	±100 to −20	.1% + .1 pF
Green	5	—	5	—	—
Blue	6	—	—	—	—
Violet	7	—	—	—	—
Gray	8	—	—	—	—
White	9	—	—	—	—
Gold	—	.1	—	—	—
Silver	—	—	10	—	—

Based upon EIA RS-153B, *Molded and Dipped Mica Capacitors*, by permission of the publisher, Electronic Industries Association.

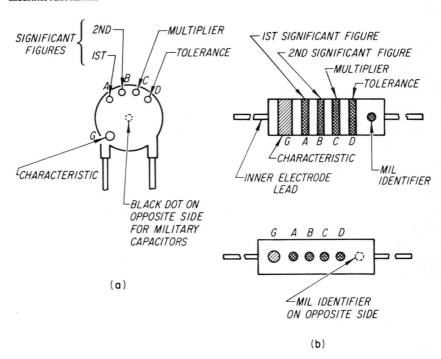

FIG. 4-3. Ceramic Capacitors and Code Markings. (a) Disc. (b) Tubular. (See Table 4-4.)

Table 4-4   Ceramic Capacitor Color Code

Color	Capacitance (pF)		Tolerances (D)		Characteristic Temp. Coeff. PPM per °C (G)
	Significant figures (A) (B)	Multiplier (C)	In per cent if greater than 10 pF	In pF if less than 10 pF	
Black	0	1	—	2.00	0
Brown	1	10	1	—	−30
Red	2	100	2	0.25	−80
Orange	3	1000	—	—	−150
Yellow	4	—	—	—	−220
Green	5	—	5	0.50	−330
Blue	6	—	—	—	−470
Violet	7	—	—	—	−750
Gray	8	0.01	—	—	—
White	9	0.1	10	1.00	—
Gold	—	—	—	—	+100

Based upon EIA RS-198B, *Ceramic Dielectric Capacitors*, by permission of the publisher, Electronic Industries Association.

percentage, when capacitance is greater than 10 pF, or on actual picofarad variation, when the capacitance is below 10 pF.

By proper selection of ceramic dielectric material, ceramic capacitors may be made with positive, zero, or negative coefficients; i.e., their capacity increases, remains constant, or decreases as the temperature increases. The temperature coefficient is given in parts per million per degree C (see the Characteristic column (G) in Table 4-4).

Miniature Capacitors

Transistors, with their low voltage requirements, brought about the development of miniature capacitors, which are high in capacitance value and small in size.

Microminiature ceramic capacitors, used for coupling and bypass applications, are available in 1000 to 100,000 pF capacity ranges, 50 WVDC, and tolerances of ±20 percent. They are made in such sizes as .2 by .2 by .1 inch thick and .3 by .3 by .1 inch thick. Capacitance and tolerance values are marked on the case.

Other capacitor sizes in low-capacity values, 100 WVDC, are available in sizes as small as .1 by .1 by .1 inch.

Miniature tantalum capacitors, a development in solid-state technology, use a semiconductor electrolyte and contain only solid materials. They are of the polarized type and are hermetically sealed in metallic cases.

They have the big advantage of combining high capacity with a much smaller size than other conventional foil capacitors. They have a temperature range of $-55°$ to $+85°C$ and a working voltage range of 3 to 35 volts dc, which makes them suitable for most semiconductor applications.

Passive Microdiscrete Capacitors

Chip capacitors are utilized in hybrid microelectronic packaging when thick or thin film deposition techniques are insufficient to provide the necessary capacitance. These capacitors come in microminiature sizes down to .050 inch square by .025 inch thick; see Fig. 4-4. They consist of ceramic dielectric, such as barium titanate, and metallic paste for the conductor plates.

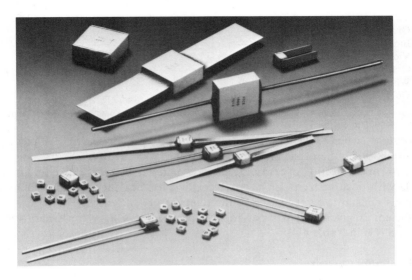

FIG. 4-4. Chip Capacitors. (Courtesy American Technical Ceramics.)

4.4 CONNECTION DEVICES

Male and female connectors are included among the various connection devices along with terminal blocks.

Connector Requirements

With new designs constantly being added, the variety of connectors almost seems endless. These components meet varied requirements: space, reliability, current-handling capacity, and so forth. They must be designed for power or RF circuits and printed circuit boards and designed so that connections may be made to them by soldering, welding, or wirewrapping.

MS Connectors

Formerly called the AN connector, this is one of the best known connector types and one of the earliest to be specified for military electronic equipment (Military Specification MIL-C-5015). Some typical connectors of this type are shown in Fig. 4-5.

The part numbers of connector shells listed in MIL-C-5015 and shown in the figure are:

MS3100   panel or wall mounting receptacle (a)

MS3101   cable receptacle (b)

MS3102   box mounting receptacle (c)

MS3106   straight plug (d)

In addition, the following suffix letters are used to designate the construction type:

A   solid shell

B   split back shell

C   pressurized

An example of the complete MS system designation code is as follows:

MS	3100	A	16	11P	
MS					Military Specification
	3100				panel mounting receptacle
		A			solid shell construction
			16		connector diameter in sixteenths of an inch
				11	insert contact pattern designation, not number of contacts
				P	plug or male contact

Female contacts are indicated by an "*S*," the abbreviation for socket.

Connectors of the "MS" type are available in the following shell sizes, which are in sixteenths of an inch: 8, 10, 12, 14, 16, 18, 20, 22, 24, 28, 32, 36, 40, 44, and 48. Thus, size 20 is $\frac{20}{16}$ or a shell diameter of $1\frac{1}{4}$ inch.

MS connectors are used in electronic equipment built for the Department of Defense and for commercial airlines. A complete MS connector consists of two units, a male or plug assembly and a receptacle assembly. Each unit has contacts to which connections are made.

There are six types of contacts, ranging in size from #20(B & S Gage) or .032 inch in diameter to #0 or .325 inch in diameter. Contact insert combinations are available from 1 to 52 in various contact sizes. The insert number indicates the details of the insert, not the number of contacts; see Fig. 4-5(e) and (f).

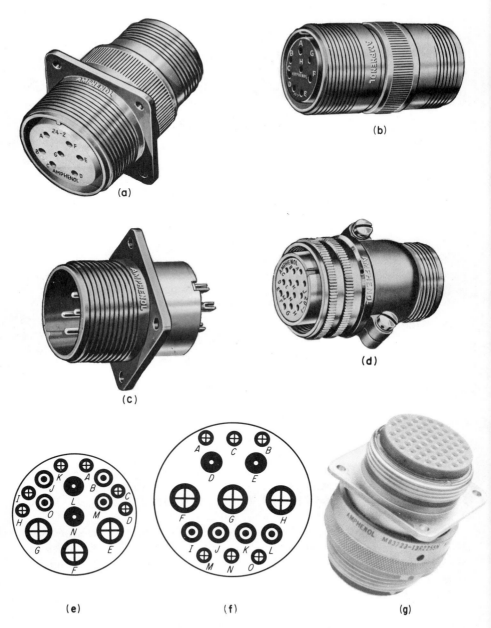

FIG. 4-5. Military Connectors: (a) through (d) Connector Shells. (e) and (f) Insert Contact Arrangements. (g) Miniature Circular Type. [Courtesy of AMPHENOL NORTH AMERICA DIVISION, Bunker Ramo Corporation.]

Another military type connector is shown in Fig. 4-5(g). This is one of the miniature circular connectors designed to meet MIL-C-26482.

Connector Varieties

In addition to the MS connectors, there are many other connector varieties that have been designed for specific tasks. These include:

1. Rack and panel connectors, which are rectangular in shape and come in a great variety of contact arrangements and sizes, Fig. 4-6(a).
2. Miniature connectors for airborne applications and computer programming, Fig. 4-6(b).
3. A connector for flexible flat cable, Fig. 4-6(c), which can also be used to provide transition from flat cable to round wire.
4. A printed-circuit-board edge card connector, Fig. 4-6(d). This connector allows connecting conventional wiring directly to the PC board.
5. Compact, rectangular printed-circuit connectors, Fig. 4-6(e). These connectors are used to join printed-circuit wiring that terminates in a number of plated contacts on the printed-circuit board. Correct printed-circuit board polarization is achieved by a key on the connector or some other means.

Terminal Blocks

These are used as terminations for equipment wiring and for other purposes. They vary greatly in size, shape, and termination arrangements, Fig. 4-7(a), although over a period of time, some have become standardized. Individual terminals are identified by suitable labels.

Being modular in construction, the number of individual terminals in the assembly can be varied to suit the requirements.

Bus Bars

In Fig. 4-7(b) is shown one form of a bus bar in use for power and ground distribution on printed-circuit cards. It is made up of a number of flat conductors insulated from each other with a dielectric material. The advantages bus bars offer are: noise reduction in high speed electronic systems; low inductance with high capacitance; and inclusion of ground shields to reduce stray electrical fields. A variety of terminations are available.

4.5 ELECTRON TUBES

Electron-Tube Elements

The various tube elements are identified and described in Chapter 9, along with their individual and combined schematic symbols.

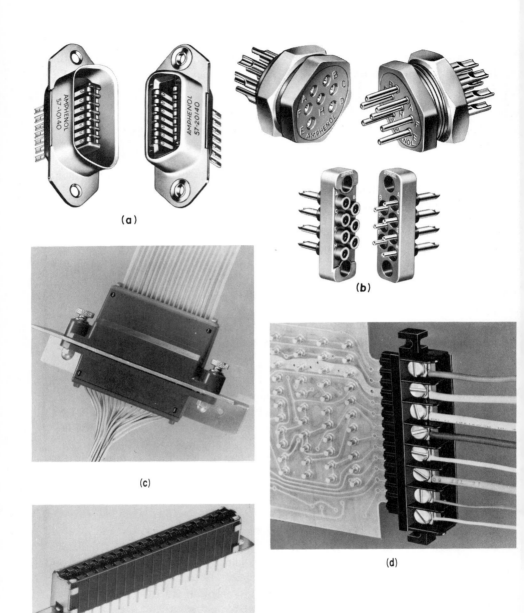

FIG. 4-6. Connector Varieties. (a) Rack and Panel Connectors. (b) Miniature. (c) Flexible Flat Cable Connector. (d) A Printed Circuit Board Edge Connector. (e) Printed Circuit Board Connector. [(a),(b), and (e) Courtesy of AMPHENOL NORTH AMERICA DIVISION, Bunker Ramo Corporation.]

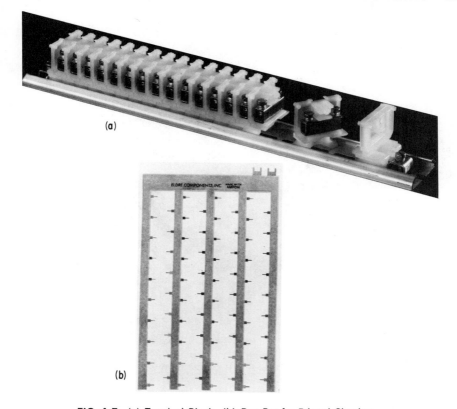

(a)

(b)

FIG. 4-7. (a) Terminal Block. (b) Bus Bar for Printed Circuitry.

Cathode-Ray Tubes

These tubes (CRT) give visual indication of a signal waveform in such electronic equipment as oscilloscopes. They are also used in radar displays, black-and-white and color television sets, and other visual displays.

One such tube (Fig. 4-8) contains an electron gun assembly, a means for deflection, and an aluminized tricolor phosphor dot screen on the inner surface of the faceplate. Three color beams are generated—blue, green, and red, which produce the multicolor images on the face of the tube.

4.6 INDICATING DEVICES

Among these may be listed such components as meters, indicator lights, and display devices.

Meters

They measure voltage—from millivolts to many thousands of volts—and current—from microamperes to many amperes. They may also measure frequency, in hertz.

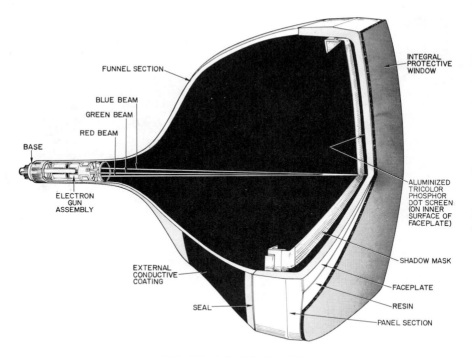

FIG. 4-8. Color Television Tube.

A newer approach for making accurate voltage, current, and other measurements is the *digital panel meter*. This is a solid-state device with 2, 3, 4, 5 or more digit display, Fig. 4-9(a), that shows the exact voltage, current, or other reading in decimal form, with far greater accuracy than the movable pointer meter.

Shown in the same figure is the printed-circuit board on which are mounted the various components that form the meter circuit.

Indicator Lights

These are also available in many styles and sizes, Fig. 4-9(b). Because they indicate circuit conditions, they are generally mounted next to the switches that control these circuits.

Lamps for indicator lights are generally tubular, designated "T," and are identified by such designations as "T1," the number indicating the diameter in eighths of an inch. To meet military requirements, the lamp must be replaceable from the front.

Display Devices

These are composed of several light-emitting diodes, with the light from each LED stretched optically to form individual segments and a decimal point. They are available in various figure sizes.

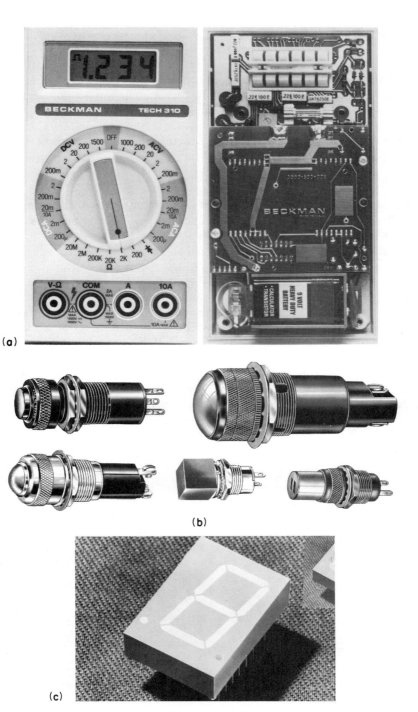

FIG. 4-9. Indicating Devices. (a) Digital Meter and Its Internal Printed Circuit Board. (b) Indicator Lights. (c) Seven-Segment LED Display.

Figure 4-9(c) shows a large seven-segment display, with an .8-inch figure (20.32 mm). It provides good visibility in bright light. It is suitable for mounting on printed-circuit boards or in standard IC sockets. The contact pins are spaced in two pin rows on .100-inch centers. Standard display colors are yellow, red, and green.

Their common use is in calculators, counters, and test equipment.

4.7 INDUCTORS

Popularly known as coils, inductors are normally associated with tuning circuits. The basic inductor is a multiturn coil that may or may not have a magnetic core to increase its inductance or to act as a tuning element.

The coil may be a single layer of wire supported on a simple coil form of polystyrene, ceramic, or plastic material. The conductor may be insulated or bare wire, flat strip, or silver-plated tubing, Fig. 4-10(a), with turns closely wound or spaced apart. It may also be wound as several "pies" on a common form, Fig. 4-10(b), for use as a radio-frequency choke (RFC). Another variation is a multiple-layer coil that is wound on a phenolic tubing form, or on a powdered iron core which acts as a magnetic core, Fig. 4-10(c); this is known as a choke coil. Coil inductance may be varied by providing taps on the coil or it may be adjusted by an iron core or ferrite "slug" that can be moved in and out of the coil winding, Fig. 4-10(d). The ferrite core or "slug" consists of metallic oxides that are compressed into various forms and fired.

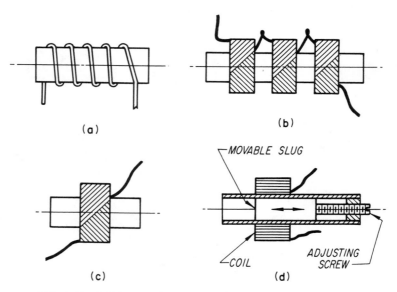

FIG. 4-10. Coil Types. (a) Antenna Coil. (b) Multi-pie RF Choke. (c) Choke Coil. (d) Adjustable Iron Slug Coil.

Generally, the following information is furnished for any given coil: its inductance, measured in microhenrys (μH) for the smaller coils with few turns of wire, or in millihenrys (mH) for the multiple-layer types; resistance, in ohms, and current-carrying capacity in milliamperes or amperes.

Inductance values are: 1 henry = 1000 millihenrys (mH) = 1,000,000 microhenrys (μH).

Iron-core coils, or choke coils, are used in power supply circuits to filter out the *ac* component or to "smooth" out the rectified current. The core consists of a number of thin "E-" and "I"-shaped laminations. These laminations are interleaved if only a nominal dc current is to pass through the choke coil; or the "E" and "I" laminations may be separated with a small air gap between them. The coil itself is multilayer, with a layer of insulating material between each winding layer. The core laminations are held together by a steel channel-shaped bracket that has its own mounting feet, Fig. 4-11.

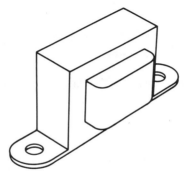

FIG. 4-11. Iron-Core Choke Coil.

The following information is generally given for iron-core choke coils: inductance in henrys at specified dc current in milliamperes, coil resistance, the dc current rating, and the breakdown or test voltage.

4.8 INTERCONNECTION COMPONENTS

In addition to the various types of wire used for interconnecting components at power and low frequencies, special means of interconnection are necessary for coaxial and waveguide work. Among these are the coaxial cables and wave-guides.

Coaxial Cables

Also known as RF transmission lines, coaxial cables consist of a single conductor, solid or stranded, which is enclosed in polyethylene, Teflon, or some other relatively thick insulation, and an outer metallic braid that serves as the ground return. The relative diameter of the inner conductor and its insulation determine the cable impedance. An insulating jacket put over the braid insulates it over the entire length except where the braid is connected to ground. Some typical cables are shown in Fig. 4-12.

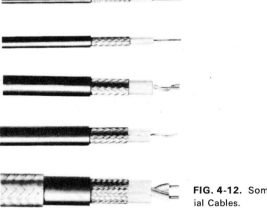

FIG. 4-12. Some Typical Coaxial Cables.

Military specification MIL-D-17E classifies the various coaxial cables and designates military numbers RG- /U for each cable.

Waveguides

A special conductor, in the form of a round or rectangular tube, is used for interconnections at frequencies above 1000 megahertz, where coaxial cables are no longer suitable. This tube, known as a waveguide, has no center conductor and acts as a boundary for electromagnetic waves. These waves are propagated from one end of the waveguide to the other by reflection from the inner surfaces of the tube.

Waveguides are made of brass, copper, silver, aluminum, and other metals. Sizes vary depending upon the frequency range to be transmitted.

Waveguide Accessories

Waveguide "plumbing," as it is popularly known, requires certain fittings to interconnect sections of waveguide to antennas, other waveguides, and so forth. Among these are the coupling flanges that correct misalignment of inner walls of the waveguide sections that are coupled together. These flanges are made in both plain and choke types. The latter has a gap deliberately added to the flange.

Rectangular waveguides have their narrow and wide sides designated as E and H planes, respectively. The various fittings, such as bends (Fig. 4-13) are classified in the same manner.

FIG. 4-13. Waveguide Bends.

4.9 RELAYS

These devices are used in electronic circuits to switch circuits electrically, Fig. 4-14. The switching operation is controlled by a coil or coils, depending upon the relay type in the conventional relays.

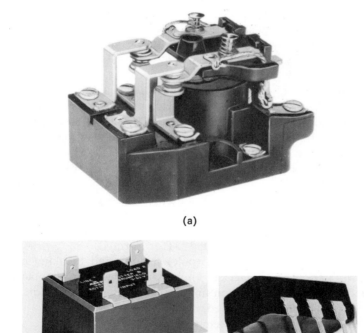

FIG. 4-14. Relays. (a) Conventional. (b) Solid State. (Both Courtesy of Potter & Brumfield, Div. AMF, Inc.) (c) Opto Isolator. (Courtesy Litronix, Inc.)

Relay Categories

Relays are frequently separated into types according to their application and size: subminiature, miniature, aircraft, printed-circuit, sealed, power, latching, and others. They are available in open, enclosed, or hermetically sealed assemblies. The latter two types are generally specified for military use.

Relays are designated as either ac or dc, and are further subdivided according to their power operation, the number of poles, the contact number and size, the coil operating voltage, and other factors.

Relay Contact Designations

To simplify the relay contact nomenclature, the National Association of Relay Manufacturers and the Department of Defense have standardized the contact arrangement by letter designations.

Each of the movable contacts on a relay is considered a pole of the relay. The relay abbreviations in common use are:

B	Break	M	Make	P	Pole
C	Closed	N	Normally	S	Single
D	Double	O	Open	T	Throw

The combination of a movable contact (pole) and a stationary contact which engage when the relay coil is energized, is known as a front, make, or normally open contact. It is abbreviated NO or Form A. The combination of a movable contact (pole) and a stationary contact which engage when the relay coil is deenergized, is known as back, break, or normally closed contacts. This is abbreviated NC or Form B. NO and NC contacts are called single-throw, which is abbreviated ST.

The combination of a movable contact (pole) and two stationary contacts, one of which is engaged by the movable contact when the relay coil is energized and the other when the coil is deenergized, is known as transfer or double-throw. This combination is abbreviated DT or Form C.

The case of a movable contact (pole) making and breaking connection between two stationary contacts is known as double-break contacts, and is abbreviated DB. This combination may be called double-make contacts when normally open contacts are involved.

Examples of relay stack-up assemblies are shown in Fig. 4-15.

Note the arrows indicating direction of motion of contact arms.

Relay contact notations are specified in the following order: (1) poles, (2) throws, (3) normal position with coil deenergized, and (4) DB or double-break or double-make contacts. For example, SPST NO DB is abbreviation for single-pole, single-throw, normally open, double-break contacts.

Relays with several sets of different contact arrangements have the contact

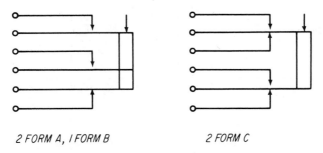

2 FORM A, I FORM B   2 FORM C

FIG. 4-15. Typical Relay Stack-up Assemblies.

forms listed in the alphabetical order of their letter symbols, as shown in Fig. 4-15. As an example, 2A1B refers to DPST NO contacts and one set of SPST NC contacts.

Figure 4-16 shows the various relay contact assemblies and their letter designations.

Solid-State Relays

These relays, Fig. 4-14(b), do not have any moving parts and have nothing to wear out. They use semiconductor components and provide reliable operation under severe shock and vibration conditions.

They are available in a SPST configuration and are capable of handling a 20-ampere load. The input side of the relay circuit can be controlled from a logic circuit to turn on a light-emitting diode (LED). The light output of the LED is sensed by a photo device which will switch on at a specific light level. This switching device action "turns on" the solid-state load switching device.

To reduce the high inrush current or severe RFI, an additional circuitry is included in the relay to ensure that the load controlled by it is turned on only near the "zero voltage" point in the ac cycle.

A separate device, called the opto isolator, Fig. 4-14(c), is available to convert the conventional solid-state relay to the relay circuitry, where there is complete isolation of the controlled output switching from the input side of the relay.

4.10 RESISTORS

The many types of resistors in use are classified by construction—fixed, continuously variable, or adjustable. Their basic designation is defined by the resistor element, which may be carbon composition, film deposit, or wirewound. Figure 9-3 shows resistor symbol variations.

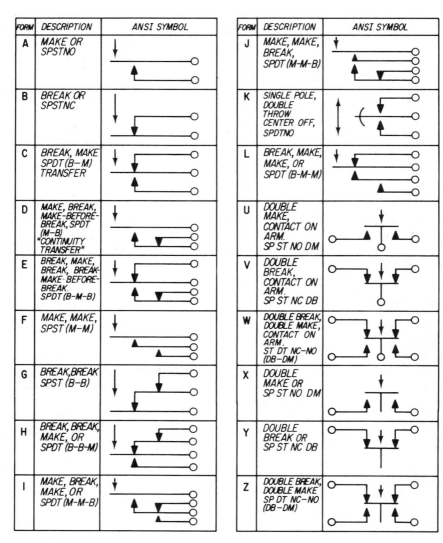

FIG. 4-16. Letters and Their Relay Contact Designations. (National Association of Relay Manufacturers.)

Resistor Characteristics

All resistors are classified on the basis of their resistance value, resistance tolerance, and power rating in watts. Additional factors are physical dimensions, terminal arrangements, mounting provisions, and effects of frequency and temperature. All of these factors must be considered in circuit and equipment design in choosing the appropriate resistor.

A major consideration in resistor application is the operating temperature, which is determined by the ambient temperature, location in equipment, available ventilation, and nearness to other components.

Resistors are generally *derated*, or operated at 10 to 50 percent of their rating, to ensure reliability and low operating temperature.

Fixed Composition Resistors

These resistors are available in $\frac{1}{10}$-, $\frac{1}{4}$-, $\frac{1}{2}$-, 1-, and 2-watt ratings. A powdered carbon mixture forms the resistance element, which is sealed in a plastic case, Fig. 4-17. The ends of the resistance element terminate in tinned copper axial leads, and colored bands on the periphery of the case indicate the value and tolerance of the resistor. This color code is given in Table 4-5.

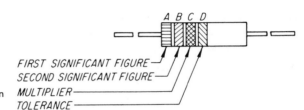

FIRST SIGNIFICANT FIGURE
SECOND SIGNIFICANT FIGURE
FIG. 4-17. Fixed Composition   MULTIPLIER
Resistor Color Coding.   TOLERANCE

Table 4-5   Color Code—Composition Resistors

Color	Significant figures (A) (B)	Multiplier (C)	Tolerance ($\pm$ %) (D)
Black	0	—	—
Brown	1	10	—
Red	2	100	—
Orange	3	1000	—
Yellow	4	10,000	—
Green	5	100,000	—
Blue	6	1,000,000	—
Violet	7	10,000,000	—
Gray	8	100,000,000	—
White	9	1,000,000,000	—
Gold	—	.1	5
Silver	—	.01	10
No color	—	—	20

Based upon RS-172B, *Fixed Composition Resistors*, by permission of the publisher, Electronic Industries Association.

Fixed insulated composition resistors may be obtained in a range of preferred values (see Table 4-1), and in tolerances of ± 5, ± 10, and ± 20 percent. The resistance range is from 10 ohms to 22 megohms (1 megohm = 1,000,000 ohms) and the tolerance in general use is ± 10 percent.

Resistance color code marking is defined in Fig. 4-17; examples are given in Fig. 4-18. In Fig. 4-18(a), reading from left to right, the first and second sig-

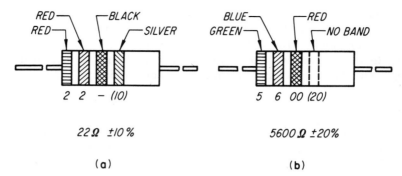

FIG. 4-18. Fixed Composition Resistors: (a) and (b) Examples of Coding.

nificant bands (Table 4-5) are colored red (2) and red (2). The multiplier band, column (C) in Table 4-5, is black or no zero, which makes the resistor value 22 ohms. The tolerance band, the fourth column, (D), is silver or ± 10 percent. The color sequence in Fig. 4-18(b) is green (5), blue (6), red (2 zeros), or 5600 ohms, with no tolerance band, or ± 20 percent.

Fixed Wire-Wound Resistors

These are made by winding one or more layers of resistance wire on a hollow ceramic tube, Fig. 4-19(a). Ribbon, such as nichrome, is used for heavier

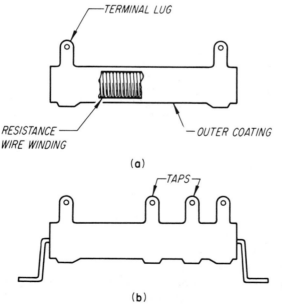

FIG. 4-19. Wirewound Resistors. (a) Typical Construction. (b) Tapped Resistor.

currents. End terminal lugs or wires are used for connections, and the resistance wire is covered with cement or vitreous enamel for protection.

Wire-wound resistors are available in a range of .1 to 200,000 ohms, with wattage ratings of 2 to 200 watts and tolerances of ± 5 and ± 10 percent. The 10 percent tolerance applies to resistors that have a value of less than 1 ohm.

The smaller wattage resistors are frequently supplied with wire axial or radial leads, which also serve for mounting purposes.

Resistors are marked along the longitudinal axis with either the trademark or the name of the manufacturer, or both; the wattage designation; and the code marking for resistor value (for military resistors) or actual value (for commercial resistors).

Tapped resistors have one or more permanently connected terminals or leads that are brought out at desired resistance values, Fig. 4-19(b). Adjustable tap resistors are made by exposing a portion of the outer coating and positioning a movable, tight-fitting band over the exposed wire to make contact at any desired point.

Precision resistors are wire-wound resistors that are available with tolerances of ± 1, $\pm .5$, $\pm .1$, $\pm .05$, and $\pm .01$ percent. They range in values from 100 ohms to .15 megohm in the miniature size and up to several megohms in the larger sizes. They are protected by encapsulation or a varnish treatment.

Military wire-wound resistors (power type) are identified as in the following example from MIL-R-26:

RW20	N	100
Style	*Characteristic*	*Resistance*

This description identifies the following:

- RW   resistor, wire-wound
- 20   rating of resistor in watts
- N   operating characteristic of the resistor explained in the specification
- 100   resistance symbol for value in ohms

The following are examples of resistance symbols for various resistance values:

$$R10 = 0.10 \text{ ohm}$$
$$1R0 = 1.0 \text{ ohm}$$
$$100 = 10 \text{ ohms}$$
$$101 = 100 \text{ ohms}$$
$$102 = 1000 \text{ ohms}$$

Wire-wound resistors are generally supported by a spring-type bracket, Fig. 4-19(b), or in the smaller sizes, by their own leads. Sufficient spacing must be allowed between the mounting surface and the resistor to secure adequate ventilation and to avoid overheating the resistor.

Passive Microdiscrete Resistors

Chip resistors are used in hybrid microelectronic packaging when it is necessary to obtain higher values in a thin-film hybrid circuit.

The film resistors are deposited on an alumina chip and come in size down to .050 inch square and .020 inch thick. They have metallized edges or beam lead terminations for connection purposes.

Variable Resistors

These consist of a circular carbon or resistive wire element mounted in a case. Contact is made to any point of this element by means of a rotating slider, Fig. 4-20.

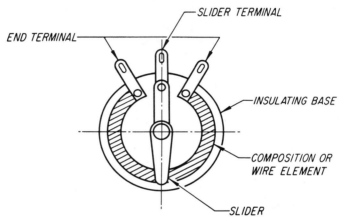

FIG. 4-20. Typical Variable Control Element.

Variable resistors are identified as *controls* when they are used on electronic equipment to control tone, volume, range, etc. They are identified as *rheostats* when they are connected in series with the controlled circuit that makes connections to one end terminal and the slider terminal. They are *potentiometers* or "pots" when they are connected across the voltage source and a variable voltage from zero to full is obtained between one end terminal and the slider.

Variable Carbon Resistors

These resistors use a carbon composition element that is formed on an insulating base and enclosed within a metallic or plastic housing. A threaded bushing is used to mount the variable resistor on the panel.

Variable composition controls are available in .1- to 2-watt power ratings and in various overall sizes, shaft diameters, and mounting bushings. Their resistance values range from 100 ohms to 5 megohms, and their tolerances are ±10 or ±20 percent. The name of the manufacturer, the resistance, and type designation are identified on the case.

Wire-Wound Variable Resistors

These are available in 1-, 2-, 3-, and 5-watt ratings, in resistance ranges of 1 to 100,000 ohms.

Subminiature potentiometers are a special style having the slider moving axially along the resistance element by turning an adjusting screw 25 or more turns. The outer housing is rectangular in shape.

Resistance Tapers

Plotting the percentage of rotation against percentage of resistance, as measured from the slider to one of the end terminals, gives a graphical relationship known as *taper*. When the resistance change is uniform in relation to rotation, such a taper is known as a *linear taper*, Fig. 4-21.

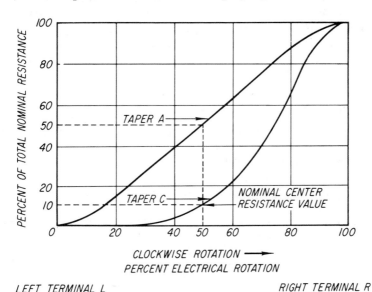

FIG. 4-21. Potentiometer Tapers, Clockwise Rotation.

The taper direction is given when the potentiometer is viewed at the shaft end. Thus, taper *A* in Fig. 4-21 has 50 percent of resistance as measured from the left to the center terminal at 50 percent of rotation. Taper *C* has only 10 percent resistance change at 50 percent of clockwise rotation.

4.11 SEMICONDUCTOR DEVICES

It is necessary to know a variety of information about such semiconductor devices as diodes, transistors, and silicon-controlled rectifiers when designing electronic equipment, laying out a chassis or a printed-circuit board, or drawing schematic or connection diagrams. This information includes the physical size, the device terminations, the orientation, and other details. Although the text and the Appendix of this book contain much of this information, the student is also advised to consult the technical manuals on semiconductor devices that are issued by the various manufacturers.

Semiconductor Manuals

These publications present a brief description of each diode, rectifier, integrated circuit, or transistor, indicating physical data, such as size, terminal arrangements, and keying in reference to terminals; base details; and other information, such as characteristic curves (see Fig. 17-18). As a rule, separate manuals deal with the various semiconductor devices, such as integrated circuits, silicon-controlled rectifiers, and transistors. These manuals may also include basic theory, construction details, typical circuits, and technical design data. Diode and transistor designations, enclosures, and base details are of particular interest to the student. The symbols for diodes and transistors are given in Figs. 9-15 and 9-16.

Semiconductor Materials

Silicon and gallium arsenide are the two basic materials used in manufacturing semiconductor devices. Silicon is used for higher power devices because it is capable of withstanding much higher operating temperatures.

Semiconductor Device Outlines

Several typical diode and transistor outlines are shown in Fig. 4-22 and 4-23. Many of the semiconductor sizes have been standardized by the Joint Electronic Devices Engineering Council, or JEDEC. Most outlines are identified by "DO" (diode outline), see Fig. 4-22(a), or "TO" (transistor outline), followed by a dash and a number, e.g., TO-5, TO-18, etc. This identfication is used for the smaller diodes and transistors. The larger power transistors and rectifiers do not follow any set standard.

Diode and Transistor Identification

Diodes and rectifiers are identified by a prefix "1N" followed by a number, while transistors are identified by a prefix "2N" followed by a number. Other transistor devices, such as insulated-gate field effect transistors (IGFET) of the

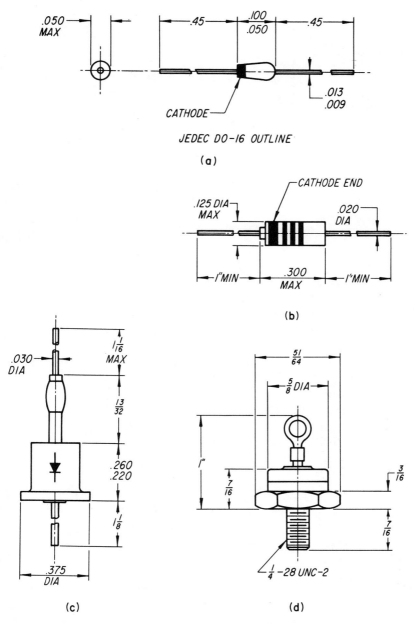

.050 MAX

.45 .100 .050 .45

CATHODE

.013 .009

JEDEC DO-16 OUTLINE

(a)

CATHODE END

.125 DIA MAX

.020 DIA

1" MIN .300 MAX 1" MIN

(b)

.030 DIA

$1\frac{1}{16}$ MAX

$\frac{13}{32}$

.260 .220

$1\frac{1}{8}$

.375 DIA

(c)

$\frac{51}{64}$

$\frac{5}{8}$ DIA

1"

$\frac{7}{16}$

$\frac{3}{16}$

$\frac{7}{16}$

$\frac{1}{4}-28$ UNC-2

(d)

FIG. 4-22. Typical Diodes and Rectifiers. (a) Microminiature Diode. (b) Subminiature Diode. (c) and (d) Rectifiers.

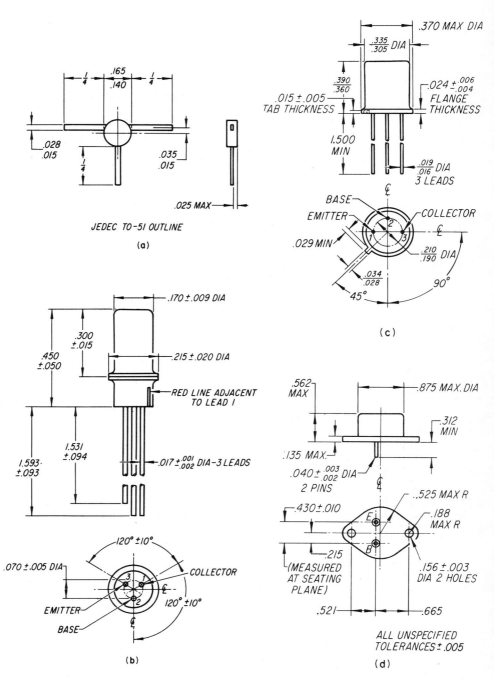

FIG. 4-23. Representative Transistor Outlines.

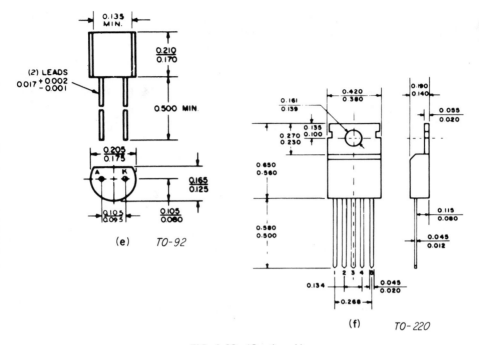

FIG. 4-23. (Continued.)

MOS type and silicon-controlled switches (SCS), are identified by a prefix "3N" followed by a number.

Diode and Transistor Connections

Diode and transistor connections are quite simple. The individual element designations may appear on their outline drawings or they may be identified under each transistor in the technical manual.

Diode and rectifier connections are generally identified by a graphic symbol placed directly on the component itself, Fig. 4-22(c).

Diode and Rectifier Design and Function

A diode, made of silicon or gallium arsenide, allows the applied current to flow easily in the forward direction but offers high resistance to the flow of current (reverse current) in the opposite direction.

A rectifier is similar to the diode in that it allows a flow of current in one direction only and is utilized for converting alternating to direct current. It is more of a power-handling device than a diode, which generally handles minute currents. Both have the same schematic symbol (see Fig. 9-15).

Typical diodes and rectifiers are shown in Fig. 4-22. A glass-encased micro-miniature diode is shown in Fig. 4-22(a), with the cathode end identified as shown. A subminiature silicon diode is shown in Fig. 4-22(b), with the wide black band used to identify the cathode end and three narrow bands next to it to identify the type number. The manufacturer's name or code appears near the anode end. The miniature-size, half-wave silicon rectifier shown in Fig. 4-22(c) has the cathode or positive connection identified by a bar.

Diodes and rectifiers of the size shown in Fig. 4-22(a) through (c) are generally supported by their leads, while the power rectifier shown in Fig. 4-22(d) depends upon its stud for mounting. It may be necessary to insulate the stud from its mounting surface. Heat sinks or some other method to adequately dissipate the heat are required in the larger rectifiers to keep the internal temperature within the specified limits.

Diode and Rectifier Types

There are many varieties of these semiconductors: tunnel diodes, zener diodes, silicon-controlled rectifiers, to name but a few. Their individual characteristics are indicated by variations in the basic graphic symbols which are illustrated in Chapter 9.

The *signal diode* is of the point-contact type. The point contact uses a fine wire against a crystal, such as silicon. This diode is suitable for operation to many thousands of megahertz and is used in microwave work. Its symbol is the same as that for the conventional diode.

The *tunnel diode*, a special diode type, reverses its conduction resistance with small impressed voltages but returns to normal operation with a further increase in voltage. This unusual characteristic, in combination with high switching speed, makes it a useful device in switching, amplifying, and oscillator circuits, even at microwave frequencies. These diodes have low power consumption and are relatively unaffected by temperature changes.

Silicon diodes or *rectifiers* are among the most widely used of semiconductor diodes. They are available in a variety of current ratings, from milliamperes to hundreds of amperes, and in voltage ratings up to several hundred volts. Heat sinks may be necessary to keep down the temperature of these diodes because of their small size. They may be operated in series for higher voltage inputs or in parallel for higher current output. Suitable resistors or capacitors are required to help distribute the voltage or current between the individual diodes.

The *zener diode*, or voltage reference diode, is a variation of the silicon rectifier diode. The reverse current of this diode remains small and then suddenly increases rapidly when the breakdown or "zener" voltage is reached. This voltage may be from one volt to several hundred depending upon the diode material and construction. This current change phenomenon is used to advantage by having the diode act as a bypass for current when the voltage input across the diode, which is connected across a voltage source, rises above a predetermined point.

The *silicon-controlled rectifier* is another special silicon diode, known generally as SCR. It can be triggered by applying a pulse to a "gate" that controls conduction in the forward direction.

The *varactor diode* is a microwave frequency junction semiconductor device. It is small in size and its junction capacity depends upon the applied voltage. Its principal use is in parametric amplifiers, subharmonic generators, and television tuners.

Diode and rectifier specifications include such information as type number, semiconductor material, power rating, input voltage and current, and allowable temperature rise. To assist in assembly and maintenance, a note should be included on assembly drawings and connection diagrams indicating the cathode connection or polarity markings of diodes and rectifiers. They should be indicated by graphic symbols on printed-circuit boards or other surfaces on which the diodes are mounted.

Transistor Types

Such crystalline elements as silicon change their electrical characteristics by controlled addition of certain impurities. The type of impurity determines whether the resultant semiconductor material is of the *P*-type (positive charges) or *N*-type (negative charges). A *PN* junction acts as a rectifier or a diode. When a second junction is added to a semiconductor diode, the resultant device is known as a transistor. A simple type may be considered as a sandwich composed of *PNP* or *NPN* sections.

The *unijunction transistor* uses an *N* silicon base with an electrical connection at each end and a *P* silicon emitter electrode between the end connections. It exhibits a negative resistance characteristic over part of its operational range and is used in oscillators, multivibrators, and tuning and switching circuits.

The *double-base junction transistor* or *tetrode* has two separate base connections on opposite sides of the center of the transistor. Its main advantage is its higher frequency response.

The *silicon-controlled switch* is a four-layer, silicon *PNPN* switch with a third junction. It is designed for digital computer and control applications, such as binary counters, shift registers, and pulse generators. It has two gates, one adjacent to the anode and the other next to the cathode.

The *field-effect* transistor is an *N*-type bar of silicon with a region of *P*-type silicon on opposite sides. This transistor type has a high input impedance and fairly high output impedance, similar to an electron tube. It is used in low signal and bridge amplifiers.

Multijunction transistors include many special transistor types which exist with three and four junctions for special applications.

Power transistors are of the conventional two-junction type and are used in a wide variety of switching and amplifier applications requiring transistors with high values of voltage, current, and dissipation.

Thyristors are a group of semiconductor devices acting as switches whose bistable action depends on PNPN regenerative feedback. They can be two-, three-, or four-terminal devices, available in both unidirectional and bidirectional devices.

The *silicon-controlled rectifier* (SCR) is the best known of all thyristor devices. In this family are included the silicon unilateral switch (SUS), the light-activated silicon-controlled rectifier (LASCR), the gate turn-off switch (GTO), and the programmable unijunction transistor (PUT).

The *silicon unilateral switch* (SUS) is a miniature SCR, having an anode gate and a built-in low voltage avalanche diode between the gate and cathode.

With applied forward voltage the junctions of a *light-activated silicon-controlled rectifier* (LASCR) are forward biased and can conduct unless one of the junctions is reverse biased and blocks the current flow. Increasing the light that enters the silicon causes the SCR to turn on.

The *gate turn-off switch* (GTO) is a four layer PNPN device, similar in construction to the SCR, and is triggered into conduction by raising its loop gain to unity.

TRIAC identifies a three-electrode semiconductor switch that will conduct by a gate signal, similar in action to an SCR. It will conduct in both directions in response to a positive or negative gate signal.

The *programmable unijunction transistor* (PUT) is a small thyristor. With the gate at a constant potential the device will remain in its off state until the anode voltage exceeds the gate voltage.

The *diac* (DIAC) is a transistor which shows a negative-resistance characteristic above a given switching current.

Transistor variations are illustrated by graphic symbols in Chapter 9.

Transistor Sizes

These devices come in a variety of shapes and sizes (see Fig. 4-23). The microminiature type in Fig. 4-23(a) has three flexible leads, with the base connection indicated by a colored dot on one of the leads.

The small type in Fig. 4-23(b) has three flexible leads, with lead #1 indicated by a reference to a red line on the metal or plastic transistor case. Transistors of this size may be connected into the circuit directly, or they may require special sockets. Another type has its leads identified as shown in Fig. 4-23(c). The reference point is the location of the index tab as related to the emitter connection. This transistor is suitable for socket or clip mounting or for mounting directly on printed-circuit boards.

A power-type transistor is shown in Fig. 4-23(d). It has a large base for dissipating heat directly or through an additional heat sink, which also serves as the collector connection.

Various transistor component symbols are given in Fig. 9-14 and complete symbols in Fig. 9-16.

Transistor specifications include the type number; material; rating in voltage, current, and power; operating characteristics; and type—PNP, NPN, or other.

Integrated Circuits

These circuits, also called molecular electronics, produce a performing electronic circuit that can be contained in a block or chip .040 to .080 inch square. The equivalent of fifty or more components may be contained within this small solid chip of material with very few external connections.

Integrated circuits are fabricated on a silicon wafer, between one and three or four inches in diameter and .015 to .020 inch thick. When the process is completed, the wafer is transformed into many hundreds of performing electronic circuits.

There are several processes used to produce these chips. Among them is the epitaxial diffused planar process. The surface of the wafer is lapped and a layer of silicon dioxide is formed by a vapor growing process. The wafer is etched and then an epitaxial layer is deposited by a vacuum process, an oxide layer is grown and the surface is coated with photoresist. A photographic mask is used to expose and develop the pattern on the coated surface. The exposed oxide layer is etched and the photoresist is removed. *P*- and *N*- type dopants are added as required by diffusion, thus forming diodes, transistor junctions, resistors, and capacitors. These are followed by photoresist coating, exposing of terminal pads, and interconnection pattern. Interconnections are formed by vaporizing aluminum. The wafer is then cut or "diced" to form the individual chips or "dice." Each die is attached to a package header or substrate to be mounted in a flat pack or TO can. Interconnecting wires are attached from the chip terminations to the header leads. A metallic or ceramic lid completes the enclosure to form a sealed package.

The integrated circuit design is first breadboarded to check its operation. Then, the circuit schematic is laid out 100 or more times larger than the chip size. This is done with a coordinatograph (see Chapter 1) on a special type film, such as Mylar®* with an opaque overlay. The overlay is removed after the desired pattern is cut out on the coordinatograph with a sharp knife. Thus, sharp, clear and opaque areas are produced. This artwork is laid out as a number of drawings, one for each circuit layer. The coordinatograph, by its rectangular coordinates, locates each point on the artwork with extreme accuracy. The artwork drawings are reduced photographically to an intermediate size to make the photomasks. These masks are moved from position to position on both the *X* and *Y* coordinates and photographed at each position to form a matrix of patterns. Finally, this matrix of patterns is reduced to chip size by microphotography.

*® duPont Company registered trademark.

Integrated Circuit Packaging

Most integrated circuits are packaged in one of the several types of packaging shown in Fig. 4-24.

The TO-type case consists of a header with sealed leads, on which the transistor or integrated circuit chip is mounted, and a hermetically sealed, cylindrically shaped step enclosure, which varies in diameter from .200 inch up, Fig. 4-24(a). The number of leads varies from three to twelve or more.

The flat pack illustrated in Fig. 4-24(b) is of the ceramic type enclosure and is used for multiple transistor assemblies or integrated circuits. The number of external connections may range from six to twenty four, and they may be on two or more sides. The shape of the flat pack makes it practical for dense packaging. Some of the flat pack sizes are $\frac{1}{4}$ by $\frac{1}{4}$, $\frac{3}{8}$ by $\frac{1}{4}$, and $\frac{1}{4}$ by $\frac{3}{16}$ inch, and all approximately .100 inch thick. Leads are identified numerically for individual flat packs in technical manuals or other literature.

The dual-in-line packages (DIP) for integrated circuits may be of the ceramic, Fig. 4-24(c), or plastic (d) types, and may have 12 to 24 flat leads on two sides of the package and bent at a right angle to it. The long flat leads make them suitable for two-sided printed circuit boards because they eliminate the need for plated-through holes and connect easily by wave soldering. Indexing marks on DIP's serve to locate the numerical identification of terminals in technical manuals.

A 14-lead, dual-in-line integrated circuit packaged in clear plastic, is shown in Fig. 4-24(e) with the chip and its lead connections clearly visible.

Typical dimensions for these three types of packaging are shown in Fig. 4-25. Note the lead numbering in relation to the various reference points.

An even larger size package is used to house large-scale integration (LSI) assemblies, Fig. 4-24(f), that may have as many as several hundred circuits on a large single chip. The terminations are brought out on two or more sides of the package, with as many as 40 terminations per side. The package itself is considerably greater in size, $\frac{3}{4}$ by $\frac{3}{4}$, 1 by 1 inch, or even larger.

Lead Arrangement

The TO-type packages may have the leads project straight through the printed-circuit board or the leads are preformed—they are spaced on a .100-inch grid around the component, Fig. 4-26(a). It is necessary to raise the package $\frac{1}{16}$ to $\frac{1}{8}$ inch above the board to allow for gradual bending of the leads. Similarly, the flat pack and DIP type packages, Fig. 4-26(b), may be offset with the same grid spacing by carefully preforming the leads.

A turned up tab concept has been used to assist in wiring the microelectronic components on the printed-circuit board. The ends of the packages enter through the holes in the board to the opposite side, where they are soldered or otherwise joined to printed wiring tabs located next to the holes.

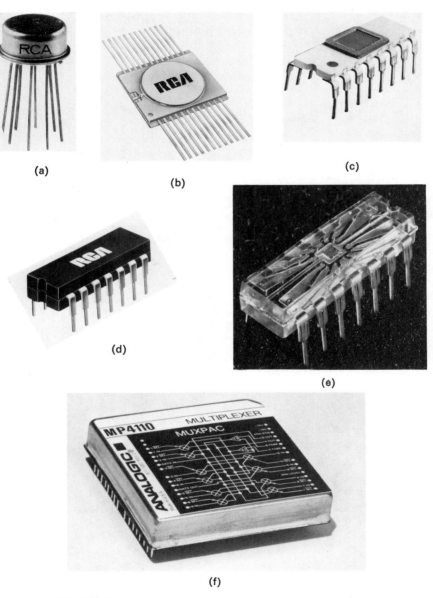

(a)

(b)

(c)

(d)

(e)

(f)

FIG. 4-24. Typical Integrated Circuit Packaging. (a) 8-Lead TO-5 Package. (b) 24-Lead Flat Pack. (c) 16-Lead Dual-In-Line Ceramic. (d) 16-Lead Dual-in-Line Plastic. (e) 14-Lead Dual-in-Line Clear Plastic—Chip and Lead Connections Visible. (f) LSI Assembly.

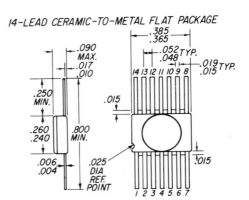

14-LEAD CERAMIC-TO-METAL FLAT PACKAGE

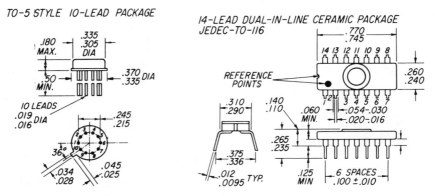

FIG. 4-25. Typical Integrated Circuit Packaging Dimensions.

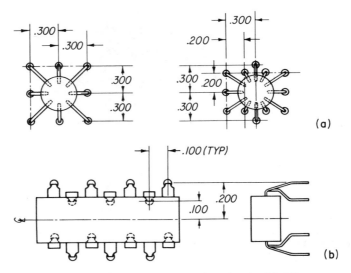

(a)

(b)

FIG. 4-26. (a) Preformed TO-Type Package Leads. (b) Offset DIP-Type Package Leads.

Chip Carriers

Although in the past the DIP type of integrated circuitry has been the standard packaging method, the rapid increase in DIP sizes that may have as many as 40 terminals or more has led to a new packaging method, namely the ceramic leadless chip carriers.

This method utilizes a square configuration, Fig. 4-27(a), as compared to the long rectangular DIP shape, and has terminations on all four sides. The illustration shows clearly the difference in the space required for each, for the same number of terminations. For example, the 24-lead DIP, center of figure (b), measures 1.25 inches by .6 inch wide, while the equivalent chip carrier is .45 by .45 inch. The interwiring of the higher density of terminations on the chip carrier can be handled readily by the multiwire method of wiring, described in Chapter 12.

Chip carriers are attached to the printed-circuit boards by the condensation reflow soldering process.

Hybrid Integrated Circuits

The hybrid circuits may be of either thin-film or thick-film types, the first being approximately .00004 inch thick, and the latter approximately .0005 inch thick. By using silk screening or other methods, various materials are deposited under vacuum on a glass or ceramic substrate.

In the thin-film process, the circuit schematic is laid out by a coordinatograph on the same type of film as used for the integrated circuits but 10 to 20 times larger than the finished size. A separate drawing is made for each circuit layer, which may include capacitor, conductor, and resistor patterns and locations for discrete components. The drawings are reduced to circuit size photographically and photomasks made. In one process, vapor deposition of conductive material on the substrate is followed by photoresist, masking, and photoetching to produce a finished pattern. Microdiscrete components are then mounted on this pattern by ultrasonic wire bonding or other means. The substrate is mounted in the package and sealed.

Integrated circuits combine or integrate many components, such as transistors, diodes, resistors, and capacitors, and all the interconnections on the same semiconductor chip. Large-scale integration has produced individual semiconductor chips with complex circuits that may contain as many as 15,000 individual elements.

To achieve greater flexibility, less complex integrated circuits are frequently combined into a compact package with such discrete devices as filter capacitors and resistor arrays as well as the semiconductor circuitry. The design of such a package can be readily altered to meet other similar requirements.

An example of a microelectronic hybrid assembly is shown in Fig. 4-28(a). This is a 44-lead digital-to-analog converter containing an input register, current switches and their regulating circuitry, and an output amplifier.

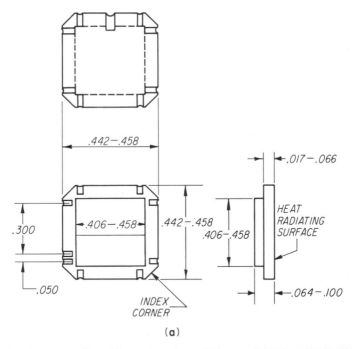

FIG. 4-27. (a) Chip Carrier Outline, 28 Terminations. (b) Size Comparison of Chip Carriers vs. DIP Packages of Equal Number of Terminations.

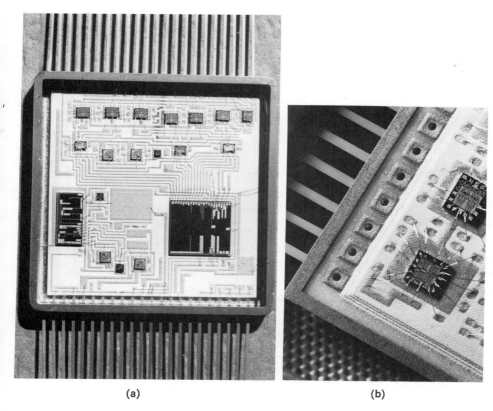

(a)   (b)

FIG. 4-28. Examples of Hybrid Microelectronic Assemblies. (a) Digital to Analog Converter. (b) Close-up View of Another Assembly.

A close-up view of a small portion of another hybrid assembly is shown in Fig. 4-28(b). The tiny interconnection leads can be readily seen.

The two hybrid assemblies shown use multiple level subtrates to provide sufficient space for the interconnections in a manner similar to multilayer printed-circuit boards.

The hybrid assemblies may range in size up to two inches square or larger. Many specialized tools are required to manufacture such an assembly and to bond its connections, which may be only .001 inch in diameter.

Microdiscrete components may be of various designs, Fig. 4-29. The pellet type of diode, capacitor, or resistor is shown in (a); in (b) a semiconductor chip such as a diode or a transistor.

Metal-Oxide Semiconductor Circuits (MOS)

This integrated-circuit type uses a simpler process than the conventional integrated circuit. It has another advantage: many more functions are possible on a given size chip. This increase in functions necessitates an increase in

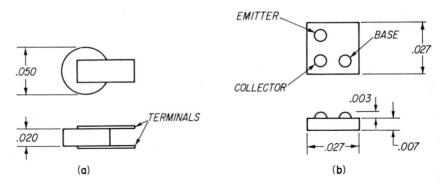

FIG. 4-29. Microdiscrete Components. (a) Pellet-Type Capacitor. (b) Semiconductor Chip.

the number of leads over the conventional integrated circuits. Thus, these leads may be located on all four sides of the package periphery.

Large-Scale Integration Circuits (LSI)

Such a circuit may be defined as one containing a large number of components performing many circuit functions as opposed to a single-function integrated circuit. For example, Fig. 4-30(a) shows a 4096-bit memory contained on a chip .088 by .094 inch.

It can be seen that extreme accuracy is required in the artwork for such complex circuitry. The ratio of the artwork to final chip size may be 200 to 500 times to ensure such accuracy.

Magnetic Bubble Memories

When an external magnetic field, with a specific intensity, is applied perpendicularly to a thin film of ferromagnetic single crystal, small cylindrical magnetic domains, normal to the film's surface, are produced within the film. They are called magnetic bubbles and can be moved within the film by changing the external magnetic field.

This technique is utilized to produce a memory function by assigning the binary digit "1" or "0" to indicate the presence or absence of a bubble at a given position. A magnetic bubble is thus obtained, Fig. 4-30(b).

Magnetic bubble memory has advantages in such applications as industrial controls, where electronic equipment must withstand dust and large temperature changes, as well as power failures.

Word processors, terminals, and airborne electronic systems are some of the applications where magnetic bubble memories are utilized. Unlike other mass-storage types, bubble memories are often packaged with the micropro-

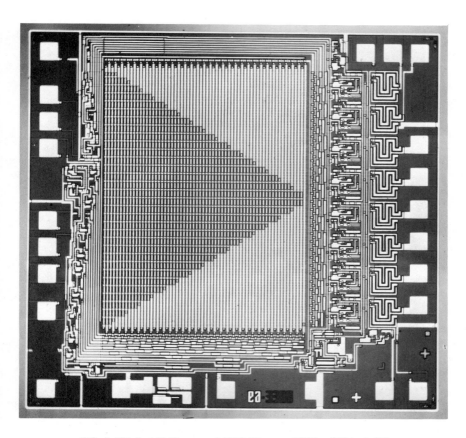

FIG. 4-30(a). LSI Memory of 4096 Bits on a .088 by .094 inch Chip.

FIG. 4-30(b). Magnetic Bubble Memory.

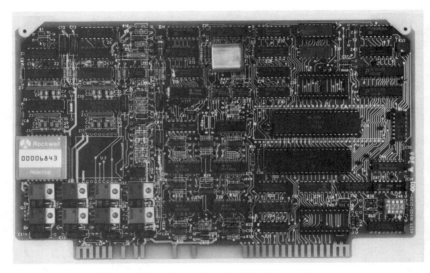

FIG. 4-30(c). Single Board Bubble Memory System.

cessor, on the same circuit board, Fig. 4-30(c). They may also be in a modular form.

Diode and Transistor Accessories

Insulators, heat sinks, sockets, and mounting pads are among the accessories that may be required in semiconductor installation.

Insulators may be either a prefabricated anodized aluminum or mica plate. The latter, .002 inch thick, isolates live rectifier terminals or transistor cases from the mounting surface and still allows rapid heat transfer from the semiconductor to its mounting, which acts as a heat sink. These insulators are available in a variety of shapes. A special nonconductive silicone grease with high thermal conductivity further decreases the thermal drop across the mica washer. Other insulators may be in the form of phenolic washers to insulate the diode or transistor terminals that extend through the mounting surface.

Heat sinks or *dissipators* are necessary to provide rapid heat dissipation from the diode, transistor, or integrated circuit and thus keep down their internal temperature. For example, power transistors mounted in TO-5 cans require the addition of suitable heat sinks to their outer housing when their power dissipation is to be increased, thus maintaining their internal temperature within a safe limit.

Figure 4-31(a) shows heat sinks or dissipators for power transistors such as triacs, while two types of dissipators for TO-5 to TO-18 type transistors are shown in Fig. 4-31(b).

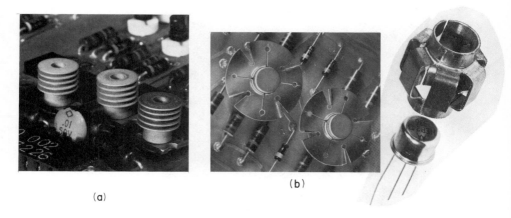

FIG. 4-31. Heat Sinks or Dissipators. (a) For Triacs. (b) For TO-5 Transistors.

The cooling capacity of a heat dissipator can be increased several fold by locating the dissipator in a forced air environment.

Sockets for transistors, flat packs, and integrated circuits, Fig. 4-32(a), are made of polysulfone or Teflon and are mounted by screws or by compression fit.

Another accessory device is the *mounting pad*, Fig. 4-32(b), which is placed between the transistor and its mounting surface.

Other Component Packaging

The packaging of semiconductors in the TO-5 case, Fig. 4-24(a), has resulted in the use of the same case size to package other components, such as relays, potentiometers, and trimmer capacitors. The advantage of this packaging tech-

FIG. 4-32. (a) Transistor Sockets.

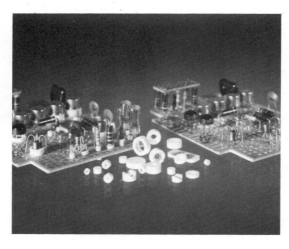

FIG. 4-32. (Continued.) (b) Transistor Mounting Pads.

nique is that it provides uniformity in the layout of printed wiring boards which are frequently used to support the components.

4.12 SWITCHES

These mechanically operated devices, available in many sizes and types, are used to make or break electrical circuits or to shift them sequentially. They are generally designated according to their application: coaxial, to switch coaxial lines; toggle, to control simple on-and-off functions; rotary, to shift circuits sequentially: and limit, to control the extent of travel of mechamical components.

Some typical switches are shown in Fig. 4-33. In (a) is shown a two-section rotary switch with terminals for printed-circuit-board use, while in (b) is shown an 8-pole rocker, dual-in-line (DIP) switch for the same application.

Reliability was always a problem in multiple keyboards, when conventional mechanically operated individual switches were used. The use of membrane-type switching, Fig. 4-33(c), can overcome this problem in situations where

FIG. 4-33. Switches. (a) Rotary Switch for Printed Circuit Boards.

(b)

(c)

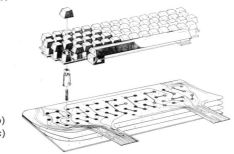

FIG. 4-33. (Continued.) (b)
Dual-in-Line DIP Switch. (c)
Membrane Type Keyboard.

each switch serves to make or break only a low value current (20 milliamperes).
It consists of two membrane pads operated by bubbles that are molded into
the top pad to close each switch circuit.

Various switch symbols are shown in Fig. 9-7.

4.13 TRANSFORMERS

These are single or multiple winding iron-core devices, with each winding except
the autotransformer isolated from another (see Fig. 9-9). They are classified
according to usable frequency, voltage input and output, type of enclosure,
mounting, capacity, commercial or military type.

Core Materials

A magnetic core is used to couple two or more transformer windings and to concentrate the magnetic flux within them. The ferrous or magnetic materials that make up such a core are alloys of iron or steel. To minimize power loss, the core is generally made of thin laminations that are insulated from each other by an outer oxide layer, shellac, or other materials. These laminations, generally of E and I shapes shown in Fig. 4-34(a), are stacked alternatively, Fig. 4-34(b), to form a complete magnetic path within the windings.

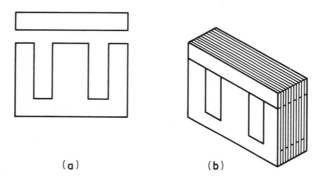

(a)   (b)

.G. 4-34. Core Materials. (a) Typical E and I Laminations. (b) Core with Alternate Stacking of E and I Laminations.

A *toroid* core is made by winding a magnetic alloy ribbon into a ring that is encased in plastic to hold the core together.

Transformer Types

Among the transformer core types are: the *shell type*, Fig. 4-35(a), which has the windings located on the center leg of the laminations, and the *core type*, Fig. 4-35(b), which has the windings placed on separate transformer core legs. The latter is used to attain good insulation between the windings of very high voltage transformers.

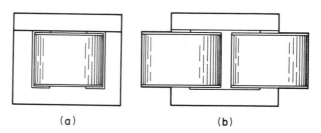

(a)   (b)

FIG. 4-35. Transformer Types. (a) Shell. (b) Core.

Power Transformers

These are used to step up or step down the input voltage to meet the requirements. A power transformer for a transmitter may have a primary input winding, a high voltage winding, and several low voltage windings. Some of these windings may be center tapped. The transformer lead color code is shown in Fig.4-36 (a).

A primary tap is often provided for a different line voltage input as in Fig. 4-36(b), while the transformer terminals may be identified numerically or by voltage notations as in Fig. 4-36(c).

A typical transformer is shown in Fig. 4-37. Such a transformer generally

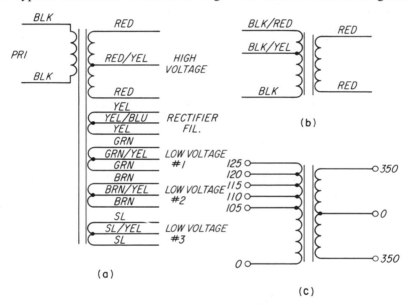

FIG. 4-36. Power Transformers. (a) Lead Color Code. (b) Tapped Primary. (c) Terminal Identification.

FIG. 4-37. Typical Power Transformer.

has a diagram showing its rating and the circuit connections. It is sealed within a metallic housing and has numbered terminals.

The following information may appear on the transformer: primary input voltage and frequency; secondary output voltages in respect to center tap, if any, and current; voltage and current ratings of the other windings and the center taps. For example: 120 V, 50–60 hertz; 350–0–350 V, 100 mA; 24 V CT, 3 AMP.

Audio Transformers

Transformers are used in audio-frequency circuits for coupling purposes between stages, or as output transformers between the output transistor stage and the speaker. The turns ratio between the windings is selected to provide correct impedance match from the input source to the output load.

The following information is generally supplied with audio transformers: primary and secondary turn ratios, impedance value of each in ohms, coil resistances in ohms, winding current in dc milliamperes, and audio power in watts.

Intermediate-Frequency Transformers

These are used in amplifier circuits that require fixed-frequency transformers to couple from the output of one intermediate-frequency stage to the input of another. Such a transformer consists of two coils on a common form which are tuned by two iron cores to the frequency selected for best gain and selectivity. The coils are assembled in an aluminum shield which is fastened to the chassis by special clips or spade lugs.

Other Transformers

Among miscellaneous transformer types are the *autotransformer*, a single-winding device that has a tap or taps brought out for a fixed voltage connection, Fig. 9-9(k). A variation of this type is the variable voltage transformer, which is wound on a toroidal core and has a movable brush that selects the output voltage from zero-to-line voltage or above, Fig. 9-9(1). The *constant voltage transformer* is especially designed to provide automatically an almost constant output voltage with wide fluctuations of line voltage.

SUMMARY

One of the outstanding characteristics of the electronics industry is the wide variety of components employed in the manufacture of electronic equipment. It ranges from mechanical hardware to integrated circuitry. The reader should become familiar with such devices, especially the semiconductor and integrated circuit components. Rapid expansion of this field requires a continuing study of current equipment and its applications.

QUESTIONS

4.1 What is the preferred value system?

4.2 Describe components referred to as "standard."

4.3 How are capacitance, inductance, and resistance values specified? List their respective designations and abbreviations.

4.4 Describe the standard color code and list ten examples, by value and code.

4.5 Describe the capacitor types and enclosures.

4.6 Name two different kinds of fixed-value capacitors and give an example of color coding for one.

4.7 Give an example of a military capacitor designation and identify the component designations.

4.8 What controls the temperature coefficient of a ceramic capacitor?

4.9 What is a "chip" capacitor?

4.10 Describe the "MS" connector types and give an example of an MS-system designation.

4.11 List three of the other connector varieties and describe each briefly.

4.12 Briefly describe the style, type, and core materials of inductors.

4.13 When are waveguides used and what are they made of?

4.14 List the applications of cathode-ray tubes.

4.15 List commonly used relay abbreviations.

4.16 Describe the resistor types and give their classifications.

4.17 Give an example of a military-type designation for a fixed composition resistor and identify the individual sections of the designation.

4.18 Name two materials used in the construction of semiconductors.

4.19 What does the abbreviation JEDEC stand for?

4.20 How does a rectifier differ from a diode?

4.21 Name two types of diodes. Describe each briefly.

4.22 How does a silicon-controlled rectifier differ from a conventional rectifier?

4.23 What is the main characteristic of a silicon-controlled switch?

4.24 List some transistor accessories and describe each briefly.

4.25 What types of components are included in an integrated circuit assembly?

4.26 Describe briefly the epitaxial planar process.

4.27 List the three types of integrated circuit packaging configurations.

4.28 Describe briefly the thin-film process for hybrid circuits.

4.29 What does the abbreviation "MOS" stand for?

4.30 How does the large-scale integration circuit differ from the conventional integrated circuit?

4.31 How are switches designated?

4.32 What information is likely to be given on an audio-transformer nameplate?

4.33 Describe the operation of the magnetic bubble memory.

4.34 What is the advantage of a membrane multiple keyboard?

4.35 What is the composition of a display device?

EXERCISES

The following exercises should help to acquire a facility in selecting components and their code designations. A component catalog, such as the Electronic Engineers Master, and other technical manuals and bulletins should be referred to for any electrical or mechanical details not given in the text.

4.1 Sketch the following relay stack-up assemblies: (a) 3 Form C; (b) 2 Form B; (c) 2 Form B, 1 Form C; (d) 2 Form A, 3 Form B, 2 Form C.

4.2 Sketch four fixed-value composition resistors, selecting a preferred value and showing color coding examples in: (a) 10 to 1000 ohm range, 5% tolerance; (b) 1000 to 10,000 ohms, 20% tolerance; (c) 100,000 to 10 megohms, 10% tolerance.

4.3 Sketch (a) a silicon diode; (b) a half-wave rectifier; (c) a power transistor; (d) a flat pack; (e) a dual-in-line package; (f) a 10-lead TO-5 transistor, including an appropriate graphic symbol for each item.

4.4 Sketch a shell-type power transformer and indicate its major parts. Include a schematic diagram of a transformer with four windings, and show the electrical values and color coding.

4.5 Make an outline drawing, twice size, of the following fixed-value resistors, showing the color bands (with colors indicated) and listing the ohmic value, the tolerance, and the military-type designations below each resistor: (a) 2 K $\pm$ 5% $\frac{1}{2}$ W; (b) 82 K $\pm$ 10% 1 W; (c) 0.51 MEG $\pm$ 10% 2 W; (d) 390 ohms $\pm$ 5% $\frac{1}{2}$ W; (e) 270 ohms $\pm$ 10% $\frac{1}{4}$ W; (f) 22 MEG $\pm$ 20% 1 W; (g) 10 K $\pm$ 10% $\frac{1}{4}$ W.

4.6 Make an outline drawing, twice size, of the following fixed-value paper dielectric capacitors, showing the color bands (with colors indicated) and listing the capacitance value, tolerance, and voltage rating below each capacitor: (a) .01 $\pm$ 10% 200WVDC; (b) .1 $\pm$ 20% 250WVDC; (c) .001 $\pm$ 10% 150WVDC.

4.7 Make an outline drawing, twice size and face view, of the following MS-type connectors, using a commercial connector catalog as a reference for the insert arrangement of the contacts: (a) MS3100A-16-10P; (b) MS3100A-18-4S; (c) MS3102A-20-8P. Identify the contacts by letter and under each list the military connector designation and size of each contact. Add the mounting and overall dimensions to flange-type connectors.

4.8 The following military designations for power-type, wire-wound resistors appear on the bill of material of an assembly drawing: (1) RW25C101; (2) RW20G103;

(3) RW50F100; (4) RW20F102; (5) RW50G100; and (6) RW25C102. List these items and identify each part of the designations.

4.9 Using Fig. 4-21 for reference, make a list of resistance values for every 20 degrees of rotation of a 5000-ohm potentiometer with a total rotation of 270 degrees for taper C.

4.10 Make an outline drawing, four times size, of the following: (a) TO-99 and TO-100 packages; (b) 8- and 10-lead integrated circuits in a TO-5 can; (c) TO-3 can; (d) 2N1017, 2N1305, and 2N1309 transistors; (e) 1N3825 and 1N3995 zener diodes; (f) 2N1481, 2N1485 and 2N2338 power transistors.

4.11 Sketch the internal circuit diagram of a seven-segment display.

5

FASTENERS, MATERIALS, AND FINISHES

Various types of fastening devices are used in electronic equipment. Some of these are conventional—screws, rivets, eyelets, and pins, while others are specialty items, specifically designed to withstand conditions in space, on the ground, aboard ship, or under water, and to meet such provisions as minimum weight and resistance to shock, vibration, and corrosion.

Metal parts may be permanently joined by riveting, soldering, welding, brazing, or crimping. Eyelets are used if solid rivets would create undue pressure on the joined parts and possibly result in early failure, i.e., switch clips assembled to ceramic base material. Rivets provide rapid assembly of parts such as brackets to the chassis. Riveting is one of the most acceptable ways to permanently join sockets, terminal strips, and many other subassemblies with plastic and metal parts, except where prohibited by military specifications.

Other fastening devices are springs, clips, or self-locking assemblies that are developed specifically for rapid production assembly and ease of replacement. Such devices as self-locking nuts, captivated screws, floating nuts, inserts, screw and lockwasher assemblies, retaining rings, and pins of various types all contribute to the goal of expediency, space and weight reduction, and ease of operation or release as required.

As with electronic and mechanical components, the electronic draftsman will find that there is a wide variety of materials and finishes entering into the manufacture of individual parts or subassemblies. A familiarity with these details will help in designing such parts or assemblies or in specifying them on engineering drawings.

The materials may be subdivided into metals, plastics, and potting compounds. Each has its own specific field of application and certain limitations.

132

Finishes may be subdivided into chemical treatments or coatings, organic or paint finishes, and electroplating.

5.1 MACHINE SCREWS

One of the most familiar threaded fasteners—machine screws—are made in six series of threads in the Unified thread form: coarse thread (UNC), fine thread (UNF), extra-fine thread (UNEF), and three special series, 8-thread (8UN), 12-thread (12UN), and 16-thread (16UN).*

The standard screw thread is the American Standard 60° V-thread type, Fig. 5-1(a). The thread series, such as 32, indicates the number of threads per

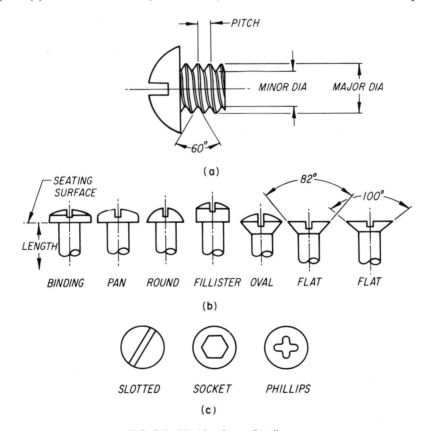

FIG. 5-1. Machine Screw Details.

*See *Screw Thread Standards for Federal Service, Handbook #28, Part* 1. National Bureau of Standards Circular #479, and ANSI Standard B1.1–1974, *Unified Inch Screw Threads* (*UN and UNR Thread Form*).

inch. The diameter of smaller screws is indicated by a number, as for instance # 10, and by the actual diameter of screws that are $\frac{3}{16}$ inch and larger.

Threads in general use are designated as Unified Coarse (UNC), while others, with more threads per inch, are designated Unified Fine (UNF). The distinction in the number of threads for the coarse and fine threads varies depending upon the thread diameter.

The smaller screws and nuts used in electronics have their diameters designated by a standard numbering system (Table 5-1) with a step increase of .013 inch between successive numbers.

Table 5-1   Machine Screw Thread Details

Size	Diameter (in.)	Threads per inch		
		Unified Coarse (UNC)	Unified Fine (UNF)	Unified Extra Fine (UNEF)
#0	.060	—	80	—
1	.073	64	72	—
2	.086	56*	64	—
3	.099	48	56*	—
4	.112	40*	48	—
5	.125	40*	44	—
6	.138	32*	40	—
8	.164	24	32*	—
10	.190	24	32*	—
12	.216	24*	28	—
1/4	.250	20*	28	32
5/16	.312	18*	24	32
3/8	.375	16	24	32*

*In common use

The relationship between two mating threads, such as nut and screw, is known as the *fit*. There are four general classifications of fits: Class 1 for rapid assembly work that does not require close tolerances; Class 2 for the majority of the work in electronic equipment; Class 3 for precision work; and Class 4 for work that requires the selection of mating threads to avoid actual interference.

In addition to the class designation, the external thread and allowance is indicated by the letter A and internal by the letter B, as in Class 2A and 2B for general electronic use.

Metric Machine Screws

These screws take the place of U.S. standard machine screws when equipment is designed utilizing metric dimensions. Such screws are listed in Table A-7 in the Appendix.

Screw Head Styles

The head types used on machine screws are shown in Fig. 5-1(b). The binding or pan head styles are commonly used because they look better and because they allow the use of larger clearance holes in mating mechanical subassemblies. Flat-head screws in countersunk holes result in flush assemblies and are available in either the 82° or 100° included angle. The latter is used in thin materials. Screw lengths are given from the seating surface shown in Fig. 5-1(b).

Screw heads are recessed in the form of a slot, socket, or a cross (phillips), Fig. 5-1(c). Since the cross recess requires a special tool, such screws are in limited use in military electronic equipment.

Machine Screw Details

Screws are identified on a drawing or on a parts list by a notation in the following sequence: thread diameter in inches or by the screw number designation; number of threads per inch; thread series; class of screw or nut thread, i.e., #6-32 UNC-2A, which is translated #6 or .138 inch in diameter, 32 threads per inch, Unified Coarse series, Class 2 screw thread. A right-hand thread is understood, and left-hand thread is specified by LH after the class designation.

The length, head type, material, and plating, if any, are also specified as part of the notation; for example: #6-32 UNC-2A × $\frac{3}{4}$ bind. hd screw, steel, cad. plated or CP. Head recess is assumed to be a slot unless otherwise specified. A simplified version, omitting the thread class, is often used; for example: #6-32 × $\frac{3}{4}$ bind. hd screw, steel, CP.

Screws are generally available in brass, steel, stainless steel, aluminum, and plastic materials, such as nylon. They may be cadmium, nickel, black nickel, chrome plated, or of anodized finish. If stainless steel, they may be left plain or passivated (see page 144).

Commercial screw lengths may be in increments of $\frac{1}{16}$ inch for the very small sizes, $\frac{1}{8}$ inch for larger sizes up to lengths of one inch, and $\frac{1}{4}$ inch for the longer ones. Most of the shorter screws are threaded up to the head.

The drill sizes required for tapped and clearance holes are listed in Table 5-2. The diameter of clearance holes is determined by the hole-spacing tolerances, which are generally greater for holes spaced far apart. It is also governed by basic part construction; i.e., sheet metal parts may require larger clearance holes.

Metric Tap and Drill Sizes

Metric threads are designated by the letter "M," followed by the *nominal size* in millimeters (mm), then by letter "x," and finally the *pitch* diameter, i.e., M2 x .40.

Table 5-2   Tap and Clearance Drill Sizes

| Screw size | Tap drill (75% thread) | | Clearance drill | | | |
| | No. | Dia | Holes up to 3″ apart | | Holes up to 6″ apart | |
			No.	Dia	No.	Dia
#2–56	50	.070	34	.111	30	.128
4–40	43	.089	29	.136	22	.157
6–32	36	.106	22	.157	14	.182
8–32	29	.136	12	.189	3	.213
10–24	25	.149	3	.213	—	1/4

There are three tolerance grades: 4, 6, and 8.

Grade 6 is the closest tolerance grade to U.S. Unified Class 2A and 2B screw threads.

See Table A-7, Metric Screw Threads and Tap Drill Sizes, in the Appendix.

5.2 OTHER FASTENING COMPONENTS

Among these may be included nuts, washers, rivets, and other fasteners.

Nuts

Like machine screws, hexagonal nuts are available in various metals and platings and come in several styles, Fig. 5-2(a). The single chamfer type is generally used for the larger sizes and the double chamfer for the smaller sizes. Nuts are specified by their width across the flats. The common nut sizes are listed in Table 5-3(a) (inch) and 5-3(b) (metric).

Table 5-3(a)   Commonly Used Hexagonal Nuts, Inch

Screw size	Dia. (in.)	Width	Thickness	Width	Thickness
# 2–56	.086	3/16	1/16	—	—
4–40	.112	1/4	5/64	1/4	3/32
6–32	.138	1/4	5/64	1/4	3/32
8–32	.164	1/4	3/32	5/16	3/32
10–32	.190	5/16	7/64	3/8	1/8

Table 5-3(b)   Commonly Used Hexagonal Nuts, Metric

Screw size (mm)	Width	Thickness	Screw size (mm)	Width	Thickness
M1.6 × 0.35	3.2	1.3	M5 × 0.8	8	4
M2 × 0.4	4	1.6	M6 × 1.0	10	5
M3 × 0.5	5.5	2.4	M8 × 1.25	13	6.5
M4 × 0.7	7	3.2	M10 × 1.5	17	8

Washers

Plain and locking washers (lockwashers) are shown in Fig. 5-2(b). The plain washer is used to cover oversize holes or slots in order to provide a bearing surface, and to protect surface finish.

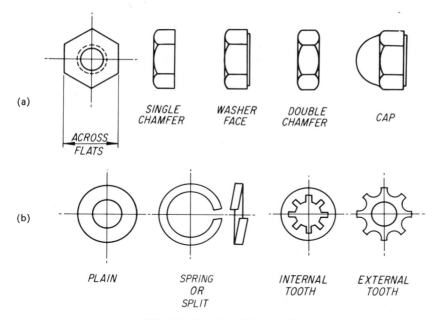

(a)

SINGLE CHAMFER   WASHER FACE   DOUBLE CHAMFER   CAP

ACROSS FLATS

(b)

PLAIN   SPRING OR SPLIT   INTERNAL TOOTH   EXTERNAL TOOTH

FIG. 5-2. Varieties of Nuts and Washers.

Locking washers or lockwashers are of several types. The *spring* or *split* washer in Fig. 5-2(b) is used with small nuts and with round, socket, or fillister head screws. The *internal tooth* lockwasher serves the same purpose and is used when appearance is important. The *external tooth* lockwasher is suitable for use under binding or pan head screws and provides maximum gripping action. Both of the tooth-type lockwashers have numerous teeth set at an angle to bite into the mounting surface and the fastener to eliminate looseness and to achieve good electrical contact.

Rivets

Riveting is one of the most inexpensive methods for joining aluminum or steel parts. The problem of corrosion can be overcome by using rivets of the same material as the parts they are joining.

Solid aluminum rivets are made of 1100, 2017, 2024, 2117, 3003, and 6061 alloys. Their physical dimensions are shown in Table 5-4. They require a clinch allowance of $1\frac{1}{2}$ times the rivet diameter. Thus a rivet $\frac{1}{8}$ inch in diameter requires a clinch allowance of $\frac{3}{16}$ inch as long as the rivet holes are only slightly larger than the rivet diameter.

Semi-tubular rivets are rivets of brass, aluminum, or steel with a hole in the end no deeper than the body diameter of the rivet. Full tubular rivets have deeper holes. The dimensions for semi-tubular rivets are given in Table 5-5(a), along with their clinch allowances.

Eyelets are variations of tubular rivets, with the hole extending completely through. They are available in odd lengths rather than the $\frac{1}{32}$ or $\frac{1}{16}$ rivet incre-ments. Those with smaller diameters are used for light assemblies and printed

137

Table 5-4 Aluminum Rivet Head Dimensions in Inches*

	BRAZIER	CSUNK	CSUNK	FLAT	ROUND	
Rivet dia "A"	.062	.093	.125	.156	.187	.250

Brazier head						
Head dia "B"	.156	.234	.312	.390	.467	.625
Head height "C"	.031	.047	.062	.078	.094	.125
Radius "R"	.114	.171	.227	.283	.339	.454

78° Countersunk head						
Head dia "B"	.114	.170	.226	.282	.338	.453
Head height "C"	.032	.047	.062	.078	.094	.125

100° Countersunk head						
Head dia "B"	.105	.170	.216	.278	.344	.467
Head height "C"	.022	.035	.042	.055	.070	.095

Flat head						
Head dia "B"	.125	.187	.250	.312	.375	.500
Head height "C"	.025	.038	.050	.062	.075	.100

Round head						
Head dia "B"	.125	.187	.250	.312	.375	.500
Head height "C"	.047	.071	.094	.117	.140	.187
Radius "R"	.065	.098	.130	.163	.195	.260

*Reynolds Metals Co.

Table 5-5(a) Oval Head Semi-Tubular Rivet Dimensions in Inches*

	Rivet body diameter						
	.059	.088	.098	.120	.143	.185	.212
Clinch allowance	.032	.046	.051	.064	.076	.098	.109
Rivet clearance hole, min	.064	.094	.104	.128	.154	.199	.234
Length							
Minimum	1/16	5/64	5/64	3/32	1/8	5/32	3/16
Increments	1/64	1/64	1/64	1/64	1/32	1/32	1/16
Head dia	.109	.147	.187	.218	.234	.312	.437
Head height	.015	.024	.029	.034	.040	.060	.068

*The Tubular and Split Rivet Council

circuitry, while those with larger diameters—such as grommets—are used to protect wiring passing through metallic partitions.

Metric Semi-tubular Rivets*

The dimensions of these rivets are all given in millimeters (mm); see Table 5-5(b). The rivet length is measured from the underside of the rivet head.

Table 5-5(b) Oval Head Semi-tubular Metric Rivet Dimensions (mm)*

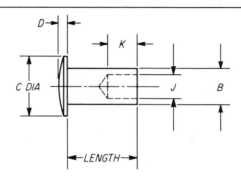

Nominal size	B	C	D	J	K
1.6	1.54–1.46	2.88	0.48	1.15	1.20
2.0	1.92–1.82	3.60	0.60	1.44	1.50
2.5	2.40–2.30	4.50	0.75	1.80	1.88
3.0	2.88–2.70	5.40	0.90	2.16	2.25
3.5	3.36–3.24	6.30	1.05	2.52	2.63
4.0	3.84–3.71	7.20	1.20	2.88	3.00
5.0	4.80–4.64	9.00	1.50	3.60	3.75
6.0	5.76–5.58	10.80	1.80	4.32	4.50
7.0	6.72–6.52	12.60	2.10	5.04	5.25
8.0	7.68–7.48	14.40	2.40	5.76	6.00

Other Fasteners

An insert is a special form of nut that acts as a tapped hole and is used to provide sufficient length of thread for fastening purposes in thin or soft sheet metal. Inserts are made in various sizes to fit different material thicknesses. Some lock in place by pressure alone.

Special fasteners, developed for severe service conditions (see Fig. 5-3) are operated by fingers or a screwdriver. Turning one element a quarter turn produces a cam-like engagement with the other part of the fastener. Turning

*Tubular Rivet and Machine Institute.

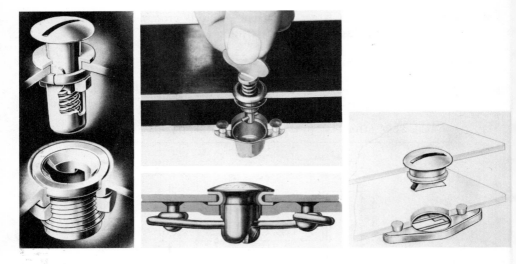

FIG. 5-3. Quick-Release Fasteners.

in opposite direction disengages the member, again with only a partial turn of the screwdriver or fingers.

Pins

Fasteners of the pin type are: the dowel, groove, roll, spiral, and the taper (Fig. 5-4). They are used to attach gears, spacers, knobs, dials, and other parts to shafts. Taper pins have a taper of $\frac{1}{4}$ inch per foot.

The pin diameter for any given shaft diameter should not exceed one third the shaft diameter; i.e., the pin selected for a shaft $\frac{3}{8}$ inch in diameter should not be larger than $\frac{1}{8}$ inch. The taper pin size selection is based on the pin diameter at the larger end.

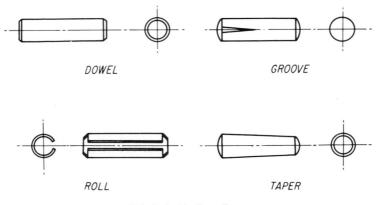

FIG. 5-4. Pin-Type Fasteners.

5.3 METALS

The lighter metals, such as aluminum, magnesium, and titanium; the ferrous metals, cold- or hot-rolled steel and stainless or corrosion-resisting steel; and the nonferrous metals, brass, copper, beryllium copper, and phosphor bronze, are all in this general subdivision. Their relative weights per cubic inch are given in Table 5-6.

Table 5-6 Relative Weights of Metals.		Table 5-7 Designations for Aluminum Alloy Groups*	
Metal	Weight per cubic inch		AA *number*
		Aluminum, 99.00% or more	1xxx
Magnesium	.065		
Aluminum	.101	*Major Alloying Element*	
Titanium	.165	Copper	2xxx
Steel	.285	Manganese	3xxx
Stainless steel	.290	Silicon	4xxx
Beryllium copper	.300	Magnesium	5xxx
Brass	.310	Magnesium and silicon	6xxx
Phosphor bronze	.320	Zinc	7xxx
Copper	.322	Other element	8xxx
		Reserved series	9xxx

*Aluminum Association

Aluminum

This metal is one of the most widely used materials in electronics. It is generally used combined with other elements to form various alloys. Among the alloying elements are copper, magnesium, silicon, and zinc. The percentage of the alloying element used is governed by the characteristics required. Although there are many aluminum alloys available, only a limited number are in use in electronics.

Aluminum alloy designations. Wrought aluminum alloys may be subdivided into two groups: non-heat-treatable and heat-treatable. The aluminum alloys in the 1000, 3000, and 5000 series are in the first class (see Table 5-7). These alloys come in various tempers, indicated by such temper designations as "O," "H12," or "H14" (see Table 5-8). The "O" designation is for the fully annealed alloys.

The heat-treatable alloys are in the 2000, 6000, and 7000 series. Their strength or hardness is increased by heating the metal to a definite elevated temperature and then quenching in cold water. This is followed by an aging process, either by heating the material to a moderate temperature or by natural aging at room temperature.

A standard designation system for wrought aluminum alloys has been adopted by the Aluminum Association. This standard has a four-digit number

Table 5-8   Temper Designations of Aluminum Alloys

H-temper *First digit designation*		*T-temper*	
H1	Strain hardened	T1	Thermally treated
H2	Strain hardened and annealed	T2	Annealed (castings only)
H3	Strain hardened and stabilized	T3	Solution heat treated and cold worked
Second digit designation		T4	Solution heat treated and naturally aged
0	Annealed	T5	Artificially aged
2	Quarter-hard	T6	Solution heat treated and artificially aged
4	Half-hard		
6	Three-quarter hard	T7	Solution heat treated and stabilized
8	Full hard	T8	Solution heat treated, cold worked, and artificially aged
9	Extra hard	T9	Solution heat treated, artificially aged, cold worked

for its basic alloy designation, with each digit indicating a definite major alloying element.

In the 2000 through 8000 series the first digit indicates the major alloying element, such as copper, 2xxx; manganese, 3xxx; and so on. The 6xxx group has more than one major alloy element.

The second digit in the alloy designation indicates alloy modification. If it is zero, the modification is the original alloy. Integers 1 through 9 indicate alloy modification.

Basic *temper designations* consist of letters, with one or more digits following. They are separated from the alloy designation by a dash, such as 5052-H34, 6061-T6, and so forth. Basic letter designations are "F" for "as fabricated"; "O" for annealed; "H" for strain hardened; "W" for solution heat treated; and "T" for thermally treated, with or without strain hardening.

The "H" temper designation is followed by a number consisting of two digits. The first digit is "1" for strain-hardened material, "2" for strain-hardened and annealed, and "3" for strain-hardened and stabilized (see Table 5-8). The second digit depends upon the degree of hardness of the material. Digit "0" is assigned to the soft or annealed state, "2" to quarter-hard, "4" to half-hard, and so on. Thus, quarter-hard material is designated by H12, H22, H32; half-hard by H14, H24, H34; three-quarter hard by H16, H26, H36. Extra-hard tempers are designated by H19, H29, and H39.

If wrought alloys are to be bent to form brackets, chassis, and other parts, it is necessary to specify the minimum bending corner radii on the drawings. Such radii depend upon the thickness of the material being bent and its composition. Table 5-9 lists the bending radii for a 90° cold bend in terms of material thickness, alloy type, and its temper. (Zero refers to the 90° sharp bend, having no radius.)

It is customary to distinguish the sheet designation from the plate desig-

Table 5-9   Bending Aluminum Alloys—Min. Radii For 90° Cold Bend*

Multiply thickness by factor shown

Alloy and temper	Thickness—t					
	.016	.032	.064	.128	.182	.258
1100–0	0	0	0	0	0	0
1100–H12	0	0	0	0	0–1t	0–1t
1100–H14	0	0	0	0	0–1t	0–1t
1100–H16	0	0	0–1t	$\frac{1}{2}$t–1$\frac{1}{2}$t	1t–2t	1$\frac{1}{2}$t–3t
1100–H18	0–1t	$\frac{1}{2}$t–1$\frac{1}{2}$t	1t–2t	1$\frac{1}{2}$t–3t	2t–4t	2t–4t
2024–0	0	0–1t	0–1t	0–1t	0–1t	0–1t
2024–T3	1$\frac{1}{2}$t–3t	2t–4t	3t–5t	4t–5t	4t–6t	5t–7t
2024–T6	2t–4t	3t–5t	4t–6t	5t–7t	5t–7t	6t–10t
3003–0	0	0	0	0	0	0
3003–H12	0	0	0	0	0–1t	0–1t
3003–H14	0	0	0	0–1t	0–1t	$\frac{1}{2}$t–1$\frac{1}{2}$t
3003–H16	0–1t	0–1t	$\frac{1}{2}$t–1$\frac{1}{2}$t	1t–2t	1$\frac{1}{2}$t–3t	2t–4t
3003–H18	$\frac{1}{2}$t–1$\frac{1}{2}$t	1t–2t	1$\frac{1}{2}$t–3t	1$\frac{1}{2}$t–3t	3t–5t	4t–6t
5052–0	0	0	0–1t	0–1t	0–1t	0–1t
5052–H32	0	0	$\frac{1}{2}$t–1$\frac{1}{2}$t	$\frac{1}{2}$t–1$\frac{1}{2}$t	$\frac{1}{2}$t–1$\frac{1}{2}$t	$\frac{1}{2}$t–1$\frac{1}{2}$t
5052–H34	0	0	$\frac{1}{2}$t–1$\frac{1}{2}$t	1$\frac{1}{2}$t–2$\frac{1}{2}$t	1$\frac{1}{2}$t–2$\frac{1}{2}$t	2t–3t
5052–H36	0–1t	$\frac{1}{2}$t–1$\frac{1}{2}$t	1t–2t	1$\frac{1}{2}$t–3t	2t–4t	2t–4t
5052–H38	$\frac{1}{2}$t–1$\frac{1}{2}$t	1t–2t	1$\frac{1}{2}$t–3t	2t–4t	3t–5t	4t–6t
6061–0 ,	0	0–1t	0–1t	0–1t	0–1t	0–1t
6061–T4	0–1t	0–1t	$\frac{1}{2}$t–1$\frac{1}{2}$t	1t–2t	1$\frac{1}{2}$t–3t	2t–4t
6061–T6	0–1t	$\frac{1}{2}$t–1$\frac{1}{2}$t	1t–2t	1$\frac{1}{2}$t–3t	2t–4t	3t–4t
7075–0	0	0–1t	0–1t	$\frac{1}{2}$t–1$\frac{1}{2}$t	1t–2t	1$\frac{1}{2}$t–3t
7075–T6	2t–4t	3t–5t	4t–6t	5t–7t	5t–7t	6t–10t

*Aluminum Company of America

nation on the basis of thickness. Sheets are classified as those which are below .250 inch thickness; plates are .250 inch or thicker.

Stainless Steel

This material, also known as corrosion-resisting steel, is used in electronic components and assemblies to meet requirements of strength and noncorrosiveness for such diverse uses as shafts, gears, supporting brackets, and other hardware. These metals must be carefully selected when nonmagnetic characteristics are required or when rotating parts could be subjected to "galling," which is an undesirable tendency that can be frequently corrected by selecting different alloy steels for parts in rubbing contact.

Some of the alloy types in use are:

Type 302. The most widely used of the nickel-chromium (8 and 18 percent, respectively) steels, this type is nonmagnetic, tough, and ductile, and has excellent corrosion resistance to salt water. It is used for springs, stampings, moldings, cables, and so on. The percentages above refer to nickel and chrome, respectively.

Type 303. A free-machining steel through addition of selenium to 8 percent nickel and 18 percent chrome—this type is used for parts fabricated by machining, grinding, and polishing, such as screw-machine products. It cannot be hardened by heat treatment, nor is it suitable for welding.

Type 304. A low-carbon, 8 and 18 percent stainless steel, it has good welding characteristics, is nonmagnetic when annealed, and can be hardened only by cold working. It can be deep-drawn, hot-forged, riveted, and upset.

Type 416. This is a free-machining stainless steel, with 12–14 percent chromium, and sulphur added for ease of machining. It is used for shafts, gears, and other parts requiring extensive machining, and it can be hardened.

Type 420. This steel requires hardening to bring out its corrosion-resisting properties. It is used for ball bearings, magnets, gears, and springs.

Passivating. Stainless steel parts become contaminated with foreign matter particles during fabrication, but the resultant rusty appearance upon exposure can be prevented by dipping the part in a 20–50 percent solution of nitric acid. This passivating procedure is a must for all fabricated parts.

Beryllium Copper

This metal is used for bellows, coaxial connector terminals, coil springs, switch contact arms, and other applications requiring a nonferrous spring contact material. It has high tensile strength, hardness, and high electrical conductivity in hardened condition, and it is readily fabricated in an annealed condition. Alloys containing 2 percent beryllium and .25 percent nickel and cobalt, 1.7 percent beryllium and .25 percent nickel and—cobalt, or .6 percent beryllium and 2.5 percent cobalt are in common use.

Beryllium copper parts have to be heat-treated after they are formed in annealed state. This is done by aging the fabricated part two or three hours at 315°C or higher, depending upon the alloy composition. This heat treatment must be specified on the drawings of parts made from this material.

Phosphor Bronze

This metal is used for lockwashers, springs, fuse clips, and other spring material applications where a nonferrous material has to be used and cost is an important consideration. It does not require any heat treatment but is cold-formed to the required shape. The alloying element is generally 5 or 8 percent tin.

Joining Metals

Metallic parts in electronic equipment are frequently joined by screws, bolts, nuts, or special fasteners if disassembly may be required. Or, they may be permanently joined by fusing the parts through brazing, spot welding, or arc welding processes.

Brazing. In this process, metal parts are joined by alloys of brass, copper, and zinc. The alloy used depends upon the metal being joined. The alloy temperatures range from 480°C upward but below the melting point of the parts being attached.

The brazing process is similar to soldering except that it requires a higher melting temperature. Flux, such as borax, is used to keep the joint surfaces clean and to protect them against oxidation during the process.

There must be a small clearance between parts to be joined by brazing to allow an even flow of the brazing material. This clearance should be specified on the detail drawing.

Brazing may be done with a torch or in a furnace, where the parts are placed in a protective atmosphere, flux is applied to the joints, and the furnace temperature is brought up to the melting point of the brazing material. This method is known as furnace brazing.

Spot welding. In this process, two or more pieces of metal, which are held between two electrodes with considerable applied pressure, are joined when a high current is passed through the electrodes. The current is turned on and off by a timing mechanism that limits the "on" time to a fraction of a second.

Aluminum parts, $\frac{1}{8}$ inch thick or thicker, can be spot-welded readily. The process is used to weld chassis corners, attach brackets and shields to chassis, and to make other kinds of permanent joints.

The limitation in aluminum spot welding is the spacing between adjacent welds. Close spacing results in poor welds due to short-circuiting effect of the adjacent welds. Edge spacing of spots is also important to prevent the edge of the parts being joined from bulging out.

Arc welding. In this process, an intense heat, such as that produced by an electric arc, is applied to the surfaces being joined, and with the addition of the proper filler material the parts are readily joined together. Aluminum and magnesium alloys, steel and copper are joined by inert-arc welding. In this case, the part being welded serves as one electrode and the arc forms between it and the tungsten electrode in an inert atmosphere of helium or argon.

5.4 PLASTICS

These synthetic organic materials are shaped into a solid finished form by application of heat and pressure. There are two basic types of plastics:

1. Thermoplastics, which, like wax, can be melted and hardened repeatedly without undergoing a permanent change in their chemical structure.
2. Thermosetting resins, which, upon application of heat and pressure, change chemically and cannot be remelted.

Examples of thermoplastics are: acrylics, nylons, polyethylenes, and polystyrenes. Examples of thermosetting plastics are: phenolics, melamine, and epoxy resins.

In addition to molding to shape, many plastic types are available in machinable sheets, rods, tubing, and other shapes. Reinforced plastics may have a filler material added to increase their strength, or their strength may be increased by laminations of sheet material, such as paper, cotton cloth, asbestos, glass fiber, or glass cloth, which are coated with thermosetting resin and bonded together with heat and pressure.

Plastics are the prime source of insulation in electronic components and subassemblies. They provide electrical insulation and mechanical mounting support against shock and vibration and seal against humidity.

Acetal Resins

These thermoplastic resins are exemplified by a duPont product called Delrin®,* one of the strongest of thermoplastics. It has a high degree of strength and rigidity and is opaque in molded parts. Its dimensional stability, resilience, and other mechanical properties approach metals by comparison. It has 15 percent elongation at room temperature, can be molded to close tolerances, and is available in extruded rod, tubing, sheeting, and shapes. It can be joined by riveting, self-tapping screws, and threading. Its coefficient of linear thermal expansion is 4.5×10^{-5} in./in./°F, or five times that of steel.

Delrin® is suitable for gears, counters, cams, knobs, and handles, and can be used as a low sliding-friction material for bearings for light loads and slow speeds and for electrical applications that require continuous operation at 80°C.

Polyamide Resins

One of these resins is nylon, another duPont product, which is a strong, flexible thermoplastic, suitable for applications up to 120°C. It has a low friction coefficient that makes it well-suited as a bearing material. It is used for such mechanical parts as gears, knobs, structural housings, and cams, and for such electrical applications as coil forms, switch parts, cable insulation, and jacketing.

*® duPont Company registered trademark.

146

It is made with such additives as molybdenum disulphide or glass to increase its bearing capabilities and tensile strength and is manufactured by injection molding, extrusion, or machining.

Nylon is sensitive to ultraviolet ray exposure and high humidity. The designation numbering of the various types available depends on the processor of the resin.

Polycarbonate Resins

A representative of these resins is Lexan, a General Electric Company product. It has a high tensile strength and heat-distortion point as well as good creep resistance and electrical properties. It is a thermoplastic, fabricated by injection molding and extrusion, and can be readily machined and adhesively bonded. It is available in nearly water-clear or opaque colors.

Its useful temperature range is up to 140°C, and it is available in film form for magnetic recording tapes and drafting films. Its applications for electrical and mechanical purposes include connectors, coil forms, dials, structural parts, and ball bearings.

Polyethylene Resins

These thermoplastic resins are translucent to opaque, manufactured by injection molding, and are suitable for temperatures up to 100°C. They are practically insoluble in all organic solvents and have extremely low water absorption. They are susceptible to ultraviolet ray exposure, however, unless they are formulated with an additive such as carbon black.

These resins are one of the lowest cost plastics and are used for packaging materials, insulation and jacketing for wire and cable, extruded film sheet, pipe, and coaxial cables.

Polyimide Resins

This resin, developed for such applications as multilayer printed-circuit boards, is a thermosetting resin, generally a glass laminate, of good mechanical properties, has an outstanding stability, and is nonflammable.

Available in sheet laminate, tubing, copper-clad laminate, bars, and other molded forms it is the product of Rhone-Poulenc, Incorporated, and has the trade name KERAMID 601.

Udel®* Polysulfone Resin

This is a high-strength thermoplastic resin which maintains its properties over a wide temperature range, −100°C to +149°C.

It is obtainable with glass fibers and spheres as reinforcement and in tube,

*® Registered trademark of Union Carbide Corporation.

rod, film, and sheet form. It is being used for printed-circuit boards operating at high temperature and high frequencies.

Polytrifluoroethylene Resins

Popularly known as Kel-F, these resins are hard, dense thermoplastics that are thermally stable from $-105°C$ to $+200°C$. They absorb almost no moisture and are transparent to translucent in color. They have high dielectric, tensile and compressive strength, are easily machined, not attacked by strong acids, and can be compression-, transfer-, or extrusion-molded in the form of sheets, tubing, and rods. They are used for making tape and films, seals, gaskets, small electronic parts, and for coating hook-up wire and rotary switches. They are also known as CTFE flourocarbons, CTFE being an abbreviation for chlorotrifluoroethylene.

Fluorocarbons

There are two fluorocarbon resins—TFE or tetrafluoroethylene, and FEP or fluorinated ethylene-propylene. Teflon, a duPont Company product, is a representative fluorocarbon resin.

TFE Teflon is suitable for gaskets, seals, electronic components, and bearings and is available in sheet, rod, and tubing. FEP Teflon is suitable for encapsulation, wire coatings, and printed-circuit boards.

TFE Teflon changes in volume when passing through a critical temperature of 18 to 25°C. The final operating temperature of precision parts should be determined to allow for this volume change. FEP Teflon does not exhibit this characteristic.

The maximum operating temperature for TFE is 260°C and 205°C for FEP. Both are suitable for temperatures below $-80°C$. They have a very low coefficient of friction and are almost completely inert chemically.

Teflon has an anti-stick property and feels slippery to the touch. It can be reinforced by such additives as glass fiber, graphite, asbestos, and quartz. The additives improve its resistance to wear by rotating shafts and lower its thermal expansion.

Fluorosint

This resin is a composition of Teflon and certain additives, which form a ceramic-like material that has certain advantages over the unmodified Teflon:

1. Cold flow is greatly reduced.
2. The coefficient of expansion is reduced to that of aluminum.
3. Bearings made of this material can operate up to 260°C and under much higher loads and speeds.

Fluorosint is available in sheets, rods, and tubing, and its electrical characteristics are similar to that of unmodified Teflon, making it suitable for stand-off insulators and other electrical parts.

Phenolics

These resins are the thermosetting type of plastics, and they have excellent dielectric strength, insulation properties, heat and water resistance, and a glossy finish. They can operate at temperatures of 150°C and in some of the special materials up to 260°C. They are molded by compression and transfer molding.

In addition to the molded phenolics, which are powders impregnated with resin, there are also various types of laminated phenolics, which consist of sheets of paper, linen, or canvas impregnated under pressure with phenolic resins. These laminated phenolics are subdivided into grades, according to the laminating material.

Paper-base laminates include (1) NEMA P and X grades, of punching stock, which have good mechanical properties, and (2) NEMA XXP and XXXP grades, of punching grade material, with good electrical characteristics. Applications are coil-form bases, terminal boards, and switch insulation.

Canvas-base laminates are classified as NEMA grade C. This is a canvas-base plastic laminate that is tough and resilient. Applications include cams, intricate machined parts, and terminal boards.

Glass-cloth laminates are (1) NEMA grade G-1 and G-2, a filament-fiber glass-cloth laminate suitable for high humidity conditions with operating temperatures up to 150°C and (2) NEMA grade G-3, a continuous filament-fiber glass-cloth laminate with excellent mechanical and electrical characteristics.

Diallyl Phthalate (DAP)

This is a thermosetting material commonly used for molding connector bodies. It is among the best of the thermosetting plastics because it has low electrical losses, high insulation and humidity resistance, and is very stable dimensionally. Its characteristics are improved further by the inclusion of glass or other additives. It is processed by compression or injection molding and is a product of Allied Chemical Corporation.

Some of the mechanical, thermal, and electrical characteristics of plastic materials are listed in Table 5-10.

5.5 POTTING AND ENCAPSULATING COMPOUNDS

The practice of potting and encapsulating components and assemblies has been increasing because it provides greater reliability to electronic components against humidity and assures against their displacement during handling,

Table 5-10   Characteristics of Plastics (Average Values at 25°C)

	Delrin	Lexan	Kel-F	Poly-ethylene	Nylon
Tensile strength (psi × 1000)	10	10	5.5	1–3	11
Elongation (per cent)	15	75	150	150	220
Flexural strength (psi × 1000)	14	12	10	1.5	14
Heat distortion (264 psi Temp. °C)	100	140	70	24	65
Expansion coefficient (in/in/°F × 10⁻⁵)	4.5	3.9	3.6	7–9.5	5.5
Dielectric strength (volts/mil)	500	400	1000	500–700	380
Dielectric constant	3.7	3.2	2.4	2.2	3.9

	Phenolics, unfilled	Teflon TFE	FEP	Diallyl Phthalate	Poly-imide	Poly-sulfone
Tensile strength (psi × 1000)	6–8	2.5–6	3	3–4	50	10
Elongation (per cent)	—	350	300	—	—	50–100
Flexural strength (psi × 1000)	9–10	—	—	7–9	50–70	15
Heat distortion (264 psi Temp. °C)	130–150	56	128	155	—	175
Expansion coefficient (in/in/°F × 10⁻⁵)	1.7–2.0	5.5	4.6	—	2	3.1
Dielectric strength (volts/mil)	200–400	1000	2100	450	750	425
Dielectric constant	4.6–6	2.1	2.1	3.6	4.5	3.0

vibration, and shock. This has also become a requirement in many military specifications.

The basic difference between potting and encapsulation is in the method of application. In potting, the component or assembly is completely encased in a shell which is then filled with a liquid that hardens or cures at a low temperature. In encapsulation, the component is coated by dipping in a viscous liquid which then hardens.

Any moisture within a component or assembly is removed by baking it under vacuum before it is encapsulated or potted.

Potting has certain disadvantages:

1. The increase in weight of the complete assembly and the inability to salvage the components unless clear, flexible compounds are used, which can be removed to make the repair.

2. The reduced ability to dissipate heat, although this can be corrected by using conducting fillers in the potting compound.

Materials Used

There are several types of materials used for potting and encapsulating.

Epoxy resins consist of the material itself in liquid form, plus a hardener. The two are mixed together and then poured into the potting container in which the component or assembly has been positioned. The container may be metallic or plastic. The curing at room temperature or higher takes from one to four hours.

Polyurethane resins are mixed with a catalyst just before they are used and are cured for several hours at 95°C or higher. These resins have good resistance to abrasion and ozone, but low heat distortion and a high curing temperature.

Silicones, or silicone rubbers, are generally mixed with a catalyst just before use. The amount of catalyst and the curing temperature control the congealing time from minutes to hours.

Other materials are used for this purpose. Some examples of clear plastic encapsulation are shown in Fig. 5-5.

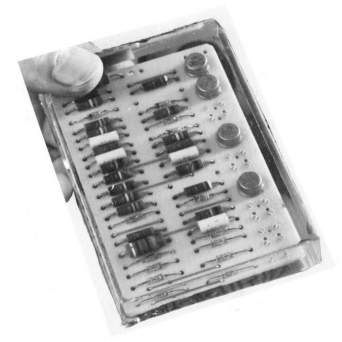

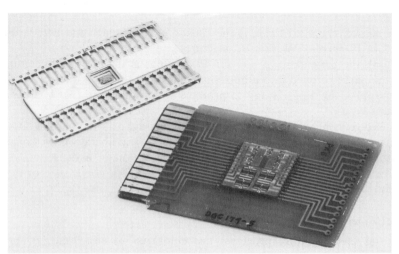

FIG. 5-5. Examples of Clear Encapsulation.

Material fillers may be either inert materials, which reduce the amount of the basic resin required, or powdered metals, such as aluminum or copper, which increase the thermal conductivity of the material.

Low-Density Potting Compounds or Foams

To reduce the weight increase associated with potted assemblies, another means is used to secure mechanical support and electrical benefits. This is in the form of foamed resins of the types listed above.

1. Epoxy resin foams are produced by adding a suitable reacting agent to the resin, which produces gas and forms a cellular structure. Temperature must be controlled to obtain a satisfactory product.

2. Silicone foams are a one-compound powder that melts upon heating and expands into foam. The temperature controls the foam density. A two-component liquid mix curable at room temperature also produces a lower density foam.

3. Polyurethane foams are produced by mixing the resin with a suitable starting mix, which results in generated gas and heat to complete the reaction. The density and, thus, the weight of the foam is controlled by choice of the starting solution.

5.6   FINISHES

There are varieties of finishes applied to the metals that are used in electronics. They include such chemical finishes as conversion coatings, anodizing, electroplating, and organic coatings or painting.

Aluminum

A common chemical finish is the caustic etch in which the part is dipped into a 2 to 10 percent solution by weight of caustic soda.

Among the better known conversion coatings is the phosphate coating, in which the part is dipped into a phosphate chemical solution, and the chromate coating applied by dip or spray to produce a chromate film that is also useful as a paint base.

An electrochemical process, known as anodizing, uses the part as an anode in an acid bath. The cathode is the tank surface. The liberation of oxygen on the surface of the aluminum results in the formation of a dense oxide film.

Surfaces prepared by any of these processes act as a good base for organic coatings, such as lacquers, paints, and varnishes.

Electroplating

Various electroplating processes are employed in electronics to protect components and subassemblies.

Steel. After parts have been cleaned and de-greased, they can be electro-plated to provide corrosion resistance and good electrical conductivity. The conductivity is improved by plating with copper, silver, or gold. Silver or gold plating is applied after the part has been copper plated.

Steel chassis are generally cadmium plated to provide a good soldering base for grounding the components as well as for corrosion resistance.

Nickel plating results in a corrosion-resistant finish that is also decorative. The thickness of the plating determines its effectiveness.

Chromium plating is used when a decorative or hard surface plating is desired. Steel is generally plated with copper, nickel, and then chromium in order to obtain the maximum resistance to exposure.

Aluminum and magnesium. These can be electroplated with any of the metals mentioned above if they are first zinc-coated by immersion or electroplating.

Nonferrous metals. Metals such as brass, bronze, phosphor bronze, and beryllium copper are frequently electroplated with silver or gold when they are used for contact purposes. They may also be rhodium plated to provide a hard wearing surface.

Electroless Plating

As its name suggests, this process does not require the use of applied poten-tial. The item to be plated and the anode of material, such as nickel or copper, are immersed in the plating bath. A very thin coating of the anode material will be deposited on the part.

One use for this type of plating is in the "additive" process of printed-circuit manufacture. In this process a thin copper coating is deposited on the substrate in the desired pattern. Details of this process are described in Chapter 13.

Organic or Paint Finishes

This is probably the most inexpensive method of finishing, especially of large-area parts, to combine decorative and corrosion-resistant features. Such finishing includes primers, lacquers, enamels, and varnishes.

Glossiness. The organic coatings vary in their degree of gloss, depending upon the application or specification requirements. Generally, electronic equipment requires a semi-gloss or dull finish to avoid possible bothersome reflection from such surfaces as front panels.

Primers. A prime coat or primer must be applied to the surfaces of metals or other materials to provide strong bonding for the organic coatings. Most of the chemical coatings listed previously for aluminum also serve as paint primers.

In addition, zinc chromate is used as the prime coat on aluminum, magnesium, and steel surfaces.

Lacquers. When used in their natural state—clear or unpigmented, lacquers protect plated surfaces against tarnishing. Any desired color may be obtained by adding pigments. Lacquers are applied in production by spraying and drying rapidly in air, thus avoiding the necessity of baking.

Enamels. These synthetic finishes furnish a heavier protective coating on electronic equipment assemblies than do lacquers. One varitey of enamels is the wrinkle type, which, through the irregular surface it produces, provides a covering for such minor surface imperfections as scratches or nicks. Another variety is the hammertone finish, which gives the appearance of a hammered metal. The enamels are generally applied in production work by spraying and require baking.

Varnishes. These are used to provide electronic equipment with moisture-, mildew-, and fungus-resistance and are applied by brushing or spraying.

SUMMARY

Many fastening devices are used in electronic equipment. Some are conventional, such as screws and rivets, and others such as quarter-turn fasteners, specialized. The draftsman should become familiar with these devices as well as the materials and processes used in electronics and manner in which they are selected.

QUESTIONS

5.1   List the methods used to join parts mechanically. When are eyelets used?

5.2   What do the following indicate: (a) $\frac{1}{4}$-20 UNC-2A; (b) #4-40 UNC-2B; (c) #8-32 UNF-3A?

5.3   What types of machine screws are commonly used in electronic work? Make a sketch of head styles, identified by name and also by types of head recesses.

5.4   Give several examples of the screw identification that might appear on a typical parts list and explain what each represents in full.

5.5   What is the difference between a tubular and a semi-tubular rivet? Sketch the various aluminum rivet heads and identify each type.

5.6   How are metric threads designated?

5.7   What are the tolerance grades in metric threads?

5.8   List four methods of joining metals.

5.9   What is the limitation in aluminum spot welding?

5.10 What types of metals are used in electronics? Give two examples of each.

5.11 How does aluminum compare in relative weight to magnesium and steel?

5.12 List the aluminum alloy series in the non-heat-treatable types. What are the temper designations for these alloys?

5.13 How is aluminum heat treated?

5.14 List at least five alloying elements in aluminum.

5.15 Give several examples of aluminum-alloy designations, together with their tempers. List examples of quarter-hard, half-hard, and three-quarter hard tempers.

5.16 What are the minimum bending radii for the following aluminum alloys: (a) 1100 half-hard, .128 inch thick; (b) 5052 annealed, .064 inch thick; (c) 6061 annealed, .258 inch thick; (d) 7075 fully heat-treated, .064 inch thick?

5.17 Describe three types of stainless steel and list their compositions, properties, and uses.

5.18 How are stainless steel parts finished after they are fabricated?

5.19 How are beryllium copper parts heat-treated after they are formed?

5.20 What is the disadvantage of using beryllium copper for spring parts?

5.21 Describe the difference between thermoplastic and thermosetting resins and give two examples of each.

5.22 What are some of the filler materials used in reinforced plastics?

5.23 Name a thermosetting resin which is nonflammable.

5.24 What are the chemical designations for the following plastics: (a) nylon; (b) Lexan; (c) Kel-F; (d) Delrin? List some of their properties.

5.25 Identify two types of fluorocarbon resins and list one outstanding characteristic of each.

5.26 What are some of the characteristics of diallyl phthalate?

5.27 Name three grades of phenolic laminates. Which grade is most suitable for use in high-humidity applications?

5.28 What is the difference between the encapsulation and potting processes?

5.29 What metals are commonly used to increase thermal conductivity of potting compounds?

5.30 List three type of foams. Describe one type of foam process.

5.31 What is anodizing? Briefly describe its formation.

5.32 What electrical process is used to protect the surface of steel chassis? What additional benefit is derived from it?

5.33 Describe electroless plating.

5.34 What has to be done to metallic surfaces before an organic finish is applied?

5.35 Make a sketch of the following screw-head styles and identify each: (a) pan; (b) oval; (c) 100° flat; (d) Phillips; (e) socket; (f) round.

6

GOVERNMENTAL
REQUIREMENTS

Since a considerable share of electronic equipment is manufactured for the
Department of Defense, the electronic designer and draftsman should be familiar with various government standards, specifications, and drawings.

The federal government, through the Department of Defense and other
governmental agencies, has issued specifications and standards that cover the
materials, components, and complete equipments purchased by the government. In some cases, these specifications also apply to the design, manufacture, rating, and testing of these components and equipment.

Although the various services of the Department of Defense and other agencies originally issued individual specifications, unified standards are issued now
by the Department of Defense and apply to the Departments of the Army,
Navy, and Air Force.

6.1 GOVERNMENT SPECIFICATIONS AND DRAWINGS

Brief descriptions and examples of these specifications, standards, and drawings
follow.

Military Specifications (MS)

These specifications are developed by the technical facilities of the Departments of the Army, Navy, and Air Force.

Military specifications cover details of cables, chemical processes, components, drawing and engineering data, semiconductors, general electronic

equipment design, hardware, instruction handbooks, insulating compounds, materials, and paints—in brief, every item that enters into the design and construction of electronic equipment.

Following is a breakdown of the coding used in the title of a military specification:

MS-3107A   *Connector, Plug, Electric, Quick Disconnect*
MS   abbreviation for Military Specification
3107   serial number within the group
A   first revision

MIL-E-3954C   *Electrical Waveguide, General Specification For*
MIL   military specification
E   first letter of military specification
3954   serial number within the group
C   third revision

Military Standards (MIL)

Military standards, also developed by the same facilities, establish standardization for codes, designations, drawing practices, electronic circuits, form factors, procedures, revision methods, symbols, test methods, and types and definitions, in order to provide uniformity in electronic equipment design, nomenclature, and appearance.

Military Standard Specifications (MIL-STD)

These requirements are specified by the Departments of the Army, Navy, and Air Force. For example:

MIL-STD-242F   *Electronic Equipment Parts, Selected Standards*
MIL   abbreviation for Military Specification
STD   abbreviation for Standard
242   numerical listing
F   sixth revision

Metric Standardization

The symbol "DOD" replaces the symbol "MIL" in new and revised specifications and standards covering "hard metric" or "hard converted" items, and other distinctly metric documents. The purpose is to provide ready identification of metric documents.

For example:

DOD-D-1000B   *Drawing, Engineering and Associated Lists*
D   metric drawing
1000   consecutive number within the group
B   second revision

Military Qualified Products Lists (QPL)

These lists have been established for materials and the products tested and approved by the Department of Defense and its subdivisions.

For example:

QPL-26   *Resistor, Fixed, Wire-Wound Type, General Specification For*
    QPL   abbreviation for qualified products list
     26   number assigned on original MIL specification

The actual Military (MIL) specification number is:

MIL-R-26E   *Resistor, Fixed, Wire-Wound, General Specification For*
    MIL   military
      R   abbreviation for resistor
     26   number originally assigned for this class of resistors
      E   fifth revision

Federal Specifications

These are prepared by the governmental agency which is directly concerned and most suited to list the requirements. For example:

QQ-M-40B   *Magnesium Alloy Forgings*
    QQ   group designation of one or more capital letters (QQ denotes metals)
      M   first letter of specification title
     40   a consecutive number within the group
      B   second revision

Drawing Types

The Department of the Navy, Naval Sea Systems Command issues the following drawing types:

MS, or Military Sheet Form Standards, which are numbered MS—in sequence. Modifications are indicated by a capital letter suffix, as in MS15527B.

USAF AN Standards include:

AN drawings of standard components and assemblies beginning with AN1.

AND drawings list design information beginning with AND 10,000.

6.2 GENERAL SPECIFICATIONS FOR MILITARY ELECTRONIC EQUIPMENT

General specifications have been issued to standardize the design, packaging, manufacture, inspection, and testing of electronic equipment purchased by the Departments of the Army, Navy, and Air Force. These specifications are based upon the ultimate use of the equipment and stress the following factors:

1. *Reliability.* This is one of the most important requirements for electronic equipment. It must operate under such difficult environmental conditions (listed in the specifications) as extremes in temperature and humidity, altitude, shock, vibration, storage in arctic and tropical regions, and explosive atmosphere. None of these conditions should affect its operation or life to any extent.

2. *Standardized parts.* To reduce the number of spare parts needed to maintain equipment, the greatest possible use of standard parts has been made mandatory in all of these general specifications. Military standard MIL-STD-242, *Electronic Equipment Parts (Selected Standards)*, lists and describes standard mechanical and electronic parts that have been approved for use in military equipment. Special approval must be obtained from the contracting authority to use nonstandard parts and, in some cases, it may be given only to reduce physical size or because approved parts are not available.

3. *Mechanical design requirements.* In many instances, electronic equipment must be installed on ships and submarines or in aircraft or missiles. Therefore, it is mandatory that the equipment be subdivided into assemblies that are small enough to pass through such restrictions as bulkheads and hatches. It must also be small and light enough to be handled, maintained, and transported easily by air to remote bases. Such mechanical production methods as modular construction and printed wiring should be fully utilized.

4. *Test and maintenance.* Built-in facilities must be provided so the equipment can be tested and serviced by relatively inexperienced personnel using standard tools and test equipment that are readily available.

A military standard, MIL-STD-454, *Standard General Requirements for Electronic Equipment*, contains many of the common requirements to be used in military specifications for electronic equipment. Each requirement is identified numerically, starting with number 1, and is referred to by its number in general equipment specifications.

The sizes of printed-wiring boards have been standardized whenever cost and technical requirements permit. They are shown in Table 6-1.

6.3 ELECTRONIC EQUIPMENT, GROUND, GENERAL REQUIREMENTS FOR, MIL-E-4158F (USAF)

Equipments of this type are purchased by the Department of the Air Force. Included in this class are aircraft landing and ground communication equipments and other apparatus of a similar nature.

Many of the specific requirements in this specification are based on MIL-STD-454, *Standard General Requirements for Electronic Equipment.*

Mechanical Details

Operating controls should be located adjacent to their associated displays. The most frequently used operating controls should be located for right-hand operation. The adjustment controls should be mounted behind access

Table 6-1 Standard Printed-Wiring-Board Sizes*

Board Number	Board Size ±.015	Hole Centers' Spacing ±.005
A1	2 × 3	1.50 × 2.50
A2	2 × 4.50	1.50 × 4.00
A3	2 × 6.50	1.50 × 6.00
A4	2 × 8.00	1.50 × 7.50
B1	5 × 3	4.50 × 2.50
B2	5 × 4.50	4.50 × 4
B3	5 × 6.50	4.50 × 6.00
B4	5 × 8	4.50 × 7.50
C1	6.50 × 3	6 × 2.50
C2	6.50 × 4.50	6 × 4
C3	6.50 × 6.50	6 × 6
C4	6.50 × 8	6 × 7.50
D1	8 × 3	7.50 × 2.50
D2	8 × 4.50	7.50 × 4
D3	8 × 8	7.50 × 7.50

*Three .125-inch-diameter extractor holes are to be located in each size board, 1/4 inch from the two adjoining edges in the upper left, upper right, and lower right-hand corners.

doors on operating panels. The characters on the dials should be large enough to be read from a distance of two feet within an included 60-degree angle, under normal viewing conditions. At least two numbers in each band on the dial must be visible at all times.

Threaded inserts or captive nuts that can be replaced readily should be used in aluminum and magnesium alloy parts. The minimum length of tapped holes should equal the screw diameter, and in highly stressed applications, the threads should be half again as long. Screws should extend at least $1\frac{1}{2}$ threads beyond the nut or any other engaging part. Whenever possible, even-numbered screws should be used.

All parts, except sealed parts, terminals, and wiring, should be accessible for circuit checking, adjustment, and maintenance with a minimum need for disassembly and special tools. Parts should be easily removed and replaced without disturbing or injuring any other parts or wiring. Rivets should not be used to mount such parts as capacitors, inductors, resistors, or transformers if parts will be removed during maintenance.

Electrical Details

Panel fuses of the extractor fuse-post type should preferably be of the indicating type.

Where practicable, insulated wires should be formed into cables and laced with twine or tape. Adequate slack should also be provided for removal of parts and units in drawers or slide-out racks.

Bushings or grommets should be used to protect wires that pass through holes in metal. Wiring and cabling should have adequate support. Wiring should be identified at the end of each lead by color coding or by marking terminals.

Any equipment with voltage sources of more than 70 volts must be protected against accidental contact when in operation. All compartment doors and covers must be equipped with interlocks if the potential within exceeds 500 volts and if they provide the access for adjustment during equipment operation.

6.4 ELECTRONIC EQUIPMENT, AIRBORNE, GENERAL SPECIFICATION FOR, MIL-E-5400T

This equipment is manufactured for all departments and agencies of the Department of Defense and includes communication, navigation, and fire-control equipments used in piloted aircraft. Aircraft electronic equipment is subdivided into four classes, based upon their altitude and temperature range:

Class 1—for operation up to altitude of 50,000 feet, and at sea level at temperatures ranging from $-54°$ to $+55°C$ ($+71°C$ intermittent operation).

Class 1A—the same as Class 1, except at a maximum altitude of 30,000 feet.

Class 1B—for operation up to 15,000 feet altitude and continuous sea level at temperatures ranging from $-40°$ to $+55°C$ ($+71°C$ intermittent operation).

Class 2—for operation up to an altitude of 70,000 feet and at sea level at temperatures ranging from $-54°$ to $+71°C$ ($+95°C$ intermittent operation).

Class 3—for operation up to an altitude of 100,000 feet and at sea level at temperatures ranging from $-54°$ to $+95°C$ ($+125°C$ intermittent operation).

Class 4—for operation up to an altitude of 100,000 feet and at sea level at temperatures ranging from $-54°C$ to $+125°C$ ($+150°C$ intermittent operation).

The addition of the letter "X" after the class number identifies equipment using auxiliary cooling from an external source.

Many of the specific requirements in this specification are also based on MIL-STD-454.

The Army, Navy, and Air Force have made it mandatory that aircraft electronic equipment be enclosed and mounted according to Military Specification MIL-C-172, *Cases; Bases, Mounting; and Mounts, Vibration*. This specification is an outgrowth of the ATR system for commercial aircraft, and some of the case sizes and mounting bases are similar in size.

Polyvinyl chloride insulating materials shall not be used in airborne applications.

Mechanical Details

Controls should be readily accessible and of rugged construction.

The heads of flathead screws may be staked or the adjoining surface metal may be upset into the screw slot in permanent assemblies.

Toggle switches should be mounted with the handle vertical, and any "off" position in the center or at the bottom. A "left-right" control may be mounted in the horizontal direction.

Screw lengths should be selected from the nearest longer standard length.

The need for special tools should be kept to a minimum. One wrench for each size and head-type of set screw, and any other required tools for operational adjustment should be mounted within the equipment. Mechanized construction, such as printed wiring, should be used as much as possible.

6.5 ELECTRONIC, INTERIOR COMMUNICATION AND NAVIGATION EQUIPMENT, NAVAL SHIP AND SHORE, GENERAL SPECIFICATION FOR, MIL-E-16400G

This specification covers the general requirements applicable to the design and construction of electronic interior communication and navigation equipment intended for naval ship or shore applications.

Unless otherwise specified, all equipment should use modular construction, utilizing the packaging techniques described in *Military Handbook*, MIL-HDBK-1239. The modular construction of the equipment must conform to the standard electronic module (SEM) program as in MIL-STD-1378B.

Naval ship and shore electronic equipment shall operate satisfactorily when subjected to the temperature ranges given in Table 6-2.

Table 6-2   Temperature Ranges (Ambient)

Range	Environmental Conditions	$C°$ Operating	Nonoperating
1	Exposed, unsheltered— ship or shore	−54 to +65	−62 to +71
2	Exposed, unsheltered— ship	−28 to +65	−62 to +71
3	Sheltered—shore	−40 to +50	−62 to +71
4	Sheltered—ship or shore	0 to +50	−62 to +71

Riveting should be used for parts not requiring removal when the equipment is serviced. Such replaceable parts as capacitors, resistors, or transformers should not be secured by riveting. Riveting and bolting should be used only if nonweldable aluminum alloys or removable panels are required; otherwise, boxes, shields, cases, etc. should be constructed by casting, drawing, bending, and welding.

Interior parts should be easily accessible for adjustment, removal, or circuit checking without the removal or displacement of wiring or assemblies to reach terminals, mounting screws, etc.

Assemblies and subassemblies should be mechanically and electrically interchangeable regardless of the manufacturer or supplier. The design should allow for the tolerances applicable to such components as transistors, resistors, connectors, printed-circuit boards, tubes, etc.

Mechanical Details

The standard 19-inch rack construction has been used in the majority of shipborne and submarine installations to mount electronic equipment. The various widths of panels and the chassis attached to them have provided the necessary dimensional variations to accommodate almost any equipment subdivision.

The new trend in this equipment is toward modular construction or subdivisioning circuitry into basic modules that can be rapidly replaced or repaired. Each module contains all the components of some basic electronic circuit, such as the shift register, amplifier, and flip-flop. This practice has been followed in the construction of such complex equipment as digital computers and is now being applied on an increasing scale to military electronic equipment design.

One of the first military applications of the modular construction was in the fire control system of the nuclear-type submarines.

Electronic assemblies should be interchangeable by the use of plug-in units, and the designs should have provisions to modernize the assemblies by adding either new or improved units in the same enclosure or space.

All nuts, except self-locking, and all machine screws except flathead type, require a lockwasher under the nut or under the screwhead if no nut is used. Aluminum parts that are secured with a lockwasher should also have a flat washer.

Any parts that will be exposed to sea water and combustion gases should be made of corrosion-resistant steel.

Screw threads should be of the coarse- or fine-thread series. Screw-thread devices that attach complete items to the supporting structure should use size 4-40, 6-32, 8-32, 10-32, or $\frac{1}{4}$-28 threads. The thread engagement in tapped parts, except nuts, should be at least as large as the diameter of the screw or bolt.

All parts should be accessible from the front of the equipment by withdrawing the chassis or by opening the access doors. The rear of each chassis should be provided with locating pins for proper alignment and stability under vibration.

Equipments for interior shipboard installation should not exceed 72 inches in height, including shock mounts. They should be capable of passage through a doorway 26 inches wide by 45 inches high, with corners of 8-inch radius, and through a hatch 30 inches by 30 inches with corners of $7\frac{1}{2}$-inch radius. See Fig. 6-1.

Equipment enclosure, exposed to the weather, shall be watertight.

Equipment, or portions thereof, weighing more than 150 pounds shall be provided with lifting eyes.

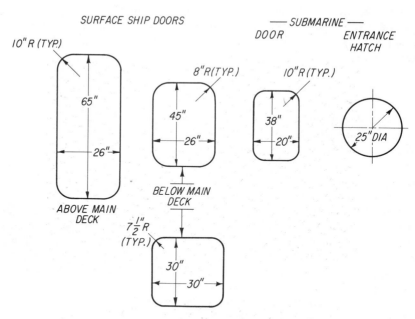

FIG. 6-1. Typical Surface Ship and Submarine Door Dimensions.

If possible, control knobs and handles should increase the final controlled effect when they are rotated clockwise.

Controls that are used only occasionally should be mounted behind access panels and operated by knobs. Controls that are used infrequently should be operated by screwdriver and be accessible when the equipment is opened for maintenance.

Electrical Details

Connectors, unless used for dynamotors, inverters, printed circuits, meters, and similar items, should have a minimum of extra unused contacts, located on outer periphery as follows:

Total Number of Pins	Extra Contacts or Contact Positions Required
1–25	2
26–100	4
Over 100	6

All panel-mounted fuse holders should be of the extractor fuse-post type and should be finger-operative.

Hook-up wire should be preferably stranded. Wherever possible, cabling should be laced with plastic cord, spirally cut plastic tubing, or other suitable material.

Connections should be provided on the face of the equipment for any test equipment that may be required for operational checkout tests.

All exposed terminals and similar points with potentials of 70 to 500 volts rms and dc should have guards or barriers to prevent accidental contact. Wherever practicable, internal wiring conductors should be combined into laced cables, using the special lacing tape available for this purpose. All conductors should be color coded, using the same coding throughout the circuit and, wherever possible, through interconnecting plugs and receptacles. All lights used to illuminate controls, switches, and dials should be dimmer-controlled from zero to full brilliance. This may be done either singly or in groups, and either optically or electrically.

Long rigid conductors or cabled flexible conductors should be securely anchored to the chassis with clamps.

SUMMARY

The design and manufacture of electronic equipment for governmental use is established by the general specifications, standards, and drawings issued by the Department of Defense and other agencies. The student should become familiar with these requirements.

QUESTIONS

6.1 Describe some of the differences between military specifications and military standards.

6.2 Discuss some of the factors that are common to all general electronic equipment specifications.

6.3 Give a short description of the mechanical requirements specified in the specification for ground electronic equipment.

6.4 What are some of the detail requirements for controls in the same specification (MIL-E-4158)?

6.5 At what voltage are interlocks required in MIL-E-4158?

6.6 How many classes of equipment are listed in specification for airborne electronic equipment? What is the maximum altitude operational range?

6.7 What replaceable components cannot be secured by riveting?

6.8   What are the connector-pin requirements for airborne electronic equipment?

6.9   List some of the practices that are required for ship-and-shore electronic equipment.

6.10 Describe some of the mechanical requirements for shipboard electronic equipment.

6.11 What is the limiting feature on submarines that determines the size of the equipment?

6.12 Give one of the requirements for lights that are used for illumination purposes on the front panels of shipboard equipment.

6.13 How should toggle switches be mounted on control panels?

6.14 What are the sizes of doors on surface ships? Make a sketch.

7

ENGINEERING STANDARDS AND SPECIFICATIONS

The standardization of methods, materials, testing, and other factors involved in the manufacture of electronic components and equipment became necessary as these factors increased in complexity, and as reliability and simplified maintenance were added to the overall requirements.

Many professional and trade organizations have compiled standards covering the various aspects of construction, identification, size, overall performance, testing, and other requirements for electronic components and equipment. The engineering standards, which represent nationwide practices, are subject to periodic revision as these requirements and procedures change. Based upon the recommendations of experts, who are members of engineering societies, technical trade orgnizations, or government standardization groups, these standards represent the best known practice at the time of issue.

The standards and specifications issued by the Department of Defense and other government agencies apply to the construction and performance of components or equipment purchased by the government.

The addition of a suffix letter to the standard's number indicates a change in the standard, so that standard's lists should be consulted to make sure that the latest issue of any given standard is used.

7.1 MILITARY STANDARDS AND SPECIFICATIONS

All equipment purchased by the Department of Defense—electronic or otherwise—must withstand more severe environmental and operating conditions than equipment designed for industrial or consumer markets.

Military components and equipment must meet such requirements as reliability (the primary and most important requirement), interchangeability, replacement, ability to withstand extremes of temperature and humidity, vibration and shock, ease of operation, and any others outlined in the contracts for such equipment. The contractor, and thus also the electronic designer and draftsman, must be familiar with these requirements and the applicable standards to check the acceptability of the equipment.

Over a period of years, the various military services have established standardization groups that have issued standards applicable to the individual services or the Department of Defense. MIL specifications and MIL standards and their numbering were described in detail in Chapter 6.

The student should become familiar with the wealth of information on materials that is included in these specifications and standards. It will enlarge his scope of knowledge and assist him in the design of equipment which must meet such specifications.

The contractor manufacturing government equipment obtains his copies of the applicable military specifications and standards from the regional offce of the service issuing the contract. The general specification listed in the contract identifies the relevant individual specifications and standards.

The DOD Index of Specifications and Standards (DODISS) is issued in two separate parts: alphabetical and numerical. Supplements are issued quarterly.

The Index is available through the Superintendent of Documents, U.S. Government Printing Office, Washington, D.C. 20402. The current price is $40 per year. However, only a limited number of the specifications listed in the Index are sold by the Printing Office.

Copies of the standards, specifications, and related standardization documents, and industry documents mandatory for Department of Defense designs may be obtained from:

Commanding Officer
Naval Publications and Forms Center
5801 Tabor Avenue
Philadelphia, Pa. 19120

They may be requisitioned by contractors doing governmental work by using a special form issued for this purpose. All items should be ordered by stating the official title listed in the Index of Specifications and Standards.

The Index lists the unclassified Federal, Military, Industry, and Departmental specifications, standards, and related standardization documents, and also those industry documents that are coordinated for Department of Defense use.

To provide ready identification of metric documents the symbol "DOD" replaces the symbol "MIL" in new and revised specifications covering "hard metric" items. Thus DOD-D-1000B identifies *Drawings, Engineering and Associated Lists.* It was originally listed as MIL-D-1000.

Table 7-1 lists some of the military standards that are of interest to the student and that relate to the general drawing practices and mandatory procedures

Table 7-1   Military Standards and Specifications

Standards

MIL–STD–8C	*Superseded by ANSI Y14.5–1973*
MIL–STD–9	*Screw Thread Conventions and Methods of Specifying*
MIL–STD–12	*Abbreviations for Use on Drawings and in Technical-Type Publications*
MIL–STD–17	*Mechanical Symbols*
MIL–STD–27	*Designations for Electric Switchgear and Control Devices*
DOD–STD–35	*Automated Engineering Document Preparation System*
MIL–STD–106	*Mathematical Symbols*
MIL–STD–108	*Definitions of and Basic Requirements for Enclosure for Electric and Electronic Equipment*
MIL–STD–167	*Mechanical Vibrations of Shipboard Equipment*
MIL–STD–196	*Joint Electronics Type Designation System*
MIL–STD–242	*Electronic Equipment Parts, Selected Standards*
MIL–STD–275	*Printed Wiring for Electronic Equipment*
MIL–STD–280	*Definition of Item Levels, Item Exchangeability, Models, and Related Terms*
MIL–STD–429	*Printed-Wiring and Printed Circuit Terms and Definitions*
MIL–STD–454	*Standard General Requirements for Electronic Equipment*
MIL–STD–455	*Alloy Designation System for Wrought Copper and Copper Alloys*
MIL–STD–681	*Identification Coding and Application of Hook-up and Lead Wire*
MIL–STD–701	*Lists of Standard Semiconductor Devices*
MIL–STD–710	*Synchros, 60 and 400 hertz*
MIL–STD–1132	*Switch and Associated Hardware, Selection and Use Of*
MIL–STD–1313	*Microelectronic Terms and Definitions*
MIL–STD–1358	*Waveguides, Rectangular Ridge and Circular, Selection Of*
MIL–STD–1378	*Requirements for Employing Standard Electronic Modules*
MIL–STD–1389	*Design Requirements for Standard Hardware Program, Electronic Modules*
MIL–STD–1472	*Human Engineering Design Criteria for Military Systems, Equipment and Facilities*
DOD–STD–1476	*Metric System, Application in New Designs*
MIL–STD–1634	*Module Descriptions for the Standard Electronic Module Program*

Specifications

MS–16662	*General Notes for Standard Electrical Drawings*
MIL–P–55110	*Printed Wiring Boards*
MIL–P–50884	*Printed Wiring, Flexible, General Specifications For*
MIL–M–83436	*Multiwire Interconnection Boards (Plated-through Hole)*

in the construction of military electronic equipment. It would be desirable for the student to have at least some of them in his technical library for reference.

The military services have issued a number of general electronic equipment specifications that describe construction and environmental conditions in detail; that list the various applicable individual component specifications; and that describe the functional, electrical, electronic, mechanical, and test requirements. These general specifications were described in detail in Chapter 6.

7.2 AMERICAN NATIONAL STANDARDS INSTITUTE STANDARDS

The American National Standards Institute is composed of more than a hundred engineering, technical, and trade associations and societies as well as government representatives. Although ANSI does not write standards, it acts as a coordinating agency in promoting uniform standards that are nationally acceptable. Use of these standards, which are sponsored by such technical societies as ASME, IEEE, and ASTM, is voluntary.

A standards price list at a nominal price and the individual standards are available from the American National Standards Institute, Inc., 1430 Broadway, New York, N.Y. 10018. Some of their standards have replaced various Military Standards, and the Department of Defense has made the use of these standards mandatory.

Table 7-2 gives examples of ANSI standards. Table 7-3 lists major metric standards from ANSI, the Department of Defense, and the ISO (International Standards Organization).

Table 7-2   ANSI Standards List

B46.1–1978	*Surface Texture*
H35.1–1978	*Alloy and Temper Designation Systems for Aluminum*
Y1.1–1972	*Abbreviations for Use on Drawings and in Text*
Y14.1 –1980	*Drawing Sheet Size and Format*
Y14.2 –1973	*Line Conventions, Sectioning, and Lettering*
Y14.3 –1975	*Multi and Sectional View Drawings*
Y14.4 –1957	*Pictorial Drawing*
Y14.5 –1973	*Dimensioning and Tolerancing*
Y14.6 –1978	*Screw Thread Representation, Engineering Drawing and Related Documentation Practice*
Y14.7.1 –1971	*Gear Drawing Standards, Part 1, for Spur, Helical, Double Helical, and Rack*
Y14.7.2 –1978	*Gear Drawing Standards, Part 2, Bevel and Hypoid Gears*
Y14.10 –1959	*Metal Stampings*
Y14.15 –1966	*Electrical and Electronics Diagrams*
Y14.15a–1970 (R1973)	*Supplement*
Y14.15b–1973	*Supplement*
Y14.36 –1978	*Surface Texture Symbols*
Y32.2 –1975	**Graphic Symbols for Electrical and Electronics Diagrams (Including Reference Designation Letters)*
Y32.14 –1973	**Graphic Symbols for Logic Diagrams (Two-State Devices)*
Y32.21 –1976	**Graphic Symbols for Grid and Mapping Used in Cable Television Systems*
Y32.16 –1975	**Reference Designations for Electrical and Electronic Parts and Equipments*
Y32E –1976	A collection adopted by the Department of Defense for mandatory use; contains items marked with an *.

Table 7-3   Metric Standards

ANSI B1.13M–1978	*Metric Screw Threads M Profile*
ANSI B1.21M–1978	*Metric Screw Threads MJ Profile*
ANSI Z210.1–1976	*Metric Practice Guide*
DOD–STD–1476	*Metric System, Application in New Design*
ISO R724	*Metric Screw Threads, Basic Dimensions*

7.3 ELECTRONIC INDUSTRIES ASSOCIATION STANDARDS

This is a national association of electronic manufacturers. It issues engineering standards at its office, 2001 Eye Street, N. W. , Washington D.C. 20006. These standards are helpful in producing interchangeable components within the industry.

7.4 INSTITUTE OF ELECTRICAL AND ELECTRONICS ENGINEERS STANDARDS

Engineering standards are available from the Institute of Electrical and Electronics Engineers, 345 East 47th Street, New York, N.Y. 10017.
Among these are:

IEEE Standard 200–1975,* *Reference Designations for Electrical and Electronics Parts and Equipments*

IEEE Standard 315–1975,† *Graphic Symbols for Electrical and Electronics Diagrams*

7.5 NATIONAL MACHINE TOOL BUILDERS' ASSOCIATION STANDARDS

This association is composed of most of the companies in the machine tool field. It is a service organization that promotes technical standards within the machine tool industry through its technical standards committee. It works closely with American National Standards Institute and similar organizations. It is located at 7401 Westpark Drive, McLean, Va. 22101. Its Joint Industrial Council (JIC) has issued a standard *JIC Electrical Standard for Mass Production Equipment* and *JIC Electrical Standards for General Purpose Machine Tools*.
This was later followed by another standard, EL-1-71, *Electronic Standards for Mass Production Equipment and General Purpose Machine Tools*.

7.6 STANDARDS ISSUED BY OTHER ORGANIZATIONS

Among the other engineering organizations that issue standards are:
Aeronautical Radio, Inc. , 2551 Riva Road, Annapolis, Md. 21401, issues specifications covering electronic equipment details for commercial aircraft.
The American Society of Mechanical Engineers (ASME), 345 East 47th Street, New York, N.Y. 10017, which has sponsored or cosponsored some of the ANSI standards.

*Same as Y32.16–1975 in Table 7-2.
†Same as Y32.2–1975 in Table 7-2.

Institute for Interconnecting and Packaging Electronic Circuits (IPC), 3451 Church Street, Evanston, Ill. 60203, sponsors standardization of printed circuitry and yearly industry meetings.

The National Electrical Manufacturers Association (NEMA), 155 East 44th Street, New York, N.Y. 10017, has among its standards a publication called *General Standards for Industrial Control and Systems*, ICS-1-1978, which is especially useful in industrial diagram work.

Radio Technical Commission for Aeronautics (RTCA), Suite 665, 1717 H Street, N.W., Washington, D.C. 20006, issues standards.

Tubular Rivet and Machine Institute, 707 Westchester Avenue, White Plains, N.Y. 10604, issues *Dimensional Standards for Semi-tubular Rivets* and *Metric Dimensional Standards for General Purpose Semi-tubular Rivets*.

7.7 SOURCES OF FEDERAL GOVERNMENT STANDARDS

The following are some of the sources of federal government standards issued for guidance in military and civilian applications:

U.S. Army Armament Research and Development Command, Dover, N.J. 07801

Federal Communications Commission (FCC), Washington, D.C. 20554

General Services Administration (GSA), Washington, D.C. 20405

National Aeronautics and Space Administration (NASA), Washington, D.C. 20546

SUMMARY

In addition to the governmental standards, the electronics industry has adopted many standards of its own. These have been issued by various industry associations. Their addresses are given in this chapter so the student may write for copies.

QUESTIONS

7.1 How is a change indicated in a standard?

7.2 What subjects do the engineering standards cover?

7.3 Discuss the various requirements that military equipment has to meet.

7.4 How are the designer and draftsman affected by these engineering standards and specifications?

7.5 What is a general electronic equipment specification?

7.6 What organizations sponsor ANSI standards?

7.7 List some of the professional organizations that issue standards.

7.8 Name the organization that issues machine tool electronic standards.

7.9 What organizations sponsor standards applicable to commercial aircraft electronic equipment?

7.10 Name some of the sources of federal government standards.

8

GENERAL ELECTRONIC
DESIGN

The designer must meet many mechanical and electrical requirements in design-ing electronic equipment, regardless of whether the equipment will be for military or for civilian use. Not only must all these requirements be satisfied concurrently, but the finished product must also be equal to or better than competitive equipment—in terms of price, reliability, meeting specifications, and ease of operation.

The mechanical design of the equipment exerts a vital influence upon its overall performance, being particularly responsible for its trouble-free operation in all types of environmental conditions. Reliability is a major factor being stressed now in both military and civilian applications, and it can be achieved only through the proper application of the various underlying electronic and mechanical principles. This is especially true of complex, modern electronic equipment which may include thousands upon thousands of electronic and electromechanical components that are integrated within a computer, a missile, a radar, or other electronic system.

Electronic equipment is subdivided into basic smaller units known as chassis and modules. A chassis provides a mounting surface or surfaces for such components as printed circuit cards, modules, transistors, resistors, and capacitors. The chassis may be fabricated from sheet metal, such as aluminum, cold-rolled steel, copper, or brass, or it may be cast. Since the chassis is metal-lic, it provides shielding against electric or magnetic fields and physical pro-tection for the components mounted on or within it. It also acts as part of the electric circuit—as a ground return, for example—and as a thermal con-ductor for such heat-dissipating components as resistors.

The chassis may serve as a thermal and electrical shield between power transistors and rectifiers, and high wattage resistors mounted on one side of the chassis and components that may be affected by their heat, such as capacitors, integrated circuits, or inductors, mounted on the opposite side.

The important factors in chassis design are size, shape, weight, material, rigidity, fabrication procedures, accessibility to components, and conformance with available manufacturing facilities.

The chassis, which is normally at ground potential, is convenient as a return for all the component grounding points. It is a common practice to limit the number of such returns, and to return all circuits common to one transistor to one ground point.

As electronic equipment has increased in complexity, it has become advantageous to subdivide it into small subunits or modules, which are much smaller than chassis subdivisions. Such modules may, for example, contain only one circuit function, such as an amplifier stage, that can be replaced by semi-skilled personnel, once the fault has been isolated.

Such construction has obvious advantages in production line manufacture. The individual modules are much easier to handle, test, and replace. When reduced to a single function, modules become "throw-aways," that is, units that are low enough in price to warrant replacement with new units rather than repairing defective ones. The modules can be completely sealed to achieve environmental protection and they can be of the plug-in type, and thus quickly replaceable. The last item is of importance in providing continuity in electronic equipment operation in military applications.

8.1 TASKS OF THE DESIGNER

These tasks involve both the mechanical and the electronic design, from the breadboard model to the prototype stages and on through to the manufacturing of the finished units.

After the particular circuit has been selected in the early development stage, the mechanical designer is called upon to decide on several major requirements: the selection of components and their placement; the physical size of the completed unit and its basic layout; the means of connecting subassemblies; compliance with applicable specifications; and a design to meet human engineering factors.

One of his first tasks is to establish the physical layout of the equipment—how to separate it into chassis or subchassis, and how to enclose the chassis in cabinets. Of necessity, his selection is based upon the space available for the equipment and its ultimate use—airborne, shipborne, submarine, consumer, or commercial. Such factors as available space in aircraft structure and the size of hatches and doors aboard ship or submarine will dictate the size and shape of the equipment. Other special applications include installations in motor vehicles and missiles where space is at a premium.

The designer's next task is to decide upon the sheet metal chassis layout, the component disposition on the chassis or such component assemblies as modules, the most suitable controls and mechanical drives, and the necessary hardware. He must develop the chassis layout, provide mounting for modules or printed-circuit boards, and also design other metal or insulating parts that will be economical to produce. Components, electronic or electromechanical, must be selected to fit the weight, power, voltage, size, and circuit needs. Their location is governed by such factors as thermal considerations and possible electrical interaction between the components.

Enclosing the chassis and its components is another task of the designer. Military equipment has to meet exacting requirements in regard to shock and vibration, as well as the various specifications for equipment types and their overall dimensions.

A thorough knowledge of the shop procedures involved in the manufacture of electronic equipment is a major asset to the mechanical designer. These procedures include not only metal fabrication but also wiring and assembly techniques that result in lower cost and a more uniform product.

Quite often, the mechanical designer is a member of a research and development group, and the electronic and project engineers rely upon his skills to solve their mechanical layout and design problems. In other cases, he may be part of the mechanical engineering department and assigned to a special task group when his services are needed. His talents will, in any case, be of considerable assistance in the mechanical design of electronic equipment for civilian or military use. Over a period of time, the electronic draftsman will accumulate a considerable amount of experience that will help him to develop the designing skills needed to meet the various requirements already mentioned. If he continues to develop his proficiency in design, he is very likely to be promoted to the design group.

8.2   DESIGN FACTORS AND REQUIREMENTS

The technical design requirements or design parameters that are likely to affect the design are mechanical, electrical, functional, and environmental.

Mechanical design requirements include, among others, such factors as: unit subdivisions, as affected by size, shape, and weight; location of components and their mounting; dimensional tolerances; shielding; and equipment marking. Electrical requirements have such aspects as: circuit functions and wiring distribution; component selection, as related to size, electrical ratings, and tolerance; and internal and external interconnections. Functional design requirements include: reliability, maintenance, accessibility, and human engineering, such as visual displays, controls, and lettering. Environmental design must consider such factors as: mechanical shock and vibration, temperature extremes, salt spray and fungus proofing, and operation in space or underwater. All of these factors are interrelated, from the initial or experimental design to the final production set of drawings.

The design of an electronic chassis, unit, or complex equipment is a continuing process that generally involves several steps. It is carried on concurrently in the laboratory, the drafting room, and the model shop. Breadboard models are constructed using a temporary chassis or other structures, Fig. 8-1(a) to

FIG. 8-1. (a) Breadboard Layout on a Perforated Plastic Base. (b) Commercial Breadboard with DIP Sockets.

mount components that may be varied or relocated during circuit development. Or they may be commercially built breadboards, Fig. 8-1(b), of various types to suit the designer's requirements.

Familiarity with the various components, particularly those described in Chapter 4, will be of considerable help in equipment layout, printed board work, chassis and panel layout, and other subassembly and detail work. The electronic draftsman is expected to progress in this work and will be given more difficult design assignments as he gains knowledge and experience.

Functional Design Factors

As previously noted, these factors include reliability, maintainability, accessibility, and human engineering. Such factors must be given appropriate consideration in equipment design.

Reliability. This is one of the most important factors in military electronic equipment design and to a lesser degree in civilian equipment. In an emergency, failure of even a minor component is likely to be of major importance in the operation of a ship or an airplane.

Problems with reliability have led the Defense Department to establish a list of standard parts (MIL-STD-242) that are required to be used in all military equipment.

Nonstandard parts require a special approval from the agency which issued the contract. Their reliability has to be demonstrated to the satisfaction of the agency involved.

Electrical reliability can be improved by de-rating—operating such frequently used components as resistors and capacitors below their rated value. In addition, components should be placed so that they will operate below their maximum temperature rating (thermal considerations). Using heat sinks for diodes and transistors is but one of the many practices followed to reduce component temperature to a safe limit. Cooling, by natural or forced means, also serves the same purpose. The use of standard parts that have proved reliable through previous extensive use is another means of reducing the rate of failure.

Mechanical reliability can be improved by embedding or encapsulating the components to protect them against moisture, high temperatures, fungus, and mechanical shock and vibration. Selecting the proper materials and methods of joining also helps to reduce the possibility of an early failure. The location of visual displays and controls also influences reliability, since they affect the ease of operation of the equipment. De-rating such mechanical devices as bearings will contribute to longer life of the equipment.

Maintainability. This design factor contributes to the ease, accuracy, rapidity, and economy of maintenance of military and civilian electronic equipment. It also contributes to the simplicity of maintaining the equipment in its normal operating condition or of quickly restoring it to that state. The prescribed maintenance must be within the capabilities of technical personnel and the testing equipment supplied.

Provisions have to be made in the design of military equipment to provide ready access to test points and to be able to utilize testing equipment to its fullest extent to quickly locate the source of trouble.

Accessibility. An electronic equipment subdivision, such as a chassis mounted within a rack, is generally mounted on chassis slides to facilitate servicing. The chassis may either be mounted vertically, so that the bottom or wiring side is exposed when the chassis is pulled out on the slides, or it may be hinged.

When access openings are provided for maintenance, they should be large enough to remove the largest part likely to need replacement and to allow tools to be manipulated through the opening. Removable access covers should have fasteners that are operated with standard tools.

Parts that are likely to fail or to require considerable time for replacement should be located in the most accessible positions on the chassis. Large, heavy parts should be located so that they can be reached and replaced without damaging other parts. The layout should also provide the visual access needed to replace removable parts.

Finally, precautions should be taken to prevent injury to personnel through electrical shock, burns, etc., when access covers are removed for part replacement.

Human engineering. In designing electronic equipment, the designer must also consider the limitations of its operator. For example, maximum arm reach should not exceed 28 inches, and the maximum hand-control force required should be limited to 25 pounds.

Among the more important design factors that are likely to affect the operator are visual displays, warning devices, and controls. He must usually coordinate them to keep the equipment in operating condition or to take the necessary corrective measures in the event of malfunction.

Visual displays. These consist of various forms of circular or semicircular scales that are movable or fixed. If the scale is fixed and the pointer is movable, the reading value should increase with the clockwise movement of the pointer and with the controlling knob rotating in the same direction. If a movable scale is partially covered, at least two major divisions should be visible at all times. If the scale is movable and the pointer is fixed, the scale should rotate counterclockwise so that the values increase in a clockwise movement.

If several dial displays appear on the control panel of the equipment, they should be arranged in horizontal or vertical rows or in a square, with all the pointers located in the same relative position. Indicator lamps or pilot lights should also be arranged in either horizontal rows or a square formation.

Controls. There are two general types of controls—linear and rotary. One linear-control type used in many electronic equipments is the toggle switch, which may be of the momentary contact or detent type. It is standard practice to have vertically operated toggle switches turn the power on in the "up" position, while the horizontally operated switch turns the power on by moving the switch to the right.

Rotary or continuous action controls can be set at any position within the movement limits. Rotary controls may be in the form of a fixed scale and a moving pointer knob, a moving scale and a fixed pointer, a rotary control knob, or a handcrank or handwheel.

Regardless of whether the control is linear or rotary, it should be located as close as possible to its indicator. Control knobs should not interfere with the observation of the visual indicator.

The number of controls on the front or control panel can be reduced considerably if they are separated by use. For example, controls that are seldom used should be located behind the front panel; controls that are frequently used, but not required for normal operation, should be located behind hinged doors; and those normally required for equipment control should be positioned on the front panel.

Knobs with pointers should be set close to the panel that carries the calibration marks. Similarly, pointers for movable scales should be set close to the scale to minimize parallax. Round knobs should have an average diameter of about $1\frac{1}{2}$ inches, with size variations used to distinguish between controls. Knobs with various handle shapes and colors, called tactile knobs, are also used for control distinction in such applications as radar equipment, which involves multiplicity of controls.

Control labelling should give specific control function rather than electrical characteristics of the circuit. Controls should be arranged from left to right in sequence of operation, not at random.

Safety. Certain mechanical design features that help protect the operator should also be considered. These include the elimination of sharp corners on cabinets, panels, doors, and other mechanical structures, and the use of a dull finish to eliminate glare. This is a requirement in most military specifications.

Lettering. Large capital letters, with a height-to-width ratio of 1:1 and colors that contrast with the background, should be used since they can be seen at a greater distance and are easier to read.

Dial calibration subdivisions should be narrower than graduation divisions and about one-half the height. Scale markings should be about $\frac{1}{16}$ inch apart for the finer subdivisions, with the major divisions extending a minimum of $\frac{1}{2}$ inch. The thickness should be from 5 to 10 percent of the graduation spacing. The pointer tips should be about $\frac{1}{32}$ inch from the scale markers and should not overlap them.

Environmental Conditions

Although not directly related to the storage or operation of the electronic equipment, environmental conditions have considerable bearing upon the ultimate design.

The ease of maintenance aboard a ship, and thus the design of the equipment, is governed to a large extent by the availability of space. For example, passageways are about 32 inches wide on a ship; the one passageway on a submarine is even narrower. Hatches and doors may even be smaller. The lack of repair space in the installation area, however, makes it almost mandatory to move the major units of equipment elsewhere for repairs. Thus, the design must reconcile these factors.

Shock and vibration. Equipment installed aboard ships or submarines must be protected against high impact or shock, and pass specified vibration tests. Equipment used on aircraft and helicopters has to withstand continuous vibration ranging from 5 to 2000 hertz, depending upon the aircraft's propulsion and construction.

Equipment subject to high impact or shock can be supported on stiff mounts, known as shock mounts, while aircraft equipment is supported on soft or vibration mounts (see Fig. 8-2). See MIL-STD-5400 and MIL-STD-16400 for more details.

Temperature extremes. Electronic equipment is expected to operate continuously over a temperature range of $-54°$ to $+125C°$ aboard aircraft and to

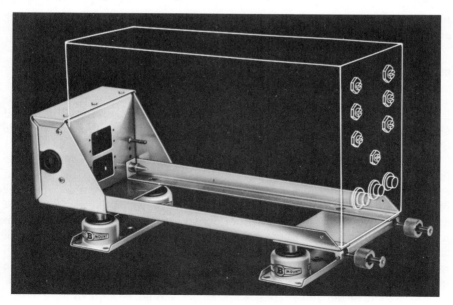

FIG. 8-2. Vibration Mount Assembly for Aircraft Electronic Equipment.

withstand storage in arctic and subarctic regions, where the temperatures may drop to extremely low levels. (Temperatures of down to $-62C°$ are specified in MIL-E-16400 for naval ship and ground equipment.) Consequently, precautions must be taken in the design of such equipment so that it may meet these temperature range limits.

Components such as semiconductors may fail to operate at the low temperatures or fail completely when the temperature within the equipment rises to the upper limit with the increase in the ambient temperature.

Low temperatures cause seizure of moving parts because of the differential contraction of materials and the possible solidification of lubricants. Some metals and plastics may also become brittle at such low temperatures. At high temperatures and the high humidities which usually accompany them, moisture may form on the insulating surfaces or components and condense in the partially sealed components because of the "breathing effect."

Metal parts corrode quickly when they come in contact with dissimilar metals. Components may become attacked by fungus under certain conditions. A solution to these problems is to enclose the complete equipment in a hermetically sealed case or to protect sensitive components by potting or coating with a fungicide material.

Other conditions. One of the best solutions for use in space appears to be sealing the equipment. Unsealed, its operation under the vacuum conditions in outer space produces damaging effects. There may be outgassing of common insulating materials, the loss of lubricating materials from the bearings of moving parts with a consequent deposit elsewhere within the equipment, arcing across terminal points, and other harmful effects.

8.3 SPACE PLANNING

To help the electronic engineers establish their circuit subdivisions, the mechanical designer should determine the approximate overall dimensions of individual chassis at an early stage of electronic equipment design. He should also decide on the general rack and cabinet layout.

Airborne, shipborne, and shore-based equipments have certain definite limitations to which the designer must adhere. In space planning, the proposed equipment is divided into a number of chassis, which are assembled into one or more enclosures or cabinets. The limitations to which the designer must adhere may involve maximum weight or dimensions of the complete equipment unit or its subdivisions. Other requirements may include the accessibility of components, interchangeability with other equipment or its subdivisions, plug-in modules, and the ability to withstand wide climatic variations or operation in space.

The weight of individual chassis assemblies should be kept to a minimum to facilitate assembly-line production, where they must be handled until their completion. Airborne equipment has strict weight limitations, and the appropriate components and mechanical design practices must be used to meet this requirement. Lift eyes have to be included when shipboard, submarine, or ground equipment exceeds a certain weight.

In airborne and mobile equipments, it may be necessary to separate the equipment into several subassemblies to meet the dimensional specifications. For example, the control panel on the equipment may be located in the instrument panel of the plane, while the bulk of equipment is placed within a rack assembly within the plane itself.

In rack-mounted equipment, the chassis subdivision system makes it possible to remove the individual chassis for inspection and repair and thus meet the accessibility requirement.

Space allowances must also be made to cool electronic equipment. This is accomplished by separating the components, by including heat sinks, or by blowers (when an excessive temperature rise is expected).

Modular Electronic Equipment

The layout of a distance-measuring equipment that locates approaching aircraft, designed for use in civilian and military planes, is an example of space planning in modern electronic equipment. The unit has a range of up to 450 miles, with an accuracy of $\pm.1$ mile.

This unit is designed to be housed in a standard ATR enclosure, Fig. 8-3. The unit and its major subdivisions are cooled by a blower mounted behind the front panel which can be tilted for ease of access.

Figure 8-4(a) and (b) show the internal construction. This is an example

FIG. 8-3. Exterior View of a Distance-Measuring Equipment.

FIG. 8-4. (a) Interior View of the Equipment in Fig. 8-3. (b) Another Interior View of the Equipment in Fig. 8-3.

of modular design where each module is tailor-made to fit within the space alloted to it. The plug-in modules mount on the I-beam chassis as shown in Fig. 8-13. This provides access to all components of the eight modules for test or repair.

Although such modular chassis construction results in high-density packaging, it is still possible to replace individual defective chassis quickly, without attempting to replace individual components. Considerable time is required to isolate the source of trouble in replacing individual components and can best be accomplished at the equipment repair base.

The goal in electronic equipment design is to utilize every cubic inch of space allotted for the equipment. At the same time, consideration must be given to such requirements as cooling such temperature-sensitive components as transistors. Test points, tuning points, and other adjustment points within the equipment must also be accessible without disassembly.

8.4 TYPES OF EQUIPMENT

Electronic equipment varies in its design, depending upon its end use. It may be roughly subdivided into three categories—consumer, commercial, and military. Consumer equipment includes radio and television sets, hi-fi, stereo, electronic organs, and high-fidelity audio equipment. Commercial equipment covers civilian aircraft radio, television transmitters, test instruments, industrial controls, and welding equipment. Military equipment is designed for the Department of Defense, as represented by the Army, the Navy, and the Air Force, and includes airborne, ground, shipborne, submarine, and missile equipment.

A single basic design cannot be applied to all equipment because it must be adapted to suit the environmental conditions in which the equipment operates, as has been previously noted.

Consumer Electronic Equipment

This equipment is designed for the mass consumer market and its basic considerations are cabinet styling, eye-appeal, and the other factors that affect its sale to the public. The mechanical designer must adapt his design to the cabinet size or the disposition of control knobs, while the size and style of the electronic chassis are governed by such large components as the picture tubes in television sets and the loudspeakers in receiving sets.

Television picture tubes may be 15-, 17-, 19-, 23-, or 25-inch in color or 12-, 16-, 19-, and 22-inch in black-and-white. The picture tube designations refer to the diagonal dimensions across the face of the tube.

The knobs on home entertainment equipment are generally large enough to allow easy operation of channel selectors and other controls that require high torque for their rotation.

As home receivers and television sets are produced in large quantities, their chassis size is not governed by a standard available chassis but instead is made to suit the cabinet requirements. A box-type chassis provides complete enclosure and thus protects against accidental contact with high-voltage components. As an additional precaution, the backs of television receivers are covered and the 110-volt input connection to the chassis is made through a special plug and receptacle, which separate upon removal of the protective back.

A back view of a color television receiver is shown in Fig. 8-5, while the modules replacing the rigid chassis previously used are shown in Fig. 8-6. The television tube is mounted on the cabinet. Interconnections between the modules consist of multiwire cables.

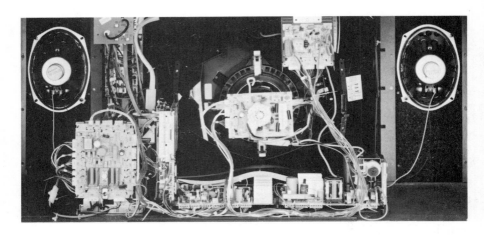

FIG. 8-5. Back View of a Color Television Receiver.

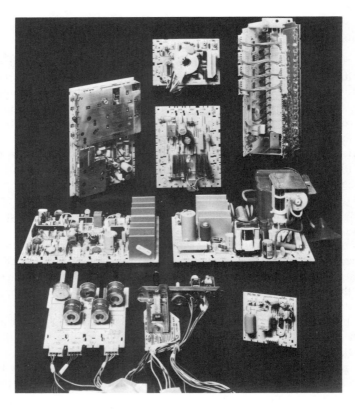

FIG. 8-6. Nine Modular Chassis of the Receiver in Fig. 8-5.

Commercial Electronic Equipment

Included in this classification is electronic equipment used in civilian aircraft, ground-based equipment, such as radio and television transmitters, among other applications.

The increasing complexity of electronic aircraft equipment has forced the airlines to establish what is now known as the Air Transport Rack or ATR system for commercial aircraft equipment.

Specification No. 404A, *Air Transport Equipment Cases and Racking*, published by Aeronautical Radio, Inc. (ARINC), a corporation formed by the commercial airlines, gives further details. Two other publications, *Guidance for Designers of Airborne Electronic Equipment*, ARINC Report No. 403, and *General Guidance for Equipment and Installation Designers*, ARINC Report No. 414, list the various requirements to be followed in designing electronic equipment for the airlines.

Specially designed mounting bases are available to mount the ATR equipment in commercial aircraft. Vibration mounts, in load ranges up to 80 pounds, protect equipment from vibration in aircraft.

Military Electronic Equipment

The construction of military equipment varies greatly depending on the environmental conditions and weight limitations encountered. Consequently, the Departments and Agencies of the Army, Navy, and the Air Force have issued general specifications (see Chapter 6) that apply specifically to equipments required by each service.

The increasing complexity of military electronic equipment has forced an increasing use of such space-saving methods as printed wiring and microminiaturization, while the use of transistors and integrated circuits has resulted in lower input power requirements and in space reductions.

8.5 PANEL MOUNTING RACKS

Adopted originally by the telephone industry for its central station equipment, the use of racks, rack panels, and chassis has become a standard practice for all large electronic equipment with both civilian and military organizations. A standard rack is a rectangular structure that consists of vertical channel members, joined at the top and bottom by other structural sections. A rack panel is a rectangular plate, at least 19 inches long and width in $1\frac{3}{4}$-inch increments, that is secured to the rack by screws or special fasteners. A chassis is a sheet-metal base of varying size and shape that supports mechanical parts and electronic components. It can be either secured to a rack panel, used separately, or provided with an enclosure for component protection. The rack

FIG. 8-7. Typical Cabinet for Spruance Class Destroyers.

may be open or in a cabinet. Such a cabinet (Fig. 8-7) may be connected to an external circulating cooling air supply. It may be "free standing," i.e., without an added external support.

Open Racks

These have been standardized at an overall width of 21 inches and a maximum overall height of 83 inches. The dimensional details are specified in such standards as EIA Standard S-310-C and military standard MIL-STD-189. Examples of such dimensional details are given in Fig. 8-8.

Rack Panels

They may be $\frac{1}{8}$, $\frac{3}{16}$, $\frac{1}{4}$, or $\frac{5}{16}$ inch thick and made of aluminum or steel. They may have open or closed-type mounting slots. Panels are mounted on military-type racks with #10-32 oval head screws and cup-type washers; on other types of racks they are mounted with #12-24 screws.

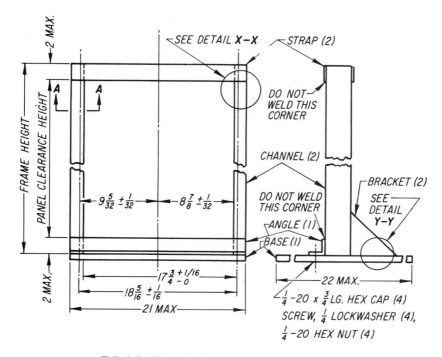

FIG. 8-8. Dimensional Details of a Standard Relay Rack.

Cabinets

The maximum height of cabinets has been established at 86 inches, with a total panel clearance height of $78\frac{1}{4}$ inches. The panel clearances vary in shorter cabinets but they are always in multiples of $1\frac{3}{4}$ inches. The maximum overall cabinet height for shipboard use is limited to 72 inches and to 60 inches for trailer use.

The panel mounting strips in cabinet racks have threaded holes, starting $\frac{5}{16}$ inch from the top cross-member of the cabinet. The horizontal mounting-hole centers are spaced $18\frac{5}{16}$ inches for the 19-inch panel. The horizontal space between the panel mounting strips is $17\frac{3}{4}$ inches, which provides a generous clearance for the 17-inch wide chassis that is generally used or for the chassis slides.

The problem of heat dissipation in a cabinet can be met by having louvres in the back, the lower front, or the side panels, where possible, to bring in and circulate surrounding air throughout the cabinet. If the heat-producing chassis is located near the top of the enclosure, it allows the warm air to rise without passing by other units within the cabinet, which could cause overheating. If a greater air flow is required to keep down the maximum temperature rise within the cabinet, forced air cooling may be used by including a centrifugal fan blower within the rack, with replaceable air filters included in the air stream.

Chassis slides (Fig. 8-9) can be used to provide ready accessibility to the chassis to be mounted within the cabinet when such accessibility is required. Handles on each panel lock the chassis panel assembly within the cabinet. In military applications, locking pins mounted on the rear of the chassis engage corresponding sockets on the cabinet to lock the chassis securely in place.

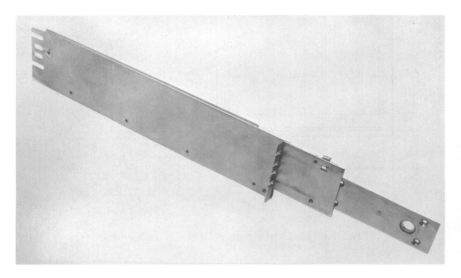

FIG. 8-9. Chassis Slide.

Miscellaneous Equipment

The Department of Defense and such commercial organizations as EIA have established dimensional standards for the design of large electronic equipment used in aircraft, land, or ship-based applications. There are other areas in which industry has followed similar standardization procedures, based upon practices and stock parts.

One equipment style in use is the console, which is a metal desk specifically designed to contain electronic equipment. It has sloping front panel cabinets above the desk top to accommodate various controls, indicator lights, meters, and so forth. Such consoles are used to control digital computers, power-station equipment, chemical or oil refinery processes, and many other similar applications.

Consoles for large equipment may not only control the equipment but may also provide meters and indicator lights for circuit checking of remotely located equipment units and indicators.

8.6 TYPES OF CHASSIS

There are two basic types of metallic chassis in use—the box and the flat plate, although variations are available in each type.

Box Chassis

This is the most prevalent type in use. It is also known as the open box chassis because the bottom (when used) is made removable, Fig. 8-10(a). The chassis is fabricated from one piece of aluminum alloy or cold-rolled steel, with corners welded and turned-in flanges that are used to attach a dust cover or shield or to mount the chassis. A similar chassis, with external mounting feet, is shown in Fig. 8-10(b). Other chassis varieties are the U-shaped or open-end chassis, Fig. 8-10(c) and (d), the latter showing mounting flanges or feet.

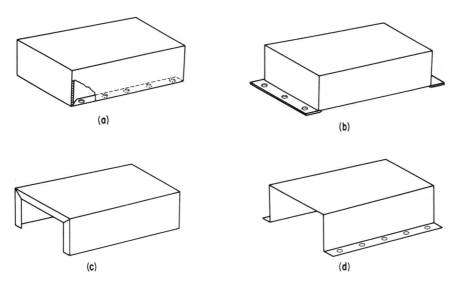

(a)

(b)

(c)

(d)

FIG. 8-10. Chassis Types. (a) Open Box. (b) End Flange Mounting Type. (c) and (d) U-Type or Open End.

Chassis corner construction varies. Some of the common methods are shown in Fig. 8-11. A spot-welded lap corner joint is shown in Fig. 8-11(a), while the corner in Fig. 8-11(b) is secured with a separate piece by spot welding. The corner in Fig. 8-11(c) is a butt joint, with the ends joined by a fillet weld. Some box chassis may be L- or U-shaped to accommodate an RF amplifier subchassis, the picture tube in a television receiver, or a subunit, such as a separate power supply.

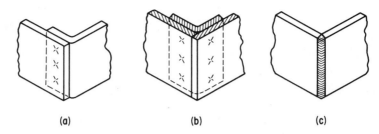

FIG. 8-11. Chassis Corner Construction. (a) Lap Corner. (b) Separate Corner Piece. (c) Fillet Weld Joint.

Flat-Plate Chassis

As its name implies, this chassis type consists of a flat plate with edges turned 90 degrees to form narrow sides that provide rigidity to the plate, Fig. 8-12(a).

Another chassis type is the panel-mounting chassis, Fig. 8-12(b), which is mounted in back of the rack panel on a standard 19-inch rack. Such chassis are available in $1\frac{3}{4}$- to 14-inch widths, in $1\frac{3}{4}$-inch increments, and in depths of approximately $5\frac{1}{4}$ inches. The chassis can be mounted vertically on the panel or in drawers that slide out of the cabinet.

A variation of the flat-plate chassis has the sides turned up to form mounting feet, which, except for its shallow depth, is similar to the box chassis in Fig.

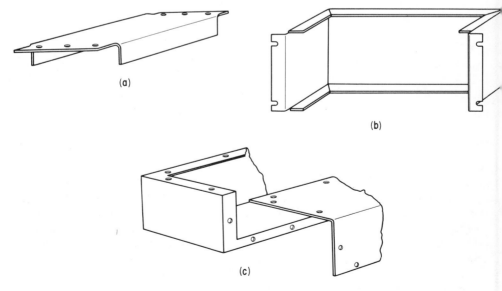

FIG. 8-12. Other Chassis Types. (a) Flat Plate. (b) Panel Mounting. (c) Split.

8-10(d). Another chassis shape may be in the form of an ell, to provide ready accessibility to all components. Two ell-shaped chassis may be used together to form a complete enclosure, as shown in Fig. 8-12(c).

An I-beam-shaped chassis is shown in Fig. 8-13. It serves as the main structure to support the various subchassis assemblies designed for commercial aircraft installations, Fig. 8-4.

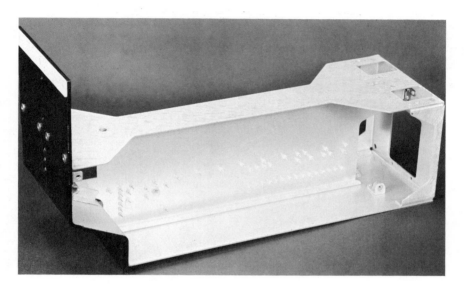

FIG. 8-13. I-Beam Chassis Construction.

8.7 CHASSIS FABRICATION

The great majority of electronic chassis are manufactured from metals, although some are made from plastics, such as phenolic materials or acrylic resins. The nonmetallic chassis are generally used for high-voltage circuits where there is a possibility of an arc over.

Materials

Metallic chassis are fabricated from steel when strength is required and from aluminum alloys for applications when weight is the prime consideration. Other materials used are copper, brass, and magnesium. Chassis are usually formed from sheet metal, although some cast chassis are also used.

Low-carbon, cold-rolled steel is generally used when steel chassis are specified. It has a smoother surface finish than hot-rolled steel and is more uniform in thickness. It has the disadvantage of requiring a protective coating by plating or painting.

Table 8-1(a), which lists the weight per square foot of various metals, was prepared to assist the electronic draftsman in estimating the weight of a chassis. Because the metals listed use two different gages, their gage numbers and equivalent thicknesses have been shown separately.

Table 8-1(b) lists the weight of various metals in kilograms per square meter.

Table 8-1(a) Weight of Materials (pounds per square foot)

| Gage | | Thickness | Aluminum | Brass | Copper | Cold-rolled | Stainless |
B&S	USS	(in.)	5052H32	½H		steel	steel
	26	.019				.77	.79
24		.020	.28	.88	.93		
22	24	.025	.35	1.12	1.17	1.01	1.05
	22	.031				1.26	1.31
20		.032	.45	1.41	1.50		
	20	.037				1.50	1.58
18		.040	.56	1.77	1.87		
16	18	.050	.70	2.24	2.35	2.02	2.10
	16	.062				2.51	2.63
14		.063	.88	2.82	2.91		
	14	.078				3.16	3.28
12		.080	1.13	3.56	3.73		
11		.090	1.26	4.00	4.20		
	13	.094				3.81	3.94
10		.100	1.40	4.49	4.66		
	11	.125	1.78	5.60	5.80	5.07	5.25

Table 8-1(b) Metric Weight of Materials (kilograms per square meter)

| Gage | | Thick-ness | Alumi-num | Brass | Copper | Cold-Rolled Steel | Stainless Steel |
B&S	USS	(mm)					
	26	4.83				3.78	3.83
24		5.08	1.36	4.27	4.51		
22	24	6.35	1.70	5.44	5.68	4.90	5.10
	22	7.87				6.12	6.36
20		8.13	2.19	6.84	7.29		
	20	9.40				7.28	7.67
18		10.16	2.72	8.59	9.08		
16	18	12.70	3.40	10.88	11.40	9.80	10.20
	16	15.75				12.18	12.77
14		16.00	4.27	13.62	14.14		
	14	19.81				15.37	15.91
12		20.32	5.48	17.27	18.08		
11		22.86	6.12	19.45	20.40		
	13	23.88				18.50	19.20
10		25.40	6.80	21.76	22.60		
	11	31.75	9.20	27.20	28.15	24.62	25.43

Methods of Fabrication

The student may find it useful to be familiar with the details of chassis fabrication in the shop. After the chassis detail drawing has been prepared, the shop personnel lay out the chassis blank in its developed or flat form. Allowances must be made in the blank size for bends and possible material shrinkage during fabrication. These allowances vary depending upon metal used for the chassis.

Once the overall blank size has been determined, the first step in chassis manufacture is to cut or shear the blank to size. The blank must be cut square and to the computed dimensions so that the bends and holes will come out in the relative positions specified on the drawing.

The developed chassis blank is normally in the form of a cross for a box chassis (Fig. 8-14). Its straight sides are cut to dimensions in a large shear, while the corners are frequently punched out in a punch press. Dies are used to blank out the entire chassis shape in one operation, if a large enough quantity is needed.

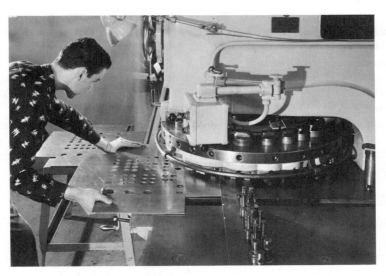

FIG. 8-14. Punching Holes in a Cross-Shaped Chassis Blank.

Punching. Chassis holes, whether round, square, or some other shape, are pierced in the chassis blank after it has been trimmed to size. This is done in several ways, depending upon the size of holes and quantity of chassis being made.

If the holes required are smaller in diameter than the thickness of the material, they may be drilled rather than punched to save needless wear on the punch.

The methods of punching holes in the chassis depend upon the chassis quantity. In large-scale production of the same chassis, special dies are used to pierce all the holes with one operation of the punch press.

In moderate production quantities, chassis holes are punched individually. One method is to use the turret-type press (Fig. 8-14). In this press, punches are located in the upper turret and the corresponding dies in the lower turret, with both turrets rotating together and locking simultaneously. The operator places the blank chassis against a back stop to locate the holes in the horizontal direction, selects the correct punch and die set by rotating the turret manually, and then positions the blank under the turret head to pierce the hole.

Another method is to use a computer-controlled turret press that automatically positions the turret and locates the chassis blank.

Numerically controlled machines are available that locate the hole positions by the X-Y or coordinate method by means of prepunched tape prepared from chassis drawings.

Forming or bending. All operations on the chassis blank—such as blanking to size or punching of holes—have to be completed before the blank is ready for bending to its final shape. Bends are made on a machine known as a brake, which may be hand- or power-operated and may range in length from 3 to 10 or more feet.

Definite bend radii should be specified on the drawing to avoid fracture of metal at the bends. They vary depending upon the material and its thickness. Recommended bend radii for aluminum were listed in Chapter 5.

Bending must also be in correct sequence, otherwise it is impossible to complete bending the chassis to shape.

A chassis is shown in Fig. 8-15 in its flat shape and after being formed to shape.

8.8 CHASSIS AND PANEL DESIGN

An electronic assembly may consist of a chassis, a panel, a printed circuit card cage, and an enclosure or cabinet which, when required, may be supported on a shock-mount base.

Some or all of the following drawings may be required for a complete set of drawings:

1. A layout of the major components on the chassis in their relative positions, mechanisms, and so on

2. A chassis detail drawing, drawn to scale to show its dimensions and the location and size of holes and cutouts

FIG. 8-15. Chassis Blank. (a) Flat. (b) Formed to Shape.

3. A chassis drawing showing the location of reference designations and other nomenclature

4. Detail drawings of such sheet metal parts as brackets and shields, such machined or cast parts as mechanical drives, such hardware as spacers, component boards, nameplates, and many other miscellaneous items

5. Subassembly and assembly drawings of the chassis and mechanical and electronic components as required

6. A layout of the panel components

7. A panel detail drawing, including lettering and finish

8. Detail and assembly drawings of printed-circuit boards

9. Detail and assembly drawings of printed-circuit card cages

10. Chassis cable or harness drawings

11. A schematic or elementary diagram

12. A wiring or connection diagram

13. Drawings of external interconnection cables

14. A layout of the cabinet or enclosure

15. A detail drawing of the cabinet or enclosure

16. A complete assembly of the chassis, panel, and cabinet, and shock-mount base, if used

17. An outline drawing

18. A parts location drawing for service manual
19. An installation drawing

Chassis Layout

The layout of a chassis is made simultaneously in the drafting room, the laboratory, and the model shop, and much time and effort can be saved by coordinating these activities. Breadboard models may be built using a temporary chassis to mount the components that are subject to value and position changes until the required results are obtained.

Component location may be indicated on a sketch submitted by the project engineer, or it may have to be determined on the basis of circuit requirements.

Chassis size may be established in two ways: (1) the equipment specification may designate a standard size chassis, fitting within a specific case, which establishes the chassis form factor as well; or (2) a definite chassis or equipment size may not be specified, which leaves it up to the designer to select the size and shape or to make part of the decision. For instance, the requirement may specify that the chassis use a standard 19-inch rack panel and may list the total volume of equipment space in cubic inches, but it still allows some variation in the form factor.

Component selection. After the circuit details have been established, it is necessary to select the appropriate electronic and mechanical components—a task that is shared by the circuit engineer and mechanical designer. A temporary list of all major components and their details should be prepared for reference, and tentative decisions should be made regarding the use of component and printed-circuit boards, since space must be allocated for them. The type and size of components will also be affected if they are to be mounted on printed-circuit boards.

Building a breadboard model. The name, *Breadboard Model*, has been carried over from the early radio working models built on a wooden base. Building such a model involves the use of an insulating or plastic material for component mounting, and specially perforated plastic boards are available for this purpose, with holes uniformly spaced on .100 inch grid pattern. These boards lend themselves to mounting mechanical and electronic components of various shapes and sizes, Fig. 8-1(a). Special terminals can be mounted in the holes to support components by their leads on one side of the board and to interwire on the other.

As these boards are at the most $\frac{1}{16}$ inch thick, it is easy to make any additional fabrication that is required, such as openings for transistor sockets, tubular capacitors, connectors, switch bushings, and other modifications.

Component location. The placement of components is determined by such considerations as location for the shortest connections, ready accessibility for replacement, freedom from interference with other parts or their movement if adjustable, ready identification, freedom from electromagnetic interference, location within the circuit path, and clearance for operating shafts or other mechanisms that are a part of the component.

Components should be located as close together as possible, without making access for maintenance difficult. The stress on miniaturization by the Department of Defense and other agencies has brought forth other difficulties, such as overheating of components because of the reduced volume available for heat dissipation and the close proximity of heat-dissipating components. These problems may, in turn, require either the addition of heat sinks or forceful cooling by a blower.

The process of arranging and rearranging components to fill a given space is known as packaging. Component location can also be determined by using cardboard or paper templates in place of the actual components on the layout, as explained in Chapter 2. This method is especially helpful when the actual components are not available.

Prototype model. Once the tentative component layout has been accepted, the next step is to build a prototype that is as near as possible to the working model, mechanically and electrically. The designer starts with a chassis built in the model shop according to the dimensions determined in the component layout. The final component placement must take into consideration such factors as weight distribution, the selection and placement of connectors, test points, and electrical clearances. The orientation of transistors is also important to keep the signal voltage leads as short as possible. Any shielding requirements should be determined so that space can be allowed for this purpose.

The designer has to make final layouts according to the component placement established by the trial method, check mechanical clearances, and start on some of his detail drawings. Corrections have to be entered on the provisional drawings made for the model shop chassis, and the material lists must be started. The preliminary drawings made for the model shop are used as a record in the preparation of regular drawings later. Layouts should be corrected if the principal dimensions have been modified or major subassemblies have been rearranged. The various alternatives that present themselves during the preliminary design should be resolved as they occur. The biggest problem faced by the designer is to keep a record of changes as they occur so the appropriate corrections can be made later.

Typical Chassis Drawing

The chassis shown in Fig. 8-16 is a representative chassis drawing. The components mounted on top of this box-type reference regulator chassis include a large capacitor; a smaller capacitor; a plug-in printed-circuit board containing semiconductor rectifiers and other components; and a dual plug-in

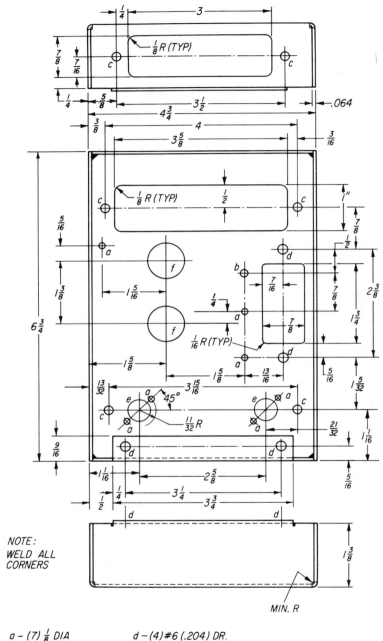

NOTE:
WELD ALL
CORNERS

a – (7) $\frac{1}{8}$ DIA   d – (4) #6 (.204) DR.
b – #22 (.156) DR.   e – (2) $\frac{15}{32}$ DIA
c – (6) #16 (.177) DR.   f – (2) $\frac{3}{4}$ DIA

FIG. 8-16. Simple Chassis Detail.

resistor unit. A small resistor and two socket receptacles are located within the chassis while a terminal board is mounted externally on one side of the chassis.

Hole conventions. Hole locations are given from two adjacent edges to maintain all the holes in their relative positions. For example, mounting holes for sockets are given relative to the socket hole to ensure that the mounting holes in the socket will match. Note the method of dimensioning the "*a*" holes relative to the socket or the "*e*" hole. By dimensioning across "*d*" holes, only one tolerance is involved. This practice of dimensioning holes in groups or clusters has become standard in chassis sheet metal work.

Another prevalent practice is to specify large clearance holes for mounting screws, components, and so on, on the drawings. Binding head screws, with a large diameter head-bearing area, have become standard items in large clearance holes in sheet-metal chassis. Table 5-2 listed the suggested sizes of screw clearance holes in sheet metal.

Another chassis drawing is shown in Fig. 8-17. Center lines are used for reference, and all holes are located relative to them. This method of laying out individual holes has the advantage of having them laid out from a straight line in actual manufacture instead of from the edge of the chassis, which may not always be straight. In addition, the center line serves as a reference from which the relative positions of other parts—such as brackets—and their holes may be determined. The hole dimensioning practice shown in Fig. 2-5(c) is followed in this crowded drawing, which also shows dimensioning of such holes as transistor socket mounting holes in clusters.

Drawing practices. Sheet metal parts are ordinarily dimensioned with fractional tolerances—generally $\pm\frac{1}{64}$—to reduce cost, while their holes may be specified with a decimal tolerance of $\pm.005$ inch if necessary.

After the various chassis views to be drawn have been selected, the space requirements for the views, dimensions, general notes, drill sizes, and so forth, are determined (see Fig. 1-22). The center lines for major components are drawn first and serve as reference points for the location of other component holes. The number of hole sizes should be kept to a minimum.

The circuit determines the arrangement of transistors or other major components. Usually, the output of one stage is located next to the input of the next, resulting in shorter leads. Power transformers and iron-core chokes are mounted close to the chassis edges whenever possible to give better support. The beginner should remember not to locate mounting holes too close to the bend. If he forgets about the radius at the corner, it will interfere with the proper seating of such fasteners as nuts.

Both sides of the chassis must be considered in chassis layout. Locating a component inside the chassis may interfere with the proposed location of a component mounted on top, thus requiring a compromise. For example, a component board location may interfere with the proposed connector location on

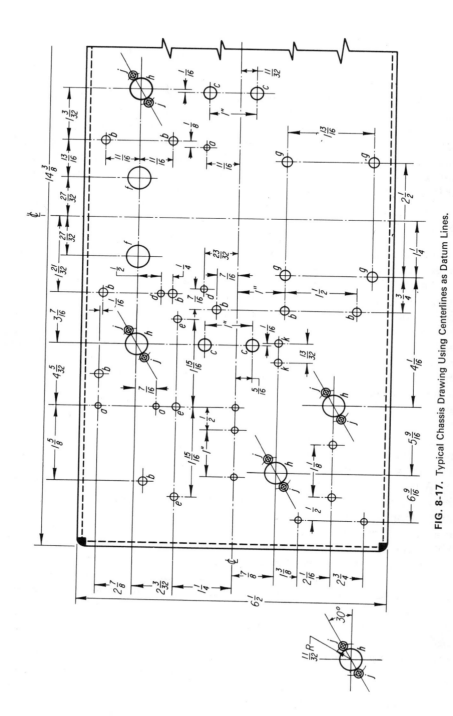

FIG. 8-17. Typical Chassis Drawing Using Centerlines as Datum Lines.

top of the chassis. Or a switch mounted on the panel may project far enough over the chassis to interfere with a printed-circuit-board assembly located on top of the chassis. Thus, chassis layout requires careful consideration of adjacent components that may create mechanical interference upon assembly. When discrepancies are discovered during the drawing of chassis or other details, they should be noted and corrections made at the earliest convenience.

Chassis reference designations. Components are generally identified on civilian electronic equipment by reference or other designations corresponding to schematic diagram designations. They must be identified in this manner on all military equipment. In the latter, reference designations must identify the transistors or other plug-in components on top of the chassis. Their sockets are identified on the opposite side. Further details of reference designation practices in military equipment are given in Chapter 10.

Reference designation marking on a chassis is shown in Fig. 8-18. A reference designation drawing is made by copying or tracing over the chassis detail drawing, showing only the chassis outline and holes but not the dimensions or hole identification letters.

The reference designations and their locating dimensions are added to this outline drawing. The designations must be carefully located so they will not be covered by assembled parts. This can be checked on the chassis assembly drawing by laying out the designations in their proposed locations. Designations should be oriented to read from the normal installed view, that is, from the panel end in panel-type equipment.

Designations are located on component center lines or at such locations where they will not be obscured by other components. Particular attention must be given to areas where components are close together. One method to indicate the identified component is to extend a short arrow from the designation.

Small Mechanical Parts

Besides the chassis itself, there are various small mechanical parts that may be required in a chassis assembly. Among these are the various types of sheet metal brackets (Fig. 8-19).

The sheet metal brackets may be the simple L-type, Fig. 8-19(a), which are sufficient for simple mounting purposes but which cannot compare to the stiff bracket, Fig. 8-19(b), or its double version in a wide bracket, Fig. 8-19(c). The clamping bracket, Fig. 8-19(d), has the long horizontal section stiffened by turning up its edges. A pull-down bracket, Fig. 8-19(e), uses riveted spade lugs that protrude through the chassis to anchor the component. The bracket shown in Fig. 8-19(f) is the familiar type used for holding choke and transformer laminations together and also for mounting purposes.

In drawing sheet metal parts, care should be exercised to avoid making some obvious drawing errors. A common mistake is showing an outer bent corner

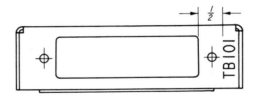

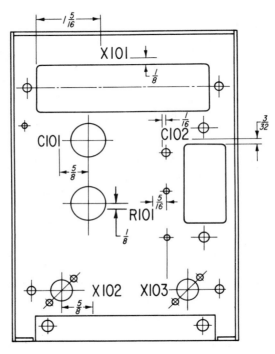

NOTES:

1. ALL CHARACTERS COMMERCIAL GOTHIC $\frac{1}{4}$ HIGH, CONDENSED

2. CHARACTERS TO BE SILK SCREEN PER SPEC. S310

3. ONE COAT CLEAR LACQUER OVER ALL SCREENED CHARACTERS, SPEC. V225

FIG. 8-18. Reference Designation Drawing.

as being square, Fig. 8-19(g), or not providing a cutout relief in the corner as in Fig. 8-19(h).

Thin sheet metal often lacks the rigidity necessary in electronic work. Some of the methods used to stiffen the material and the lightening holes are illustrated in Fig. 8-20.

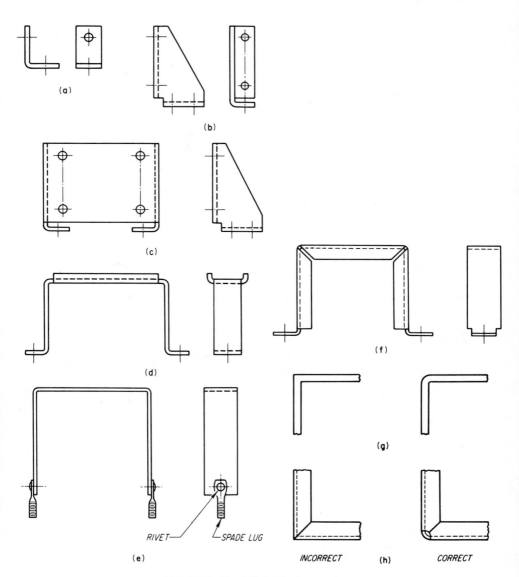

FIG. 8-19. Sheet Metal Brackets.

Chassis Subassemblies and Assemblies

In electronic equipment, drawings of subassemblies and assemblies are used to show all the parts and components in their relative positions. A subassembly is generally one of the smaller subdivisions, which contains a number of parts that can be separated into individual components. An assembly is a larger unit,

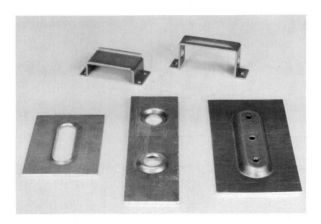

FIG. 8-20. Methods for Stiffening Sheet Metal.

such as a chassis or a panel assembly, that may include a number of sub-assemblies such as component and printed-circuit boards that are part of the overall assembly. The overall assembly is generally reserved for the larger or final assembly drawing.

The subassembly or assembly drawings are usually full scale, or enlarged if necessary, and are developed from the detail drawings of individual parts. This procedure helps to locate errors or discrepancies in the detail drawings. Thus, the assembly drawing is also a useful checking tool.

End and bottom views of the chassis may be required for a chassis assembly (Fig. 8-21). If a front panel is attached to the chassis, a third or panel view may also be necessary to identify the various panel components and hardware. Unimportant details such as those that would normally appear on the sub-assembly drawing are omitted on the assembly drawing; these include minor features of the individual components that do not contribute to understanding the drawing. However, certain details that are helpful for shop assembly should be included, as for example, including terminal numbering on the drawing to indicate the orientation of sockets (Fig. 8-21) or indicating sufficient components or other items on subassemblies if they could possibly be installed in more than one position.

All parts in the assembly should be identified by item numbers on the parts list and shown on the drawing. Such hardware items as a screw-lock washer-nut assembly do not have to be shown in full detail. Installed, they may be indicated only by a series of joined balloons or circles as in Fig. 8-21.

Dimensions may be included on the assembly drawing, if necessary, to clarify details shown on it or to serve as information for an installation drawing.

On complex assemblies with some of the components hidden by others in the view shown, it may be necessary to include auxiliary views, enlarged if necessary, to bring out details of the equipment. Such a view is of considerable help to shop personnel who are assigned to assembly or wiring.

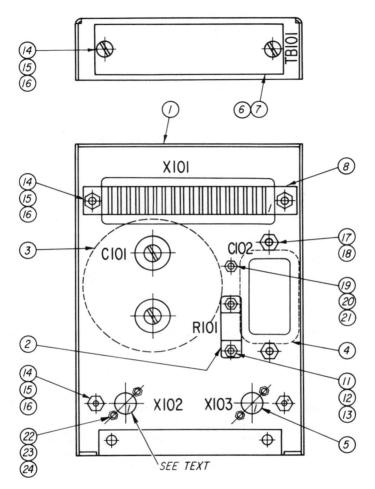

FIG. 8-21. Chassis Assembly Drawing.

The main assembly drawing should include an itemized list of the drawing numbers of schematic and connection diagrams, the cable or harness assemblies, the interconnection diagrams, and any other information that relates to the chassis or unit shown on the assembly drawing.

Panel Drawings

Panels are laid out in a manner similar to the chassis layout, and detail drawings are made from this layout. These drawings show the location, size, and shape of holes for panel-mounted components and for fastening the panel to the chassis.

Other detail drawings show the markings on the panel face and reference designations on the back side. The markings on the front may be done by silk-screen process, by engraving the panel surface and filling with a contrasting color, by etching with raised characters on a painted background, or by other commercial processes.

Other identification methods used include metallic labels with adhesive backing and decalcomanias.

Nameplates are generally attached on the front panels as part of equipment identification and are a requirement for military electronic equipment. Separate specifications have been issued for details of nameplate nomenclature, as for example MIL-P-15024/2, *Plate, Identification, Unit or Plug-in Assembly.*

Cases or Enclosures

Their dimensions are dictated in two ways—by the established standard-ized sizes such as ATR and MIL-C-172, which in turn use standardized chassis, or by the equipment specifications, which determine the chassis size. In the latter case, it is necessary to make a detail drawing of the enclosure according to the specifications and to include such applicable details as provisions for shock mounts, external plug connections, locating pins, and so forth.

It is also necessary to specify on the drawing the enclosure finish, inside and out, and any marking that may be required. A separate assembly drawing has to be made if a number of parts are to be included as part of the enclosure.

Parts Identification Drawing

This type of drawing, which indicates the location of individual parts or components on an electronic assembly, is frequently used in service manuals. It may be in the form of an outline drawing that identifies the components and controls by number designation and function (Fig. 8-22) or it may be a photograph of a completed assembly to which numbered balloons have been added. These balloons correspond to the parts list or identify the parts by reference designations. Such identification methods are helpful in servicing, in test measurements, and in repair work.

8.9 MODULAR CONSTRUCTION

The size of the module depends on whether conventional discrete components are used or whether the components are of microminiature size. The discrete components may be conventional size resistors, capacitors, transformers, relays, switches, and so forth. The microminiature components may be chip capa-citors, subminiature trimmer capacitors, miniature resistors, or semiconductors such as transistors, diodes, flat packs, dual-in-line packages, or integrated circuit packages of very large-scale integration (VLSI) type, large-scale inte-gration (LSI) type, medium-scale integration (MSI) type, or hybrid circuit

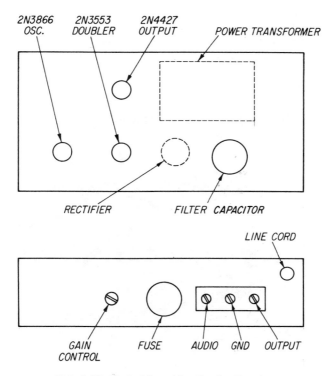

FIG. 8-22. Typical Parts Identification Drawing.

packages. The LSI packages are likely to contain more than one function within the package.

Module Varieties

There are several basic types of module construction:

1. The modules of the Standard Electronic Module Program (SEM) have been developed by the Navy and the program is listed as standard by the Department of Defense.

MIL-STD-1378B, *Requirements for Employing Standard Electronic Modules,* and MIL-STD-1389A, *Design Requirements for Standard Electronic Modules,* have been approved for use by all departments and agencies of the Department of Defense. Another is specification MIL-M-28787A, *Standard Electronic Modules Program, General Specification For.*

These modules are part of a military electronic standardization program and supersede the previous Standard Hardware Program (SHP). The newer very large-scale integration (VLSI) functions require a different packaging approach. They are of a standardized mechanical packaging design and provide selected electronic functions for use in such military equipment as radar, navigation, automatic testing, etc.

The basic SEM module is a single-span, single-thickness module shown in Fig. 8-23. Its basic overall dimensions are: 2.740 inches wide, 1.85 inches high, and .275 inch thick, with a projecting fin on top.

Larger SEM modules, Fig. 8-24, are based on increasing the basic width in 3-inch steps.

The projecting fin structure on each SEM module serves as a heat dissipator, extracting means, and an identification marker. Each module also has two

FIG. 8-23. Single Width SEM Module.

FIG. 8-24. Larger SEM Module.

special pins to key it radially. They prevent the module from being inserted in the wrong location within the assembly, from being plugged in reverse, or from damaging the connector contacts.

The module connector has two rows of contacts spaced on a .050-inch grid system. A maximum of 40 pins is standard.

Pin functions on a module are as follows:

Pin No.	Function
1, 21	+5 volts
2, 22	Clock line
3, 23	Positive supply voltage other than +5 volts
10, 30	Signal ground—zero volts
11, 31	Frame ground (chassis)
19, 39	Clear line or not used in the case of multiple negative voltages
20, 40	Most negative or least positive supply voltage
All remaining pins	Their use is optional

Originally designed to use discrete components, these modules now use microminiature components, Fig. 8-25. This results in more efficient space and circuitry utilization. The modules are cooled by natural airflow or forced-air cooling against the top fins.

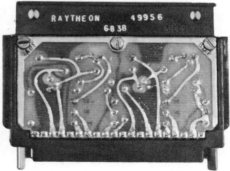

FIG. 8-25. SEM Modules with Microminiature Components.

2. Another example is the digital computer module of the all semiconductor type. Such modules are plug-in, etched printed wiring boards or cards on which resistors, capacitors, transistors, and other components are secured by wave soldering. These cards are all identical in shape and size. Each card contains all the components for a basic electronic circuit—amplifier, shift register, flip-flop, etc. Thousands of these cards may be used in the computer assembly and spares kept on hand for quick replacement of defective cards.

Such modules cannot be considered as standardized, however, because they may vary in size, shape, means of connection, and other details from one manufacturer to another.

3. The number of individual circuits that can be packed on one card has increased tremendously with the development of LSI and VLSI integration. Thus, the size of the modules or the number required has decreased.

Such modules may be of the plug-in type, designed for a specific application such as the modules shown in Fig. 8-4(a), test equipment, power supply, amplifier, D/A conversion, etc. The variety is endless and increasing every year.

Module Connectors

Connectors on the flat modules may be in either of two forms: a multiterminal connector mounted on one end of the insulation board forming the module, or the connections may be plated tabs on one or both sides of the end of the board. Such tabs are actually an extension of the printed wiring that interconnects the board components (Fig. 8-26).

FIG. 8-26. Plated Tabs Form the Connector on This Module.

Module Assemblies

Modules, such as SEM's and printed-circuit boards, require mechanical mounting so they can withstand environmental conditions, be readily interconnected, and provide physical protection to the modules and their connectors.

In an assembly, the SEM module connectors are interconnected by various methods, such as wire wrap, termi-point, soft soldering, and multilayer printed circuitry.

In many applications that employ diodes and transistors to a great extent—such as computers—the desirability of having component groups, such as logic circuits, easily replaceable has resulted in the use of small mechanical assemblies, or "cards," which in turn plug into the larger mechanical assemblies that compose the equipment. Many of these card assemblies are printed-circuit boards with a set of contacts on one end that serves as both a connector to the external circuitry and also as a mechanical mounting (see Fig. 8-27). Circuit alterations can be easily made or a defective assembly replaced by plugging in a different card assembly.

FIG. 8-27. Card Cage for PC Boards.

Card materials vary, depending upon their end use. The better grade cards, $\frac{1}{32}$ to $\frac{3}{32}$ inch thick, are of glass cloth impregnated with silicone or melamine resin. Small cards are generally mounted by their connectors, while the larger cards may have additional support from side members with grooves just wide enough to admit the cards. This method is known as the "card cage."

Spacing of Semiconductor Devices

The TO-5 type components should be spaced approximately $\frac{3}{4}$ inch center-to-center to allow sufficient space for printed wiring conductors on a printed-circuit board. Adjacent flat packs can be spaced $\frac{1}{2}$ inch center-to-center and one inch center-to-center in adjoining rows.

Modular Test Equipment

The application of plug-in modular construction has been extended to test instruments for electronic equipment. By using plug-in units in oscilloscopes (Fig. 8-28), it has become possible to extend their frequency range, increase the number of traces, and provide other features, thus increasing the usefulness of the instrument and also allowing for future technological changes.

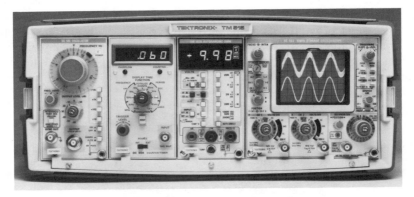

FIG. 8-28. Oscillator Counter/Timer with Plug-in Modules.

Military Modular Equipment

The adoption by the Department of Defense of the previously described SEM modules for its electronic equipment construction has resulted in several desirable advantages. Among these are cost reduction, lower design and development costs, improved reliability, and simplified logistics support in regard to spare parts and components. To these can be added simpler testing procedures and mass productions of limited types of modules, selected on the basis of functional requirements.

In the Trident modular system, used on nuclear submarine fire control equipment, there are more than 15,000 modules (Fig. 8-29), standardized by circuitry into over 160 varieties, SEM type. The modules, in turn, plug into panel module assemblies that support the modules and provide solderless wire-wrapped interconnections between them (Fig. 8-30). The module assemblies, in turn, are mounted on equipment rack doors six feet high, Fig. 8-31(a), that swing open from the rack frame to provide access to the modules for testing or replacement.

The intricacy of interconnections was resolved by using numerically controlled wire-wrap machines to make all the connections on the module assemblies. The modular design makes it possible to assemble the complete system at the manufacturing facility and to disassemble and reassemble it aboard the submarine.

In other Navy equipment the modules are mounted in "free-standing"

FIG. 8-29. SEM Modules Installed in Trident Equipment.

FIG. 8-30. LOC-4-SHOCK Assembly Structure for Trident SEM Modules.

FIG. 8-31. (a) Trident Fire-Control Equipment Rack Doors Swing Open to Facilitate Maintenance. (b) Aegis "Free-Standing" Cabinet.

cabinets, Fig. 8-31(b). These cabinets are anchored only on the bottom or deck surface.

Metric Design

Military Standard, DOD-STD-176, *Metric System in New Design*, provides guide lines when a decision has been made to design an item, equipment, or system in metric units.

Until the metric system is adopted on a national basis, it will be necessary to use existing components, thus creating a mix of straight inch-pound, hybrid, and metric designs.

The word "METRIC," preferably enclosed in a rectangle, should be placed on drawings near the title block, with lettering the same size as the drawing number. On other technical documentation the metric identifier should be placed near the document number.

8.10 OTHER DESIGN ELEMENTS

The designer has to consider such additional factors as shielding and wiring techniques in the layout of electronic equipment. It is essential to be familiar with them to make the necessary allowance in layout arrangement and space requirements.

Shielding

Two types of shielding may be required in electronic equipment—electrostatic or electromagnetic, or both.

Electrostatic shielding. This type may be required to reduce or eliminate the escape of radio-frequency energy into surrounding equipment or components. The source of this energy may be an RF oscillator that has an access cover with insufficient points of contact or overlap to contain the energy within the enclosure.

Shielding may take one of the following forms: a metal spring, solid or with fingers, made from phosphor bronze, beryllium copper, or other copper-bearing spring material; a fine wire-mesh screen of copper or other low-resistance material; or a formed wire-mesh gasket. These shielding means are used to prevent RF radiation through cooling air openings, cabinet lids, waveguide flanges, and other couplings.

To avoid the development of RF leakage, the design should provide for good metal-to-metal contact at all chassis seams and joints. Screws used to assemble parts should be spaced no more than $1\frac{1}{2}$ to 2 inches apart. The parts themselves should be clean and free of paint or other insulating finishes at the points of contact.

Metal springs, made in the form of a series of fingers, are generally satisfactory on flat surfaces, such as sealing doors or small chassis work where

RF shielding is necessary. Beryllium copper springs must be properly heat-treated to maintain adequate spring pressure.

Wire-mesh gaskets are generally specified for irregular or cast surfaces, where the shielding means must conform to the metal surface. These consist of Monel or some other corrosion-resistant material that is formed into knitted wire-mesh of round, square, or rectangular cross-section to become a gasket that fits in the opening. Such metal-conductive gaskets are used in sheet metal or cast cabinets, waveguide flanges, or other openings that require prevention of RF leakage. The gasket in waveguide flanges should project above the flange groove to make good electrical contact with the upper flange and be inside the bolt circle to prevent possible leakage at the bolt holes.

Shielding against EM/RF interference in meter and readout devices has been accomplished by including shielded windows that have knitted wire-mesh laminated within the window material.

Magnetic shielding. Such components as cathode-ray tubes must be shielded against magnetic effects, either by shielding the tube itself or by shielding the sources of low-frequency magnetic radiation such as power transformers and chokes. In the majority of cases, shielding of tube itself is the most satisfactory method because the shield protects it from all possible sources of such radiation. If a cathode-ray tube is not properly shielded, the external magnetic fields are likely to affect its response.

8.11 COMPONENT BOARDS

The method of mounting the individual components in electronic equipment is determined to a large extent by whether it will be used for commercial or military purposes. Components may be wired directly across terminals as part of point-to-point wiring, be assembled together on an insulating board or a printed-circuit board, or be a part of a modular assembly. The commercial type of equipment is more likely to utilize point-to-point wiring.

To avoid having a dense concentration of components on multisection switches, transistor sockets, and other major components, component boards are often utilized to provide an orderly component mounting and interwiring. Such boards can be built as separate assemblies, including their outgoing wires; they provide better component mounting and a place for component identification.

Component board materials vary depending upon the application. General-use boards are made of a phenolic paper or fabric base laminate. The better-grade boards are of glass cloth impregnated with silicone or melamine resin—a material that is more resistant to moisture and has better mechanical strength and electrical properties than the phenolic laminates.

Commercial component boards are available in $\frac{1}{16}$- and $\frac{3}{32}$-inch thicknesses, depending upon the board size. Such boards are made in several sizes and styles.

Terminals

Many varieties of terminals are available for component boards. Some of the styles shown in Fig. 8-32 are identified as being of standard, midget, or subminiature size, with shank lengths suitable for mounting on board thicknesses from $\frac{1}{32}$ to $\frac{1}{4}$ inch. Terminals are made of brass and are silver-plated or have a solder-dip finish. They are mounted in board holes by rolling over the shank with a special tool made for this purpose.

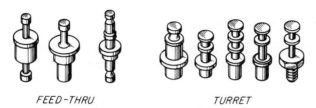

FEED-THRU   TURRET

FIG. 8-32. Terminal Styles for Component Boards.

Board Mounting Methods

The board location determines its mounting method. Boards may be mounted against the inside surfaces of the chassis sides or at right angles to the chassis surfaces. Two mounting techniques are shown in Fig. 8-33. A common method uses spacers between the board and its mounting surface, Fig. 8-33(a). A plain L-shaped bracket may also be used, although it is not recommended if vibration is likely to be encountered. Boards may also be mounted flat against the mounting surface, Fig. 8-33(b), by using such insulation material as phenolic

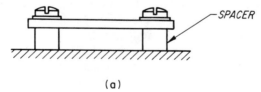

(a)

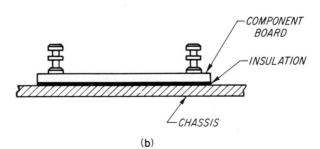

(b)

FIG. 8-33. Board Mounting Methods. (a) Spacer. (b) Insulation Plate.

or glass laminates, as long as there are no sharp protrusions on the side of the board that bears against the insulation piece.

Boards should be placed in accessible locations where they do not block hardware (such as screws and other fasteners) or cover other components or their mounting screws. It should not be necessary to remove the board to replace other parts, and it should be possible to replace the board assembly or to disconnect its leads without other disassembly.

Laying Out a Component Board

Boards used in electronic equipment are not restricted to any definite shape or size. They may be of any shape and terminal location required by the application.

The circuit shown in Fig. 8-34 is an example of a component board layout. All of the components shown, with exception of test points, are to be mounted on a rectangular component board located within a module assembly (Fig. 8-35). The board is to be mounted flat within the module. Because the module size is fixed, the board dimensions are limited to a maximum size of $1\frac{7}{8}$ by 5 inches long.

If possible, the circuit layout should be followed when the relative placement of components is laid out on the board. Components with common connections should be noted, as for example C1 and R4 and R10, R8, and R9.

A tentative component disposition is selected and sketched (Fig. 8-36), together with the interconnections and outgoing leads to the remainder of the circuit. A layout with a minimum of crossovers and long interconnections is preferred.

Component sizes, consistent with their value and voltage or wattage rating, can be determined from catalogs. Terminal spacings have to be selected to determine the length and width of the board. In the majority of cases, a $\frac{3}{8}$-inch spacing between adjacent rows of terminals and $1\frac{1}{4}$ inch across is adequate for such small components as resistors, capacitors, diodes, and inductors. Discrete components more than $\frac{3}{8}$ inch wide or more than 1 inch long are likely to require additional support by clamps or other means to prevent possible lead failure under vibration conditions. Component leads should project from $\frac{3}{16}$ to $\frac{3}{8}$ inch beyond the body of the component to avoid possible internal damage to the component from soldering or bending the leads. Long component boards should be supported by brackets every 4 to 5 inches.

After the component size, spacing, and layout have been decided, the next step is to lay out the board and to indicate all the mounting, terminal, and other holes that are required (Fig. 8-37). By placing the board on spacers, it is possible to use double-ended terminals and to have the interwiring and outgoing leads on the bottom side.

Reference designations are also shown on the detail drawing, next to the components or their terminals. They may be stencilled or silk-screened and protected with clear lacquer.

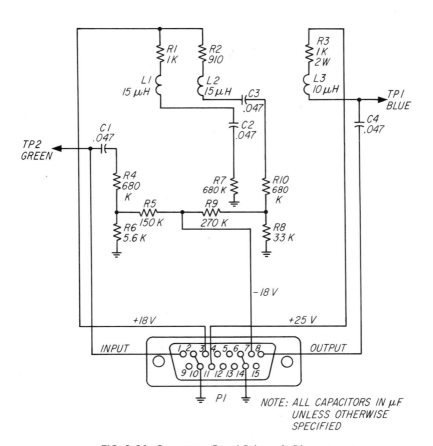

FIG. 8-34. Component Board Schematic Diagram.

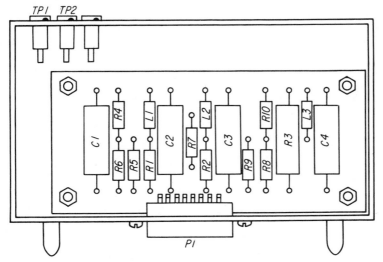

FIG. 8-35. Component Layout for Fig. 8-34.

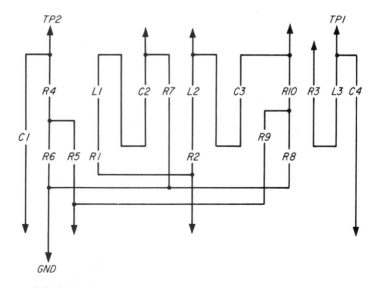

FIG. 8-36. Component Board Arrangement of Circuit in Fig. 8-34.

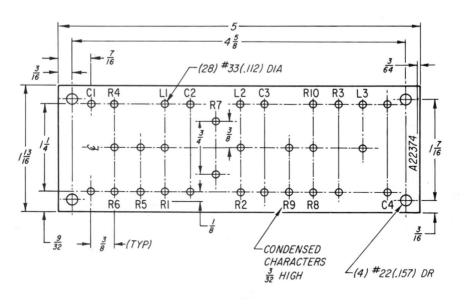

NOTES:

    I. CENTER PART NUMBER ON BOARD

    2. PROTECT DESIGNATIONS WITH CLEAR LACQUER

FIG. 8-37. Component Board Detail of Circuit in Fig. 8-34.

Several drawings are usually required for a component board assembly, starting with the board detail drawing. These drawings may include the following:

1. The board detail and reference designations
2. The assembly of terminals and mounting, if part of board assembly
3. The interwiring and lead assembly
4. The complete board assembly.

A complete board assembly is shown in Fig. 8-38. The leads and interwiring are listed on a previous assembly (see item 3 above).

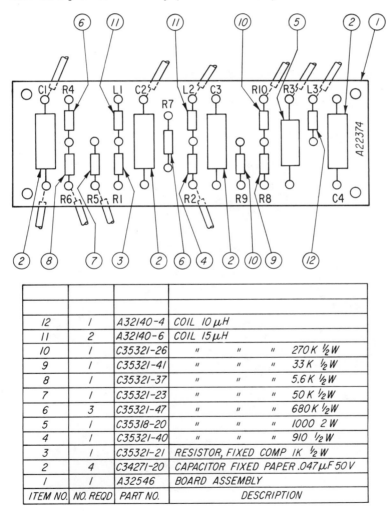

12	1	A32140-4	COIL 10 μH
11	2	A32140-6	COIL 15 μH
10	1	C35321-26	" " " 270K ½W
9	1	C35321-41	" " " 33K ½W
8	1	C35321-37	" " " 5.6K ½W
7	1	C35321-23	" " " 50K ½W
6	3	C35321-47	" " " 680K ½W
5	1	C35318-20	" " " 1000 2W
4	1	C35321-40	" " " 910 ½W
3	1	C35321-21	RESISTOR, FIXED COMP 1K ½W
2	4	C34271-20	CAPACITOR FIXED PAPER .047μF 50V
1	1	A32546	BOARD ASSEMBLY
ITEM NO.	NO. REQD	PART NO.	DESCRIPTION

FIG. 8-38. Component Board Assembly Drawing.

8.12 MINIATURE AND SUBMINIATURE COMPONENTS

The growing trend toward miniaturization of electronic equipment, accelerated by the increasing use of semiconductor devices, has resulted in the development of many new components, miniature and subminiature in size. In some cases, miniaturization has been possible because transistors have lower voltage requirements. In others, technological advances have resulted in a decrease in the component size.

Component miniaturization has sometimes been accomplished by scaling down the mechanical parts of the components (such as screws and shaft diameters), by using plastics to a greater extent, or by using very fine windings. In other instances, the TO-type transistor package has been established as the governing dimension for such components as relays, potentiometers, and so forth.

Among the recent developments that have resulted from the vastly increased density of semiconductor circuitry within a given area can be cited such examples as a desktop computer, Fig. 8-39(a), and a slide-rule calculator (b).

These are but a few examples of the many others where such circuit density not only decreased the item's size but also resulted in improved reliability due

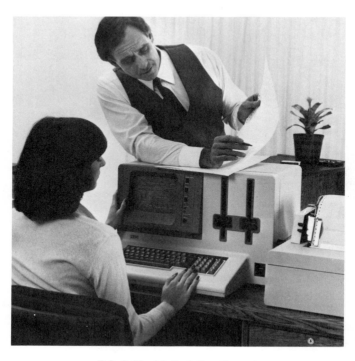

FIG. 8-39. (a) Desk-Top Computer.

FIG. 8-39. (Continued.) (b)
Slide-Rule Calculator.

to wiring elimination between individual components and better mechanical stability.

Another example of reduction in equipment size is the transistor switching power supply shown in Fig. 8-40. The reduction here has been accomplished by increasing the switching frequency to 20 kilohertz, thus increasing the overall efficiency. This high switching frequency became possible with the increased reliability of the power transistors employed in this equipment.

FIG. 8-40. Transistor Switching Power Supply.

SUMMARY

The mechanical design of electronic equipment, which may range from a tiny printed-circuit board to a computer housed in many large racks requires the team effort of many persons to produce an acceptable set of drawings, suitable to manufacture, test, and install such equipment. The draftsman should become familiar with the many factors involved in designing such equipment, such as cost, producibility, wiring techniques required, and so forth.

QUESTIONS

8.1 List some of the decisions that a mechanical designer has to make in designing new equipment.

8.2 What are some of the factors that govern the chassis layout?

8.3 Why is a knowledge of shop fabrication procedures important to the electronic designer and draftsman?

8.4 Name the four functional design requirements with which an electronic designer should be familiar. Discuss each briefly.

8.5 What are some of the functional factors that affect electronic design, and how do they contribute to better electronic equipment design?

8.6 What are some of the human engineering factors that must be considered?

8.7 How are controls segregated and where are they located?

8.8 What are some of the environmental conditions that affect the design of electronic equipment? List some of the difficulties caused by low and high temperatures.

8.9 What are some of the limitations in the space planning of electronic equipment?

8.10 List the various methods used to cool electronic equipment.

8.11 Define briefly a standard rack, a panel, and a chassis.

8.12 What are the three principal end-use categories that affect the design of electronic equipment? List some of the applications in each.

8.13 Name some of the factors that affect the design of consumer electronic equipment. What is the principal component that governs the size of a TV receiver?

8.14 What are some of the important factors in chassis design?

8.15 What are some of the materials used for chassis and how are they selected?

8.16 Describe briefly chassis fabrication in the shop, and list the machines and methods of fabrication.

8.17 What important detail should appear on a drawing of a formed chassis?

8.18 What are some of the drawings required for a typical unit or chassis assembly?

8.19 What are some of the steps to be followed in a trial chassis layout?

8.20 What is a breadboard model?

8.21 What is packaging? Describe some of the methods to be used in component arrangement.

8.22 What is a prototype model?

8.23 Describe some of the practices followed on drawings of sheet metal parts.

8.24 What are some of the common errors in chassis layout? Describe some of the ways to avoid them.

8.25 How are the various components identified on a chassis? How are socket-mounted components identified in military electronic equipment? Make a sketch illustrating this method.

8.26 Describe the methods of chassis marking in commercial production.

8.27 Describe the details that should appear on a typical assembly drawing and make a sketch of a simple parts location drawing.

8.28 Briefly describe modular design, and list its advantages.

8.29 Describe briefly two basic types of modules.

8.30 What is the size of a single-span, single-thickness SEM module?

8.31 What are the standard contact arrangements on SEM connectors?

8.32 Give a brief description of a digital computer module.

8.33 How do the modules in the Trident fire-control system connect into the interconnection network? Describe the main features of the system.

8.34 Describe the industrial applications of modular construction.

8.35 What is a component board? List some of the construction features and advantages.

8.36 What drawing types might be required for a production component board?

8.37 Describe shielding details as they apply to electronic equipment.

8.38 List some of the subminiature components and state their dimensions. Use an electronic catalog to obtain this data.

8.39 What military standard establishes a list of standard parts and what is required for using nonstandard parts?

8.40 List six pin numbers and their functions on a SEM module.

8.41 How is metric system identified on drawings and technical documentation?

EXERCISES

Make preliminary sketches before starting finished drawings and check them carefully. On finished drawings include such data as call-outs, reference designations, specifications, typical notes, and any other necessary information. Consult technical manuals, individual catalogs, and such combined catalogs as Electronic Engineers Master *for any detailed mechanical and electrical data required. Particular care should be taken to develop a layout that is feasible and economical for shop fabrication. Panel slots are $\frac{1}{4}$ inch wide, $\frac{1}{2}$ inch deep, and $1\frac{31}{64}$ inch to center from each horizontal edge.*

8.1 Sketch a panel with a number of dial displays. Show pointers, toggle switches, and indicator lamps in preferred positions.

8.2 Calculate the approximate weight of a steel chassis, 13 by 17 by 3 inches high and

0.062-inch thick, with two mounting lips, $\frac{3}{4}$-inch wide, along the length of the chassis. Use Table 8-1(a) for reference.

8.3   Calculate the approximate weight of a $20\frac{31}{32}$-inch wide, 19-inch aluminum rack panel, $\frac{1}{4}$ inch thick. Base your calculations on an aluminum weight of 0.10 pound per cubic inch. Do not allow for slots in weight calculations.

8.4   Make sketches of: (a) a box chassis; (b) a lap joint; (c) a flat-plate chassis; (d) a panel-mounting chassis; (e) an I-beam chassis.

8.5   Make a sketch showing some of the common drawing errors in sheet metal work and the necessary corrections.

8.6   Make a working drawing of a rack panel in Exercise 8.3 and locate the following components in a symmetrical arrangement: (a) three digital meters; (b) a three-position, $1\frac{1}{2}$-inch diameter, rotary switch, single section, bushing-mounting type, located below the center meter; (c) 2 two-position toggle switches, rated 5 A 125 V, with two indicator lights above the switches, mounted below the left and right meters; (d) suitable panel handles. In making the layout, allow for a chassis 2 inches high, 17 inches long, and 10 inches wide, to be assembled to the rack panel one inch from the bottom of the panel. Include holes for mounting the chassis and for chassis support brackets. Select components from a catalog such as *Electronic Engineers Master*. Include the following engravings on the panel: above the meters, from left to right, "Voltage," "Current," and "Frequency"; above the rotary switch, "M1," "M2," and "M3," and below it, "Meter Selector Switch," on the left toggle switch, "Power" below, and "On" above; and on the right toggle switch "60 hertz" below and "On" above. All lettering is to be in vertical, upper-case condensed characters.

8.7   Make an assembly drawing of the rack panel and blank chassis in Exercise 8.6 including call-outs for all components, hardware, and so forth. Make a parts list as part of the drawing and identify all components by their commercial designations. Include notes as required.

8.8   Make a side-view drawing of the rack and panel assembly in Exercises 8.6 and 8.7. Add chassis slides to the assembly and dimension the extra mounting holes.

8.9   An amplifier assembly uses an open-ended chassis, $3\frac{3}{4}$ inches wide, 6 inches long, and $1\frac{1}{2}$ inches high, with $\frac{5}{16}$-inch end flanges and $\frac{3}{8}$-inch-wide internal mounting flanges. Four 14-contact DIP sockets (select from electronic catalog) are mounted on the top surface, while a volume control with a shaft $\frac{1}{4}$ inch in diameter and $1\frac{1}{4}$ inches long, is mounted on one side and a miniature screw-type fuse holder is mounted on the opposite side. A six-terminal barrier terminal block is mounted near one end of the top chassis surface, two phone jacks are mounted at the opposite end. The chassis is made of #20 gage cadmium-plated steel. Make a suitable layout and prepare the following drawings: (a) chassis detail; and (b) an assembly drawing using local notes to identify the various mechanical and electronic components.

8.10 A standard 19-inch rack panel, $6\frac{31}{32}$ inches wide, is to be used for mounting the indicating, control, and output means of a regulated power supply. Make a front view drawing arranging the following components in a symmetrical layout in outline form: (1) two round flange meters, $2\frac{1}{2}$ inches in diameter, with three #6 screw mounting holes; (2) an instrument removable-type fuse; (3) a two-position, on-off toggle switch, with a $\frac{1}{2}$-inch diameter jewel above; (4) a two-position rotary switch, $1\frac{1}{2}$-inch maximum diameter, operated by a skirted knob $1\frac{3}{8}$ inches in diameter; (5) two standard round panel handles mounted on $4\frac{1}{2}$-inch centers; and (6) two rectangular connectors, one male for input connections and one female for output. Include the major dimensions and locate panel mounting slots.

8.11 Part of a voltage regulator circuit is mounted on a glass epoxy board, $2\frac{1}{2}$ by $3\frac{1}{2}$ by $\frac{1}{16}$ inch thick. The following components are mounted on it: one silicon diode 1N964; two NPN-type switching transistors 2N2193A; $\frac{1}{4}$-watt, $\pm 5\%$ tolerance, fixed-composition resistors with ohmic values of 430, 270, 270, 130, and 100; one variable wire-wound resistor, $\frac{1}{4}$-inch square and $1\frac{1}{4}$ inches long; one fixed-composition resistor, 150 ohms, 2 watts, $\pm 10\%$ tolerance. Seven single-ended terminal studs are to be mounted for external connections on the long end of the board, and four mounting holes for #6 screws are to be located $\frac{5}{16}$ inch from the edge.

Lay out a board with all fixed resistors and the diode in one row and component terminations supported by small single-ended turret terminals. Terminals are to be spaced $\frac{7}{32}$ inch apart except for the 2-watt resistor terminals. Make the following finished drawings: (a) board detail; (b) board assembly with all components and call-outs shown. Assign reference designations and show them on board detail.

8.12 Make a working drawing of a panel-mounted chassis with $\frac{3}{4}$-inch internal flanges and $\frac{3}{4}$-inch mounting flanges, three mounting slots on each side, welded corners, and outside dimensions of $10\frac{7}{16}$ inches wide and $4\frac{1}{4}$ inches high. Chassis is to be of 5052 H32 aluminum, .064 thick, and to have iridite finish.

8.13 Make a reference designation drawing of the back of the rack panel in Exercise 8.6, and include general notes. All designation characters are to be $\frac{3}{16}$ inch high. Sufficient space should be allowed so that the designations may be readily seen when the components are mounted in place. The dimension locations of designations should be from the center of component mounting holes.

8.14 Make drawings of the module shown in Fig. 8-35. The component board assembly is shown in Fig. 8-38. The module external dimensions are $3\frac{1}{2}$ by 6 by $1\frac{1}{4}$ inches, .064 aluminum alloy. Within the module are: three miniature test jacks, two of which are insulated from the module; four riveted component board spacers, $\frac{3}{8}$ inch high; two riveted back-guide pins; and a 15-pin connector of the shape shown in Fig. 8-34. Use Fig. 8-35 as a reference in locating the various components, and make the following drawings: (a) module detail; (b) reference designations, $\frac{1}{8}$-inch characters, shown on inside of the module; (c) complete module assembly, including the component board assembly, a parts list, and such notes as required.

8.15 Make a reference designation drawing of the chassis shown in Fig. 8-22. Black gothic-type characters, $\frac{1}{8}$-inch high, are to be silk-screened on top and inside the chassis to identify the dotted components shown. The transistors are to be numbered from left to right. The rectifier is a four-diode bridge, and the chassis is 6 inches long. Locate all designations in reference to edge of chassis or the center line of component mounting holes.

8.16 A cable tester is housed in an aluminum sloping front housing measuring $4\frac{3}{4}$ inches wide, $4\frac{1}{4}$ inches deep, and $3\frac{1}{2}$ inches high in the front and $5\frac{1}{2}$ inches high in the rear. It has a panel opening with $\frac{1}{2}$-inch flanges extending into it from four sides. The panel is made of red and white plastic laminate and is $4\frac{3}{4}$ inches wide by 4 inches long. It is secured to the flanges by six #4-40 mounting screws that screw into tapped holes in the flanges. The following components are mounted in three horizontal rows on this panel: first row, one red indicator light, body diameter $\frac{1}{2}$ inch, on the left; press-to-test pushbutton switch, same body diameter, in center; one green indicator light, same body diameter, on the right; second row, two BNC-type RF connectors; third row, two microminiature coaxial connectors.

Make a symmetrical layout and identify the various components on the front panel as follows: red indicator light, "Cable Shorted"; green light, "Cable OK"; pushbutton switch, "Cable Tester." Use upper-case condensed characters, $\frac{1}{4}$ inch high. Make the following drawings: (a) housing detail: (b) panel detail; (c) complete assembly with call-outs.

8.17 A 3mHz generator assembly uses a chassis $1\frac{3}{8}$ inches wide, 9 inches long, and 1 inch deep. The following components are mounted on top of the chassis in the order listed: (1) a BNC coaxial connector for input connection; (2) a GE-51 transistor; (3) a double-coil, capacitor-tuned assembly in a housing that measures $2\frac{1}{4}$ inches long, 1 inch wide, and $2\frac{3}{4}$ inches high and that has semicircular ends; (4) 2N1711A transistor; (5) a tuned coil in an aluminum can 1 by 1 by $1\frac{1}{4}$ inches high; (6) a BNC coaxial connector for the output; and (7) a three-terminal block for external power connections that is mounted on the side of the chassis. Both transistors use sockets. Lay out the chassis and make the following drawings: (a) chassis detail; (b) chassis assembly with call-outs for all components.

8.18 Sketch a component board layout on cross-section paper including the following: (a) two $\frac{1}{2}$-watt resistors; (b) three small, paper tubular capacitors; (c) a silicon diode; (d) four 2-watt resistors. Include jumper wires, insulation sleeving, terminals, and component board in the materials list. Include dimensions, and identify all parts with reference designations.

8.19 Make a detailed drawing of the component board in Fig. 8-37 in METRIC dimensions and use metric drill sizes for the holes.

9

GRAPHIC SYMBOLS

Because of the complexity of schematic and wiring or connection diagrams, and also because of the need to conserve space, graphic symbols are used in electronics to denote the various electronic, electrical, and mechanical components. To further clarify each component, its various parts or elements are also portrayed in an unmistakable manner. In many cases, these graphic symbols, such as the one used for a switch, resemble the actual component or its elements, thus helping to make the diagram easier to read.

The currently accepted symbols reflect a considerable evolution since the early days of the electrical industry. At that time, the symbol used by the manufacturers had to bear a strong resemblance to the component itself. In some instances, however, the manufacturers disagreed as to the choice of graphic symbols. The confusion caused by this lack of uniformity was further complicated by the emergence of the electronics industry, which, in many instances, requires the use of complex symbols for delineation of components.

9.1 GRAPHIC STANDARDS

The present graphic standards for electronic communications equipment, issued by the American National Standards Institute (ANSI), have been adopted by the Department of Defense for mandatory use. Among other groups that have contributed to the establishment of uniform graphic symbols are the Institute of Electrical and Electronics Engineers (IEEE), and the American Society of Mechanical Engineers (ASME).

The graphic symbol standard adopted by the industry and the Department of Defense is ANSI Y32.2-1975, *Graphic Symbols for Electrical and Electronics Diagrams,* which is sponsored by the American Society of Mechanical Engineers and the Institute of Electrical and Electronics Engineers. Many representatives from both industry and government served on the committee that developed this standard.

As most electrical circuits require symbols that seldom appear in electronic communication schematic diagrams, the industrial control industry still maintains its own set of symbols. These symbols and their applications are discussed in Chapter 14.

9.2 SYMBOL COMPOSITION

A graphic symbol is formed from a number of simple elements: arcs, angles, circles, rectangles, semicircles, short lines, squares, tees, triangles, and zigzag lines.

Some examples of graphic symbols are given in Fig. 9-1. The general symbol for an antenna is a tee with lines drawn from each side to the base to resemble a flat-top antenna with a lead-in. The symbol for a single cell of a battery consists of two horizontal tees, with a long line always on the positive side and a shorter line on the negative side. A capacitor symbol is composed of an arc with a line at its center to indicate connection and a tee to indicate the opposite conduct-

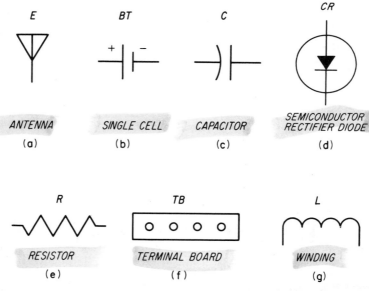

FIG. 9-1. Typical Graphic Symbols

ing plate to which connection is made. A semiconductor rectifier diode consists of a large circle, a filled equilateral triangle, and a horizontal bar which represents the cathode. The resistor is another frequently used symbol. This is composed of three zigzag elements in a row. A terminal board consists of a number of small circles—one for each terminal—enclosed in a large rectangle that signifies the outer boundary of the terminal board. A coil winding is symbolized by a series of connected semicircles.

These examples illustrate that the graphic symbols, in many instances, have a marked resemblance to the physical form of the component. In addition, the function of the component also has a direct bearing on its graphic representation.

Since a graphic symbol consists of one or more basic elements, these elements must be in their proper relationship to each other to make the symbol clear and readily identifiable.

Orientation

The meaning of a symbol is not altered by its orientation. However, by convention, many symbols—such as the ground symbol—are almost always drawn in one particular direction.

Line Weight

Although heavier lines are sometimes used for emphasis, line width does not alter the meaning of a symbol. Nevertheless, it is a good drawing practice to keep all symbols and connection lines on the diagram the same width, as is done on mechanical drawings.

Symbol Size

The ultimate use of a diagram should be considered when it is drawn. In many cases, it will be reduced photographically for use in an instruction book or for other purposes. Therefore, the draftsman should make the symbols, the line weights, the spacing of interconnection lines, and the reference designations as well as other lettered material large enough to be legible after reduction.

Since templates and other mechanical devices are generally used to draw graphic symbols, these devices determine the size of the symbols. It is preferable, however, to use the same size for symbols of the same type and no more than two sizes on any given circuit drawing.

Arrowheads

Unless otherwise specified, arrowheads may be drawn either open ($\longrightarrow$) or closed ($\longrightarrow$).

Terminals

The standard symbol for a terminal, a small circle, may be drawn at the point where connection lines are attached to any symbol, but it is not considered part of the graphic symbol.

Separation of Graphic Symbol Elements

When required by complexity of the diagram, the component parts of such devices as relays, switches, and contactors may be separated throughout the diagram as long as the correlation of these parts is indicated. The details of such separations are explained in Chapter 11.

Reference Designations

It is necessary to add such information as value, part designation, impedance, and type to identify the various symbols. In addition, reference designations consisting of a symbol abbreviation letter and number are used throughout an electronic diagram to identify similar components. These designations are required on all military diagrams, and their applications are discussed in greater detail in Chapter 10.

Reference designations on industrial diagrams follow a different pattern, which is explained in Chapter 14.

9.3 GRAPHIC SYMBOLS

In the graphic symbol standard, the symbols are arranged sectionally in family groups by general type.

For easy reference to the standard, the graphic symbols that follow have been separated into the standard family groups. With the exception of the first two groups, the remaining symbols have been separated into alphabetically arranged family groups.

Two sets of graphic symbols are shown in ANSI Y32.2-1975. One set is for use on single-line or abbreviated diagrams. The other is a set of complete symbols with all connection wires shown. The complete symbols are used in this text with the exception of the waveguide symbols.

In most cases, the corresponding reference designation for civilian and military circuit diagrams has been included above each graphic symbol as a further aid.

Symbol Modifiers

These symbols serve to modify some of the basic graphic symbols. Figure 9-2(a) shows the three adjustability modifiers—adjustable, linear, and nonlinear, respectively. They should be drawn at a 45-degree angle across the body of the basic symbol.

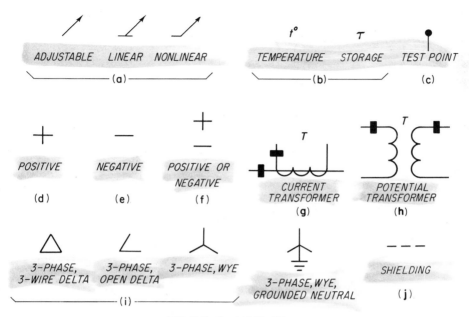

FIG. 9-2. Symbol Modifiers.

Two of the special property indicators are shown in Fig. 9-2(b), temperature and storage (as in storage diode), respectively, the latter being the Greek letter Tau.

Equipment test points are identified on graphic diagrams by a symbol as in Fig. 9-2(c).

The well-known positive and negative symbols, Fig. 9-2(d) and (e), are used on diagrams to indicate polarity. The combined symbol, Fig. 9-2(f), is used when polarity reversal occurs periodically.

Current transformers have a polarity mark as in Fig. 9-2(g) and potential transformers utilize the polarity mark shown in Fig. 9-2(h).

Interconnection of the transformer windings is identified by placing one of the symbols in Fig. 9-2(i) next to the transformer symbol.

Shielding is shown in Fig. 9-2(j).

Basic Items

Among these symbols are resistors, capacitors, antennas, batteries, and others not included in other sections of the standard.

The standard *resistor* symbol shown in the upper part of Fig. 9-3(a) has three peaks on each side and lines drawn at 60 degrees from the horizontal. The lower symbol in Fig. 9-3(a) through (d) is a rectangle, a simplified form used in industrial circuit diagrams. The tendency, however, is to use the zigzag symbol in all electronic circuits to eliminate possible confusion with other symbols.

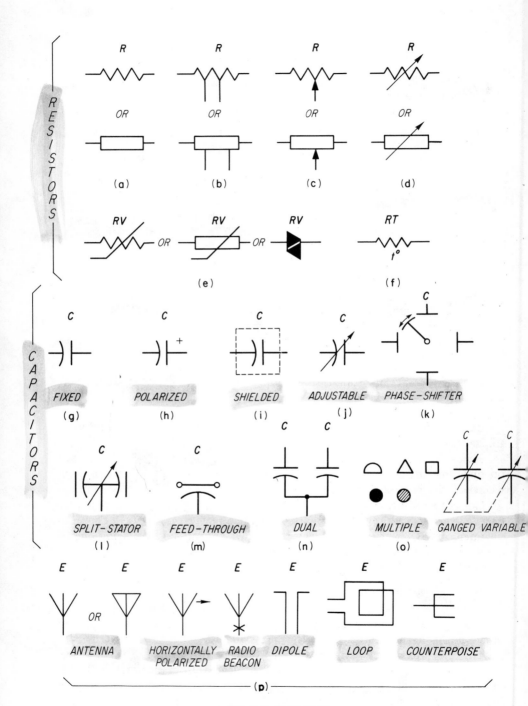

FIG. 9-3. Basic Items.

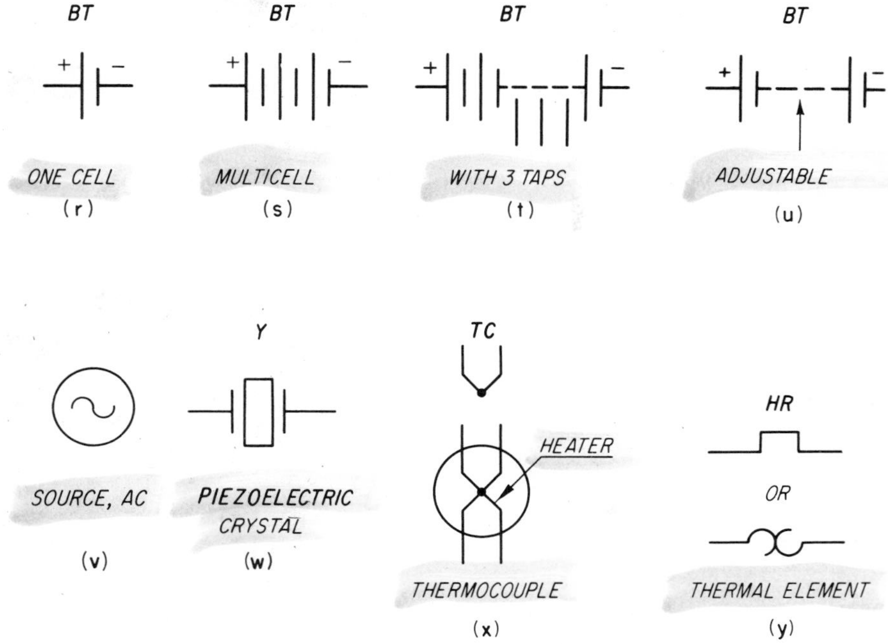

ONE CELL (r) MULTICELL (s) WITH 3 TAPS (t) ADJUSTABLE (u)

SOURCE, AC (v) PIEZOELECTRIC CRYSTAL (w) THERMOCOUPLE (x) THERMAL ELEMENT (y)

FIG. 9-3. (Continued)

The symbols for a tapped resistor appear in Fig. 9-3(b). Although the normal symbol has three peaks, it is sometimes necessary to increase it to four or more to accommodate the taps.

A potentiometer symbol, Fig. 9-3(c), normally has three terminals, two for the ends of the resistor element and one for the sliding contact, which is represented by the arrowhead. Rheostats with two terminal connections use either the symbol in Fig. 9-3(c) or (d) with one connection unused.

A voltage sensitive resistor or varistor, whose resistance is nonlinear, is indicated by the special symbol in Fig. 9-3(e). Another special resistor, known as a thermistor, Fig. 9-3(f), has the designation $t°$ included next to the resistor symbol to indicate that the resistance of the resistor element fluctuates with temperature changes.

Information on the color code system for resistor identification values and other data was covered in Chapter 4.

Some of the common capacitor symbols are shown in Fig. 9-3(g) through (n). The curved element represents the outside electrode in fixed paper and ceramic dielectric capacitors, the moving element in adjustable and variable capacitors, and the low-potential element in feed-through capacitors.

Electrolytic capacitors are generally polarized, and therefore their polarity must be indicated as shown in Fig. 9-3(h).

An adjustable or variable capacitor is shown in Fig. 9-3(j), while a capacitor with a split-stator assembly is illustated in Fig. 9-3(l). Dual or multiple unit capacitors that have a common positive or negative connection or capacitors with a case, which serves as the common terminal, are illustrated in Fig. 9-3(n). The various leads or terminals on dual or multiple unit capacitors are identified with colored dots or special markings as in Fig. 9-3(o).

There are several symbols used to represent various *antennas*, depending upon their form, Fig. 9-3(p). Qualifying symbols may be added next to the antenna symbol such as an arrow horizontally or vertically to indicate horizontal or vertical polarization.

The conventional dry *cell* or one cell of a storage battery, Fig. 9-3(r), may be listed in the cell category. Two or more such cells constitute a battery, Fig. 9-3(s). A number of connected cells—or a battery—with three taps is shown in Fig. 9-3(t), and a multicell assembly with an adjustable cell tap symbol in Fig. 9-3(u).

An oscillator or a generalized alternating current source is represented by the symbol shown in Fig. 9-3(v).

Piezoelectric crystal units that control the frequency of a circuit use the symbol shown in Fig. 9-3(w). The rectangle represents the *thin* slice of quartz or other frequency-controlling material, which is mounted between two conducting plates, indicated by the tees.

The upper part of Fig. 9-3(x) shows a temperature-measuring thermocouple. This is a welded junction, composed of such metals as iron and constantan, which produce a voltage of magnitude related to the junction temperature. The lower half of Fig. 9-3(x) is a thermocouple symbol with an integral heater.

The general symbol for a temperature-actuated device or thermal element is given in Fig. 9-3(y). The two three-quarter circles in the lower symbol represent a bimetallic strip that changes shape with temperature change.

Acoustic Devices

These provide an audible response to an electric current that is circulated through their actuating windings. Their symbols are shown in Fig. 9-4(a) through (c). The two horizontal connections in Fig. 9-4(d) go to the voice coil of an electromagnetic type loudspeaker, while the two vertical connections go to the energizing coil. Voice coil leads should be identified.

The symbols for a single receiver and a double headset are shown in Fig. 9-4(e) and (f). Notice the resemblance between the symbols and their actual physical appearance.

Circuit Protectors

These may be fuses that melt upon sudden increases in current or under overload conditions, or they may be automatic, such as circuit breakers or overload relays.

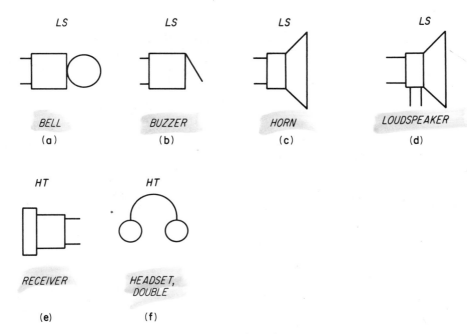

FIG. 9-4. Acoustic Devices.

Fuses, indicated by four alternate symbols, Fig. 9-5(a), are generally used as a means of protection because of their small size.

Symbols for the circuit breaker contacts are shown in Fig. 9-5(b) and (c). An additional symbol for a thermal overload device is included in Fig. 9-5(c).

The arrester also has several symbols, Fig. 9-5(d), depending upon its construction.

Composite Assemblies

Such assemblies may consist of a number of components connected together mechanically or electrically. The use of a composite symbol saves space and avoids repetition.

When a rectangle is used to represent a circuit element, Fig. 9-6(a), some of the following letter combinations may be used within the rectangle to identify the device:

EQ	equalizer		PS	power supply
FL	filter		RG	recording unit
FL-BP	filter, bandpass		TTY	teletypewriter
FL-LP	filter, low-pass			

The amplifier triangle in Fig. 9-6(b) points in the direction of current flow.

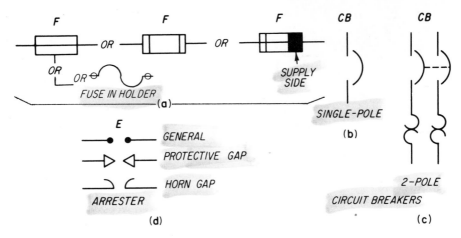

FIG. 9-5. Circuit Protectors.

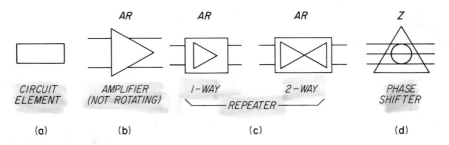

FIG. 9-6. Composite Assemblies.

While this is also true of the one-way repeater, Fig. 9-6(c), used in telephone communications, the triangles in the two-way repeater point toward each other.

The symbol for a 3-wire, or 3-phase, shifting network is shown in Fig. 9-6(d).

Contacts, Contactors, Switches, and Relays

Contact symbols may be in various forms; some of the basic contact elements are given in Fig. 9-7(a).

Basic contact assemblies take the forms shown in Fig. 9-7(b) through (d). The left-hand symbols in (b), (c), and (d) are generally used in industrial work for contactor representation. The length of their parallel lines should be $1\frac{1}{4}$ times the width of the gap. The symbols in (b) represent closed contact assembly, in (c) they indicate open contact assembly, and in (d) they indicate transfer contact assembly.

The several coil symbols for an operating coil of a relay or other device are

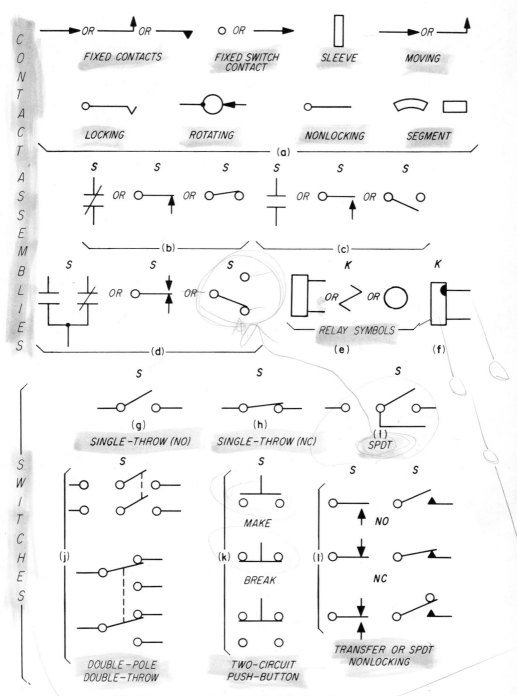

CONTACT ASSEMBLIES

SWITCHES

FIXED CONTACTS   FIXED SWITCH CONTACT   SLEEVE   MOVING

LOCKING   ROTATING   NONLOCKING   SEGMENT

(a)

(b)   (c)

(d)   RELAY SYMBOLS   (e)   (f)

(g) SINGLE-THROW (NO)   (h) SINGLE-THROW (NC)   (i) SPDT

(j)   (k) MAKE / BREAK   (l) NO / NC

DOUBLE-POLE DOUBLE-THROW   TWO-CIRCUIT PUSH-BUTTON   TRANSFER OR SPDT NONLOCKING

FIG. 9-7. Contacts, Contactors, Switches, and Relays.

241

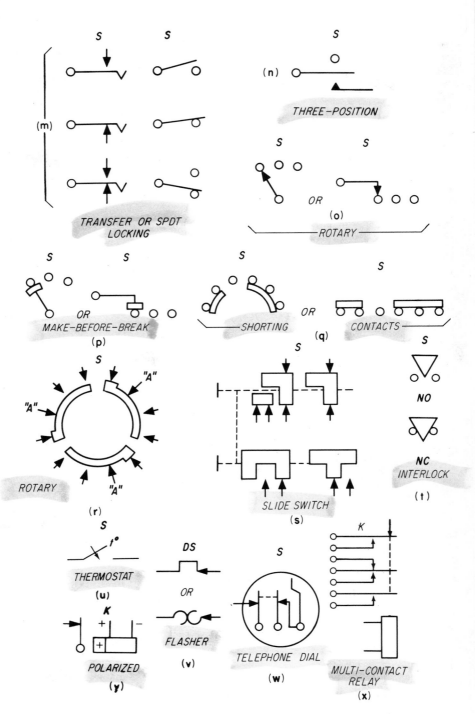

FIG. 9-7. (Continued)

shown in Fig. 9-7(e). A semicircular dot is used, Fig. 9-7(f), when it is desirable to indicate the inner end of a relay winding.

Switch symbols are normally drawn in a position that has no operating force applied. However, a note may be needed on the diagram to explain the operation of a switch that can be in two or more positions without the application of an operating force or that is operated by a mechanical device.

With the switch shown in its normal position, its contacts may be either open or separated (normally open or *NO*) or closed together (normally closed or *NC*), see Fig. 9-7(g) and (h). The switch in Fig. 9-7(g) is also referred to as "ON-OFF," or "single-pole, single-throw."

A transfer switch or "single-pole, double-throw" is shown in Fig. 9-7(i). Alternate arrangements of a "double-pole, double-throw" switch, commonly abbreviated *DPDT*, are illustrated in Fig. 9-7(j). The switch form—lever or limit—is not indicated on the diagram. However, this information is cross-referenced, by its reference designation, to the mechanical assembly that lists the designated switch.

Pushbutton switch symbols are illustrated in Fig. 9-7(k).

Nonlocking or momentary contact buildups in relays, jacks, key, and toggle switches are shown in Fig. 9-7(l). The symbols shown on the right are normally used for toggle switches.

Locking switch symbols for relays, jacks, and toggle switches are shown in Fig. 9-7(m), with the symbols on the right-hand side normally used for toggle switches.

The combination of locking and nonlocking contacts for toggle switches is shown in Fig. 9-7(n), which illustrates the following circuit states: closing, off, and momentary close, or make.

The rotary or selector switch, Fig. 9-7(o), has its own symbol arrangements, which are closely related to its mechanical construction. In its simplest form, the stationary contacts or terminals are represented by the conventional terminal circles with the moving element or contact arm shown by an arrow-headed line pivoted at the center. Connection to the arm is made at the center circle. Normally, the contact arm is drawn at the first (off) or starting switch position. The variation at the right shows the arm projecting along the line of contacts (circles). The contact arm in Fig. 9-7(o) is also known as break-before-make or non-bridging type because the contact between the arm and the stationary contact is broken before the arm makes a connection with the next contact.

The make-before-break type of bridging contact, where adjacent contacts are momentarily shorted by the moving arm segment of the switch, is shown in Fig. 9-7(p). This type of circuit closure is frequently used in power and control circuit work.

A segmental contact, not connected to any external circuit, is shown in Fig. 9-7(q). This type of contact shorts two or more switch contacts together as the switch shaft is rotated.

In drawing rotary switches, the electrical contact sequence is presented clockwise as viewed from the front or knob end of the switch. When switches have more than one section, section one is considered nearest to the control knob. When contacts are on both sides of the switch wafer, the front contacts are considered nearest to the control knob and are so indicated on the diagram.

The rotary switch symbol, Fig. 9-7(r), shows the actual shape of the rotating segments, with their projections making sequential contact to the switch contacts that extend from the stationary part of the switch wafer. The common poles A (contact arms) make continuous contact to the switch segments throughout the switch rotation. Some applications of switch symbols will be discussed in Chapter 11.

A typical three-position slide switch is shown in Fig. 9-7(s). This is a pushbutton slide-type switch, and the slides are normally shown in the released position. As the switch moves from left to right, the movable sections, which resemble their actual physical shapes, make sequential contact to the various arrowheads, which represent the switch terminals.

The safety interlock switches have a special identification symbol as shown in Fig. 9-7(t). This figure illustrates two forms of the switches often used in electronic equipment as protection against accidental contact with high voltage circuits.

Some examples of specialty switches that have their own symbols are limit, liquid-level, flow-actuated, pressure- or vacuum-actuated, temperature-actuated, and foot-operated switches. The symbols are given in Chapter 14.

A temperature-actuated switch or thermostat is shown in Fig. 9-7(u). The conventional switch symbol has the contact motion indicated by an arrow and the $t°$ symbol added to indicate that this is a temperature-actuated switch. A flasher unit symbol is presented in Fig. 9-7(v).

Another specialized form of a switch is a telephone dial mechanism shown in Fig. 9-7(w).

A relay is another switching type of device. The various relay contact arrangements were explained in Chapter 4. An example of a multicontact build-up, consisting of two forms A, or make assemblies, and one form C, or transfer contacts, is given in Fig. 9-7(x).

As relay operating coils and their contact assemblies are often separated on the diagrams to clarify and simplify wiring, care should be taken to properly identify such relay segments.

Electron Tubes

The various electron tube elements are shown in Fig. 9-8(a) through (d). The direct or filament electrode appears in Fig. 9-8(a), and various cathode types are shown in Fig. 9-8(b). On the diagram, leads may be shown connected to either end of the indirectly heated cathode or connected to both ends. Controlling electrode types are given in Fig. 9-8(c) and the anode symbols in (d).

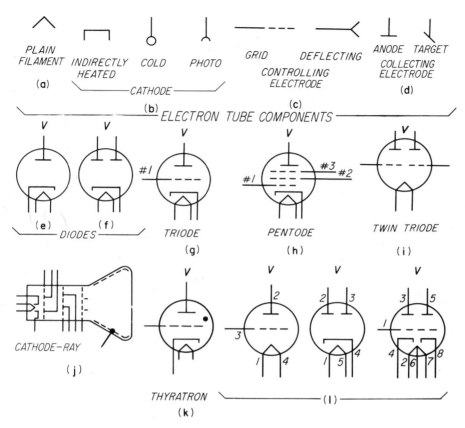

FIG. 9-8. Electron Tubes.

Various electron tube types are shown in Fig. 9-8(e) through (1). The simplest type of electron tube is a *diode*, Fig. 9-8(e), which consists of an indirectly-heated cathode, Fig. 9-8(b), that emits electrons. When the anode or plate is made positive in respect to the cathode, the electrons flow toward the plate until it is made negative. Thus, the diode acts as a valve or a rectifier. Figure 9-8(e) represents what is known as a half-wave rectifier, while the extra plate in Fig. 9-8(f) provides full-wave rectification.

By inserting a control element, known as the control grid (or grid #1), between the cathode and the plate, as illustrated in Fig. 9-8(g), it is possible to control the electron flow by varying the voltage on the grid. Such a tube is known as a *triode*.

The *pentode*, or five-element tube, in Fig. 9-8(h) has an additional screen grid (grid #2) and a suppressor grid (grid #3) added between the grid and the plate.

Some of the specialized tubes are shown in Fig. 9-8(i) through (k). An elongated tube envelope may be used if necessary to avoid crowding the tube elements within the graphic symbol. The symbol may also be split by placing the tube elements close to their functional components on the diagram.

Tubes such as in Fig. 9-8(i) combine two or more sets of tube elements within one envelope; they were developed to reduce cost, space, and weight in electronic equipment.

One special tube, illustrated symbolically in Fig. 9-8(j), is the well-known cathode-ray tube (CRT), of the three-gun, electromagnetic deflection type, used in color TV receivers.

Another special tube type known as the thyratron, Fig. 9-8(k), utilizes a grid element for control. Once the control grid starts the ionization in this tube, the grid has no further control over the conduction of the tube. Acting then as a rectifier, the tube conducts until the anode-cathode voltage is reduced below the critical value.

Tube electrode numbering. To identify the socket connections with the various tube elements, it has become customary to number the connections around the periphery of the electron tube symbol, as shown in Fig. 9-8(1). The numbering is based on the numerical identification of the tube pins as viewed from the bottom of the tube, going in clockwise rotation from the tube base key or other reference point.

The following general rules should be followed in drawing the electron tube graphic symbols:

1. Show only one filament or heater symbol with tubes having more than one filament or heater, unless extra connections are provided for the extra filaments or heaters. Show the filament or heater tap connected at the vertex of the symbol whenever it is brought out to a tube terminal or connected internally to another tube filament.

2. Draw two or more grids, tied internally within the tube, as separate symbols. However, control grids in the form of two windings should be drawn as one grid.

3. Draw only one place element for tubes that have an internally connected grid next to the plate.

Inductors, Transformers, and Windings

Although both are acceptable, the semicircle symbol at the left in Fig. 9-9(a), representing an air core winding or inductor, is easier to draw than the symbol at the right, which was used earlier. The ends of semicircles should be drawn in a straight line, and although three or four are usually used, the number is determined by the specific application.

The symbols in Fig. 9-9(a) are used for inductors or windings of all types

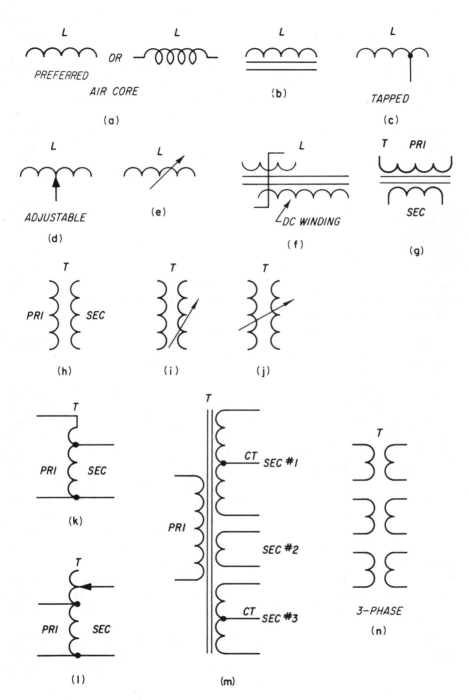

FIG. 9-9. Inductors, Transformers, and Windings.

and for windings in rotating machines, although in the latter case the magnetic core is not shown.

If the inductor has a magnetic core, it is indicated by two parallel lines next to the inductor symbol, Fig. 9-9(b).

Note: Although the core symbol in Fig. 9-7(b) and in other illustrations in the book is shown *above* the winding symbol, it may be drawn *below* it. The choice depends upon the space available or selected for placement of reference designations or other identification.

Figure 9-9(c), (d), and (e) are symbols to be used for inductors that are, respectively, tapped, adjustable, or continuously adjustable.

In some applications, inductors or coils are used as "chokes" to decrease the flow of the alternating-current component in a circuit that also carries direct current. The "choke" effect depends upon the alternating-current frequency and increases with the number of turns composing the inductor and the introduction of a magnetic core.

By adding a separate winding on the core that already has an alternating-circuit coil and passing direct current through this winding, variations in this current will also cause variations in the core magnetization and thus vary the "choking" or reactance of the alternating-current coil. The symbol for such a saturable inductor or reactor is shown in Fig. 9-9(f).

The transformer symbol in Fig. 9-9(g) consists of two or more independent windings coupled by a common core.

Transformers that operate at low frequencies require a magnetic core to reduce the number of turns that would otherwise be required. An air core, Fig. 9-9(h), is sufficient at high frequencies. Note that the designations "PRI," or primary, and "SEC," or secondary, are used to distinguish the primary or input side and the secondary or output side. Although not part of the symbol, these designations are helpful in distinguishing the various transformer windings.

A transformer with an adjustable inductor winding is represented in Fig. 9-9(i), while variable coupling between the windings is indicated by Fig. 9-9(j).

An autotransformer has a single winding with a tap, Fig. 9-9(k). However, if the autotransformer is adjustable, the symbol shown in Fig. 9-9(l) is used.

A multiple-winding transformer is shown in Fig. 9-9(m). Note that the number of semicircles in the primary winding has been increased because of the overall length of the symbol. Taps at the center of the secondary windings #1 and #3 also use the identification "CT." The symbol should also indicate whether the taps are to one side of the center of the winding. The magnetic core symbol of two straight lines should extend up to the maximum length of the windings.

Multiple-phase or polyphase transformers used in power line work are represented by a separate set of windings for each phase, Fig. 9-9(n). The interconnection of these windings is indicated by one of the symbols in Fig. 9-2(i) placed next to the transformer symbol.

Lamps and Visual-Signaling Devices

There are two basic types of lamps—fluorescent and incandescent. The various lamp symbols are shown in Fig. 9-10. In addition, the special ballast lamp symbol, Fig. 9-10(a), is for a device varying nonlinearly with the temperature change of the resistor element.

Fluorescent lamp symbols are shown in Fig. 9-10(b), while the choice of symbol for the cold-cathode lamps depends on whether the lamps are of the ac or dc type, Fig. 9-10(c).

The incandescent lamp symbol is indicated in Fig. 9-10(d).

Among the visual signaling devices, the symbols in Fig. 9-10(e) represent an

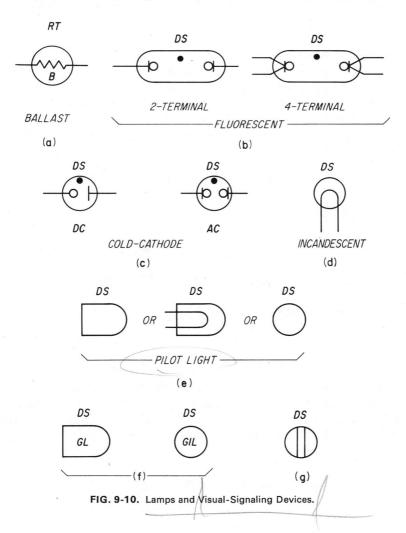

FIG. 9-10. Lamps and Visual-Signaling Devices.

indicating, pilot, or signaling lamp. The following abbreviations, placed within the lamp symbol or to next it, serve to identify the light color or its type:

LIGHT TYPE OR COLOR ABBREVIATIONS

A	Amber	NE	Neon	R	Red
B	Blue	O	Orange	W	White
C	Clear	OP	Opalescent	Y	Yellow
G	Green	P	Purple		

To avoid confusion with the meter or relay symbols, a suffix "L" or "IL", should be added to the abbreviations, see Fig. 9-10(f). A jeweled signal light symbol is shown in Fig. 9-10(g).

Mechanical Functions

Since it is often necessary to identify such mechanical details as linkage or motion on circuit diagrams, some special symbols have been assigned for this purpose. Figure 9-11(a) shows the symbol for mechanical connection or linkage. It consists of a series of short dashes that connect such linked mechanical devices as switch sections, which are separated on the circuit diagram. These dashes should be interrupted, however, if they cross connection lines.

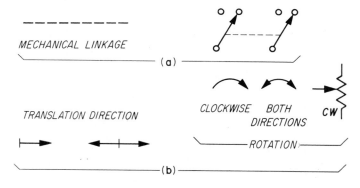

FIG. 9-11. Mechanical Functions.

The symbols in Fig. 9-11(b) indicate the translation direction of power or signal motion, and rotation. The abbreviation "CW" gives the position of the adjustable contact at the limit of clockwise travel viewed from knob or actuator end. Similarly, "CCW" indicates the counterclockwise position.

Readout Devices

Meters and electromagnetically operated counters fall under this classification.

The general symbol for a meter is a circle, Fig. 9-12(a). Meter function is

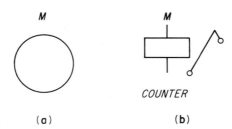

FIG. 9-12. Readout Devices.   (a)   (b)

indicated by a letter combination placed within the meter circle. Some of these combinations are:

A	Ammeter	OHM	Ohmmeter
AH	Ampere-hour	PF	Power factor
DB	Decibel meter	V	Voltmeter
F	Frequency meter	VA	Voltammeter
μA or uA	Microammeter	W	Wattmeter
mA	Milliammeter	WA	Watt-hour meter

The symbol for an electromagnetically operated counter is given in Fig. 9-12(b).

Rotating Machinery

Generators, motors, synchros, and other machines are in this classification. Designations that identify the type of machine may be added to the basic symbol which is a circle, Fig. 9-13(a).

The basic circle symbol may be used to identify generators or motors, with their abbreviation underlined by a straight line for direct current machines or a hertz symbol for alternating-current machines, see Fig. 9-13(b). The circle symbol may also be used in conjunction with the various field symbols, Fig. 9-13(c), to form a composite symbol for direct-current machines. Motor and generator winding connections may be shown in the basic circle as in Fig. 9-13(d). Alternating-current machines may be shown in the simplified form, Fig. 9-13(e).

Semiconductor Devices

Silicon is the material most commonly used for semiconductor devices. Adding controlled impurities produces semiconductor material of the *P* or *N* type. A *PN* junction of these materials will act as a diode or rectifier.

A transistor is a solid-state device that may be thought of as a sandwich composed of *NPN* or *PNP* sections. In a junction-type transistor, the three sections are composed of an emitter, a center section or base, and a collector.

The transistor elements are symbolized by a base or bar, Fig. 9-14(a); a base with two ohmic connections, Fig. 9-14(b); a rectifying junction *P* region on *N*

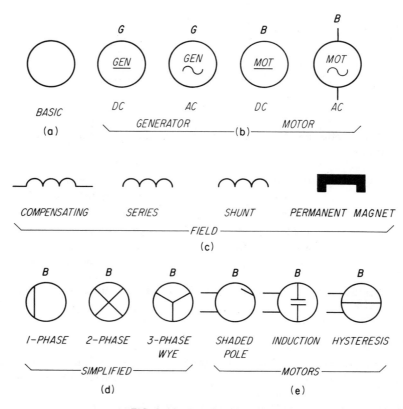

FIG. 9-13. Rotating Machinery.

region, Fig. 9-14(c); and a rectifying junction N region on P region, Fig. 9-14(d). The acute angle arrowheads at the top of symbols (c) and (d) should be filled and should touch the base symbol. They should be positioned about half the length of the arrow from the base symbol. The arrowheads at the bottom of symbols (c) and (d) are filled equilateral triangles.

Figure 9-14(e) represents the enhancement-type semiconductor region commonly associated with the FET's (field effect transistors), consisting of a number of ohmic connections and a rectifying junction.

A P-type emitter, an N-type emitter, and a collector symbol are shown in Fig. 9-14(f), (g), and (h), respectively. Note that the arrows point toward the base symbol in the P-type emitter (PNP) and away from it in the N-type emitter (NPN). Emitter and collector symbols are drawn at 60-degree angles from the base symbol.

Insulated gate symbols are shown in Fig. 9-14(i) and (j), the first representing a single gate and the second multiple gates.

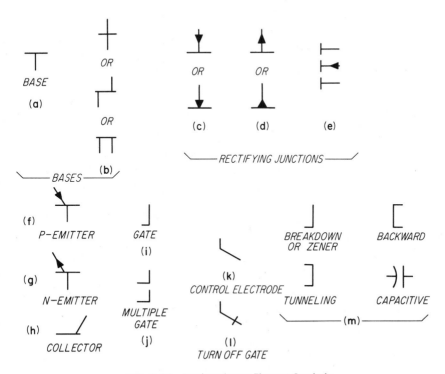

FIG. 9-14. Semiconductor Element Symbols.

A gate or control electrode is represented by the symbol in Fig. 9-14(k) while (1) shows a turn-off gate.

Special property symbols, Fig. 9-14(m), are added to the basic semiconductor symbols. These additional symbols are placed either next to the main symbol or within the envelope.

The two-terminal semiconductor devices are shown in Fig. 9-15. The envelope symbol may be omitted if no confusion is likely to result or if none of the elements are connected to it.

The symbol shown in Fig. 9-15(a) indicates any solid state or semiconductor that permits the flow of electrons in only one direction of the flow of current, while the bar represents the cathode or positive end of the rectifier.

The alternate symbols for a capacitive diode or varactor are shown in Fig. 9-15(b) while (c) and (d) represent temperature dependent and photosensitive diodes, respectively.

Photodiodes may be of the photoemissive type, Fig. 9-15(e), while (f) shows the symbol for a storage diode as denoted by the Greek letter Tau within the envelope.

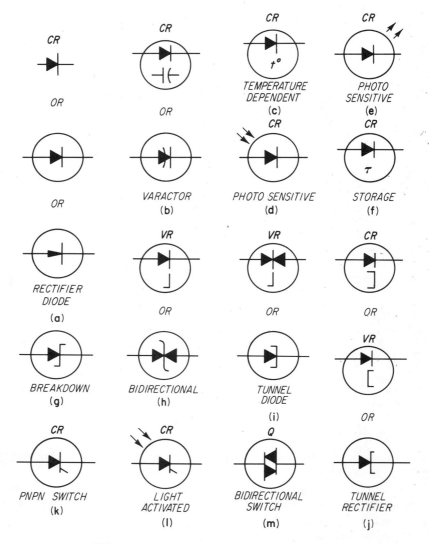

FIG. 9-15. Two-Terminal Semiconductor Devices.

The complete symbol for a breakdown diode is shown in Fig. 9-15(g). This device is used as a regulator to control the voltage of the circuit to which it is connected. The symbol in (h) is for a bidirectional diode or trigger diac.

The tunnel diode shown in Fig. 9-15(i) is a diode in which the current increases, decreases, and then increases again with progressive increases in input voltage. A tunnel rectifier or backward diode is depicted in (j).

Several thyristor symbols are represented in Fig. 9-15(k), (l), and (m). The symbol in (k) is for a diode *PNPN*-type switch, in (l) is for a light-activated diode switch, and in (m) is for a bidirectional diode switch.

The three-or-more terminal semiconductor devices are shown in Fig. 9-16. The basic transistor symbols in Fig. 9-16(a) and (b) show the build-up of the

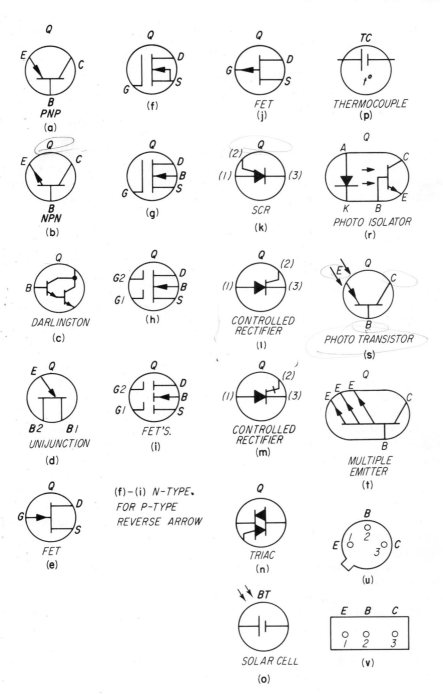

FIG. 9-16. Three or More Terminal Semiconductor Devices.

elements given in Fig. 9-14. The symbol in Fig. 9-16(a) is a single-junction, *PNP*-type transistor, while the symbol in Fig. 9-16(b) is of the single-junction *NPN* type.

The symbol in Fig. 9-16(c) is for an *NPN*-type Darlington transistor.

The unijunction transistor, Fig. 9-16(d), is a single-junction device with a double-base connection diode.

There are several symbols which represent the field-effect transistors (FET's). The symbols in Fig. 9-16(e) through (i) are for *N*-channel junction gate and insulated gate. The symbols for the *P*-channel junction gate and insulated gate would be the same except that the arrowheads would be reversed as, for example, Fig. 9-16(j) compared with Fig. 9-16(e). The letters on the symbols are abbreviations as follows: *G*-gate, *D*-drain, *S*-source, *B*-bulk or substrate. They are for reference only and are not a part of the symbol.

The symbol in Fig. 9-16(e) is for an *N*-channel junction gate FET, three terminal device, while an *N*-channel insulated gate, depletion-type single-gate, active substrate internally connected to source FET is shown in Fig. 9-16(f).

A four-terminal FET, Fig. 9-16(g), is an *N*-channel insulated gate, depletion type, single-gate, active substrate device with substrate connection brought out. A variation is shown in (h) which is similar to (g) with the exception that there are two gates and the FET has five terminals.

Another five-terminal FET is the *N*-channel insulated gate, enhancement type, two-gate, active substrate externally terminated, Fig. 9-16(i).

As mentioned previously, Fig. 9-16(j) represents an FET of *P* channel type, thus the reversed arrowhead. Thyristors are another type of semiconductor device with three or more terminals. They include a silicon-controlled rectifier, commonly known as *SCR*, and the silicon-controlled switch (*SCS*).

The symbol for a semiconductor controlled rectifier, *N*-type gate is shown in Fig. 9-16(k). In (l) is shown a controlled rectifier, *P*-type gate, while the symbol in (m) is for a controlled rectifier, *P*-type gate of the turn-off variety.

The symbol for a triac or gated switch is given in Fig. 9-16(n). This device can handle large currents and has found use in such application as zero-current crossover switching devices to minimize RF interference.

Other semiconductor devices are photosensitive cells, such as barrier photocells or solar cells shown symbolically in Fig. 9-16(o), and temperature measuring thermocouple symbols in Fig. 9-16(p).

The symbol for a photon-coupled isolator of the photoemissive-phototransistor type is shown in Fig. 9-16(r). This is used to isolate the output of a solid-state relay from the circuit it controls.

The symbol for a phototransistor of the *NPN* type is given in Fig. 9-16(s). A multiple emitter *NPN* transistor symbol is illustrated in Fig. 9-16(t).

Transistor element numbering. Several forms are used to bring the transistor element connections out to the header. When there are three connections in a semicircle, they read clockwise in the following order, as viewed from the bottom of the transistor, Fig. 9-16(u): #1 or emitter next to the index tab, #2 or base, and #3 or collector. When the connections are brought out in a straight line, the two closest together are #1 or emitter on the outside, #2 or base, and #3 or collector spaced farthest away, Fig. 9-16(v). These are some of the common lead arrangements. The individual transistor specifications in transistor manuals give other lead arrangements to which the student may wish to refer.

Hybrid circuits. The trend toward microminiaturization has brought about the development of multiple diode-and-transistor assemblies. These circuits are a mixture of etched circuitry elements to which have been added discrete elements such as resistors and capacitors.

They generally use the TO-5 type package or the dual-in-line (DIP) type, which may have as many as 80 pins.

Integrated circuits. Since the integrated circuit package may have hundreds of transistors, diodes, and other components, it became necessary to represent such a package in a simplified manner.

There are two types of symbols employed to represent such a package. One is a rectangle and the other an equatorial triangle. The triangle has been reserved to represent an amplifier; see Fig. 11-12(m). Most integrated circuits are represented by a rectangle, which identifies the pin numbers, selected by the manufacturer, numerically or by function or both. An example is shown in Fig. 9-17.

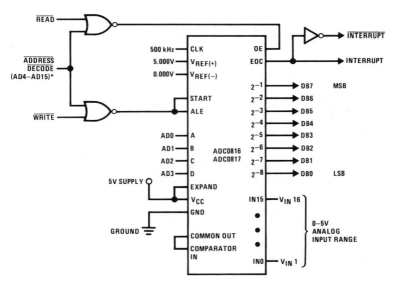

FIG. 9-17. Typical Integrated Circuit Package. (Copyright 1981 National Semiconductor Corporation.)

Reference designations. Hybrid assemblies and integrated circuit packages are identified by letter "U" on schematic diagrams.

Symbol practices. The following general rules should be followed in drawing graphic symbols for semiconductor devices:

1. Whenever possible, use existing component symbols to form new symbols.
2. Draw diode symbols either with or without the enclosing circle, and preferably use a circle for all transistor symbols.
3. Draw the base symbol approximately one-third of the distance from the bottom of the circle, except for unijunction and field-effect transistors. In these cases the base element symbol is located in the center of the circle.
4. Use a dot to indicate collector or base element grounded to the enclosing case.
5. Symbol connections should be located at intersections of a modular grid, see Fig. 9-18.

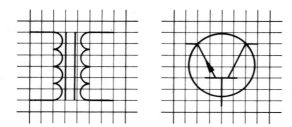

FIG. 9-18. Symbol Connection Points at Grid Intersections.

6. Indicate the index tab or other reference point on hybrid assembly symbols.
7. Show the lead reference point on integrated circuit assemblies.

Terminals and Connectors

A circle is used to indicate a single terminal, while alternate multiple-terminal assemblies are shown in Fig. 9-19(a). Terminal identification is generally governed by circuit convenience rather than sequence.

When necessary for simplification, parts of a given terminal board may be separated throughout the diagram and identified as in Fig. 9-19(b), with the reference designation added.

Due to the many connecting devices used in electronic equipment, it is necessary to identify the various types through the use of graphic symbols. The symbols also distinguish the fixed part of a connector from its mating or removable counterpart. The pin (male or plug) is a single-lead connector shown in Fig. 9-19(c) and mates with the socket (female or receptacle). Note that the lines forming the arrowheads are drawn at a 90-degree angle. The circuit wiring determines whether or not the pin or socket terminals of a connector assembly are

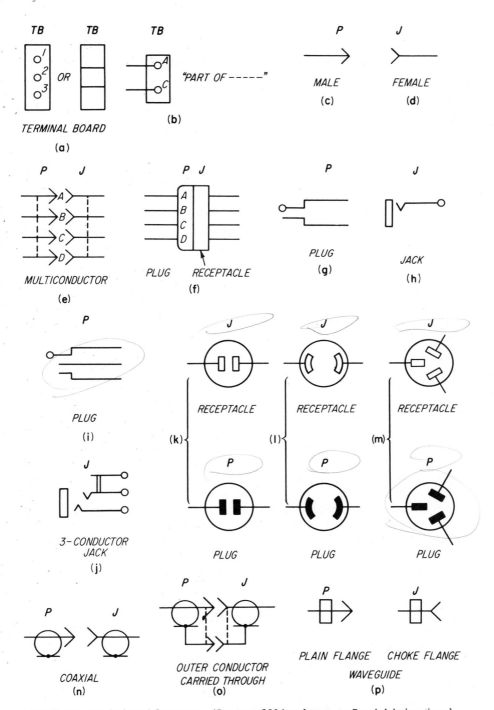

FIG. 9-19. Terminals and Connectors. (See page 260 in reference to *P* and *J* designations.)

attached to the chassis. The male or pin contacts are not wired to a high volt-age circuit if it would be dangerous to have them exposed when they are sepa-rated from the mating receptacle.

Multiple-plug and receptacle assemblies are shown in Fig. 9-19(d), (e), and (f). Terminal and connector identification in Fig. 9-19(d) and (e) show the actual disposition of pin and socket connections, while Fig. 9-19(f) illustrates a sim-plified version.

ANSI Standard Y32.16-1975, *Reference Designations for Electrical and Elec-tronics Parts and Equipments*, specifies that connector reference designations shall be assigned as follows:

1. The movable (less fixed) connector of a mating pair shall be designated *P*.
2. The stationary (more fixed) connector of a mating pair shall be designated *J*.

No mention is made as to the type of contacts on either part.

Thus, the plug or *P* part of the connector may carry all male contacts or all female contacts or a combination of both, and is therefore not limited to all male contacts. Similarly, the same applies to the *J* part of the connector. In most instances, the connector or *P* part consists of all male contacts, as shown in Fig. 19(d), and the *J* part all female contacts.

Another type of connection device, the plug-and-jack combination, is shown in Fig. 9-19(g) through (j). This device may be of the two-wire variety shown in Fig. 9-19(g) and (h) or the three-wire variety, shown in Fig. 9-19(i) and (j). In each case, a small circle is used on the plug to indicate the tip, and on the jack to indicate a terminal. The rectangle on the jack symbol indicates the sleeve. The middle element in Fig. 9-19(i) is connected to the ring part of the plug. This connection corresponds to the lower spring element in Fig. 9-19(j).

For power-supply purposes, such as a two-wire convenience outlet, the con-nector symbol is somewhat different. The contacts are shown as viewed from the mating face semipictorially, with the male contact outline filled, and the female outline open, Fig. 9-19(k) and (l). Polarized or nonreversible connectors, where the third contact is frequently used for grounding purposes, are shown in Fig. 9-19(m) and (n).

It is a general practice to connect the bottom of lowest terminal, as shown in Fig. 9-19(m), to ground.

The coaxial connector, which connects the high-frequency leads between electronic equipment units, is also frequently used. The symbol for the coaxial line—a circle with a base line underneath—is added to the conventional pin and receptacle symbol as shown in Fig. 9-19(o). When it is desired to show the outer conductor carried through on the coaxial line, the symbol is drawn as in Fig. 9-19(p).

Waveguide sections are coupled together mechanically by flanges, Fig. 9-19(q). There are two types of flanges in use—plain and choke. The latter has a groove of definite size cut in the face of the flange, which eliminates the need for good electrical contact within the guide and still maintains the desired electrical characteristics. Note that the waveguide shape is shown next to the flange symbol. The symbols are similar to the connector symbol, except that they are larger and have the waveguide shape added.

Transmission Path

This may be a conductive path or a connection line, Fig. 9-20(a), on a schematic diagram, a common bus, or ground connection.

The following letters may be used to indicate the type of transmission:

F	telephony	V	video (television)
T	telegraphy or transmission of data	S	sound (television)

The crossover loop is no longer used when connection lines cross on a diagram, whether at 90 degrees or less, Fig. 9-20(b). Junction is indicated by a dot as in Fig. 9-20(c) and (d). The staggered type of junction shown in upper part of Fig. 9-20(d) is preferred to the straight line junction shown at the bottom of the figure.

The introduction of a bus bar for power connections in complex electronic assemblies has resulted in the adoption of a special symbol, Fig. 9-20(e).

When conductors are associated with each other, the symbols shown in Fig. 9-20 are used to denote the various conditions. The conductors in Fig. 9-20(f) and (g) are twisted together unless otherwise noted. Conductors that are cabled together are shown in Fig. 9-20(h), (i), and (j), the first showing how shielded conductors are separated on the diagram for convenience.

Associated or future wiring is shown by short dash lines, Fig. 9-20(l).

Recognition symbols for coaxial and waveguide transmission paths are identified as in Fig. 9-20(m) and (n).

Strip-type transmission line symbols are of two types, the strip line at the left part of Fig. 9-20(o) or the sandwich line on the right.

Coaxial and waveguide lines are terminated by the various methods shown in Fig. 9-20(p).

There are three symbols in general use for circuit return connections. The first, or ground connection, Fig. 9-20(r), is a direct connection to earth or to a structure serving the function of earth ground. The second (s) is a chassis or frame connection, and the third (t) is a common return of same potential not necessarily at ground potential.

UHF, VHF, and SHF Circuits

RF transistors and special means for conducting or guiding the radio-frequency energy are required with microwave frequencies of 1000 mHz to 300,000 mHz.

A waveguide, Fig. 9-20(n), is used for this purpose. This is a hollow tube, either circular or rectangular, which varies in size, depending upon the operating frequency.

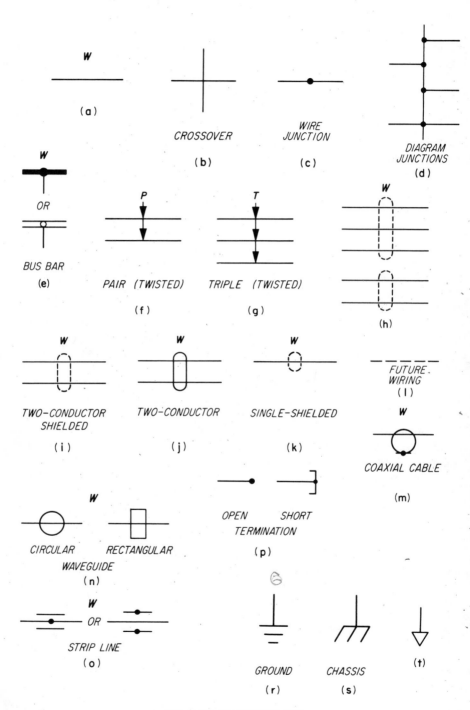

FIG. 9-20. Transmission Path.

The narrow side of the rectangular waveguide is referred to as an E-plane and the wide side as an H-plane, Fig. 9-21(a).

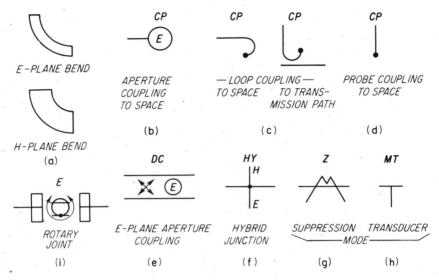

FIG. 9-21. UHF, VHF, SHF Circuits.

Waveguide sections are coupled by means of special flanges, as shown in Fig. 9-19(p).

A coupling aperture is an opening in the wall of the waveguide that transfers energy to an external circuit, Fig. 9-21(b). The designation "E" or "H" is part of the symbol in this and other symbols. Additional transmission paths are indicated on the symbol when necessary.

Two other symbols in common use are coupling by loop and coupling by probe, Fig. 9-21(c) and (d), respectively. A directional coupler, Fig. 9-21(e), is a junction of two waveguides that are coupled to induce the energy to travel in the same direction.

A hybrid junction is indicated by a cross with a dot at the center and "E" and "H" planes indicated as shown in Fig. 9-21(f). The junction consists of four terminated waveguide branches that transfer energy from any of the four branches into any two of the remaining three.

Waveguide propagation is in the form of "modes" of various field patterns. Mode suppression, Fig. 9-21(g), eliminates an undesired field pattern in a waveguide, and a mode transducer or transformer, Fig. 9-21(h), converts a field pattern from one mode to another.

A rotary joint, Fig. 9-21(i), is a device used to connect radio-frequency output from a waveguide to a rotating antenna, which is indicated by a circular waveguide.

Cable Television Systems

The increasing use of cable television systems (CATV) has prompted the inclusion in this book of some of the graphic symbols that are used in mapping and grid diagrams of these systems, Fig. 9-22.

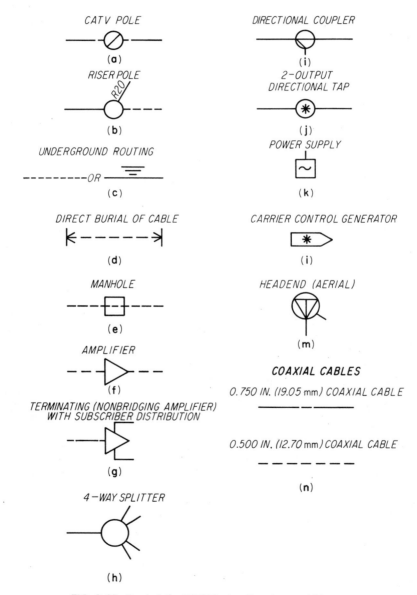

FIG. 9-22. Symbols for CATV System Drawings and Diagrams.

For convenience, the symbols included are identified directly under each symbol. ANSI Standard Y32.21-1976, *Graphic Symbols for Grid and Mapping Diagrams Used in Cable Television Systems,* shows all of the approved symbols. This standard does not use reference designations or class designation letters.

Waveform Symbols

In some schematic diagrams, waveform shapes should be shown at various circuit points to assist in servicing. Some of the simpler shapes are given in Fig. 9-23. The complex waveforms are recorded by photographing the waveform image appearing on the oscilloscope connected to the circuit and transferring the image to cross-section or graph paper.

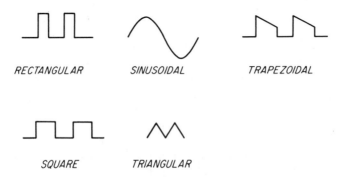

RECTANGULAR   SINUSOIDAL   TRAPEZOIDAL

SQUARE   TRIANGULAR

FIG. 9-23. Waveform Symbols.

9.4 DETAILS OF SYMBOL DRAWING

The special templates available for drawing schematic diagrams were described in Chapter 1. Symbol size is obviously determined by the templates available.

Special care should be taken in using templates, as some of the symbols must be composed from the individual elements outlined on the template. When composing such symbols, it is necessary not only to maintain uniform spacing but also to select the proper elements, since many of the elements are similar in appearance.

Transistor Symbols

A transistor envelope diameter of $\frac{5}{8}$ inch is large enough to accommodate the various transistor element combinations.

The symbols are drawn on grid paper with $\frac{1}{8}$-inch spacing (see Fig. 9-24). The transistor details have been described on page 258. The external connections to the emitter and collector are made at the point where their lines intersect with the envelope.

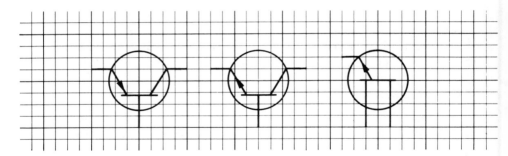

FIG. 9-24. Transistor Symbols.

Other Symbols

Some of the other commonly used symbols are shown on grid paper in Fig. 9-25. The resistor symbol, Fig. 9-25(a), is about $\frac{5}{32}$ inch high. The three half-loops on the inductor symbol, Fig. 9-25(b), are about the same length as the resistor. The capacitor elements, Fig. 9-25(c), are separated by $\frac{3}{32}$ inch, and their length is equal to about $2\frac{1}{2}$ spaces. The other symbol proportions may be measured against the grid background.

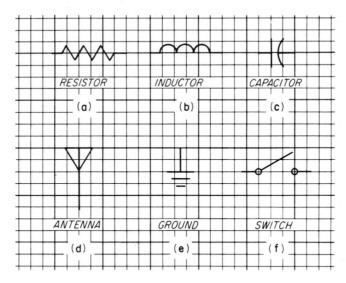

FIG. 9-25. Commonly Used Symbols.

Similar Symbols

Some of the symbols that might be confused because of their similarity or size in construction are shown in Fig. 9-26. The difference in envelope size between the signal light and the transistor, Fig. 9-26(i), is noticeable.

The numbers indicate their relative diameter sizes. Figure 9-26(b) shows the various arrowhead details; Fig. 9-26(d) the relative size of waveguide flanges and

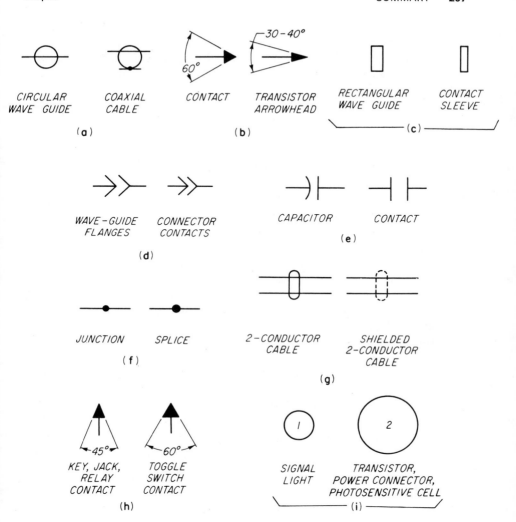

FIG. 9-26. Similarity in Symbols.

connector contacts; Fig. 9-26(e) the difference in spacing between capacitor and contact elements; and Fig. 9-26(f) the relative size of wire junction and splice dots.

SUMMARY

The various electronic devices in a diagrammatic circuit are represented by standard graphic symbols. Although these symbols are generally the same size throughout the drawing, no standards have yet been established as to which size each symbol is to be drawn. There are special commercial templates available to draw circuit symbols. Transistor symbols may be drawn with or without the circle envelope. In addition, the location of graphic symbols on the drawing does not necessarily correspond with their actual location in the equipment.

QUESTIONS

Sketches should be included with answers if necessary.

9.1 What is a graphic symbol?

9.2 What government and industry standard on graphic symbols is available to guide the electronic draftsman?

9.3 List six geometric forms that are used to compose a graphic symbol.

9.4 What is the significance of symbol orientation? Name one symbol that is generally drawn in only one direction.

9.5 What is the most important factor in composing a graphic symbol?

9.6 What are some of the factors that determine the size of the symbol?

9.7 What is the purpose of reference designations on circuit diagrams? Name the elements they are composed of.

9.8 Name the two types of graphic symbols shown in the standard covering military diagrams.

9.9 How are connections indicated at junction points? Which method is preferred? Why?

9.10 What is a symbol modifier? Give two examples.

9.11 List four basic symbols and identify each by a sketch.

9.12 When are more than three peaks used for a resistor symbol?

9.13 How is the movable element in a variable capacitor represented symbolically?

9.14 What does the rectangle represent in a piezoelectric crystal unit symbol?

9.15 What constitutes a composite assembly?

9.16 In what position are switch and relay contacts normally shown?

9.17 Name the two types of contact arms used in rotary switches.

9.18 Describe the operation of the saturable reactor.

9.19 When is a magnetic core required in a transformer?

9.20 What is one of the disadvantages of an autotransformer?

9.21 List six abbreviations for lamp colors.

9.22 Name the three elements of a transistor.

9.23 Describe the action of a tunnel diode.

9.24 List the abbreviations used on field-effect transistor symbols.

9.25 What are some of the general rules for drawing semiconductor symbols?

9.26 In what way does the choke waveguide flange differ from the plain type flange?

9.27 Describe and illustrate the two types of junctions used on schematic diagrams.

9.28 Describe a waveguide.

9.29 Describe any three of the following: (a) unijunction transistor; (b) field-effect transistor; (c) silicon-controlled rectifier; (d) triac. Draw the symbol for each type selected.

9.30 What components are represented by the following letter designations: (a) T; (b) R; (c) F; (d) A; (e) BT; (f) S; (g) X; (h) Q; (i) HY; (j) CR; (k) C; (l) Y; (m) J; (n) L; (p) U; (r) W; (s) RT; (t) AR; (u) G; (v) K; (w) Z; (x) LS; (y) DS; (z) M.

9.31 What semiconductor devices are classified as thyristors?

9.32 Describe a photon-coupled isolator.

9.33 How are connector reference designations assigned?

9.34 What type of contacts are generally found on the connector plug?

EXERCISES

Use opaque cross-section paper with 4, 8, and 10 divisions per inch to draw symbols. Draw single-stroke symbols with 2H or 3H pencil in sizes proportionate to a transistor envelope $\frac{5}{8}$ inch in diameter. See Figs. 9-25 and 9-26 for drawing details. Use symbol templates wherever possible.

9.1 Make a sketch of two symbols to indicate transformer interconnections.

9.2 Sketch four antenna symbols and identify each symbol.

9.3 Draw and identify the four alternate symbols for fuses.

9.4 Sketch the basic contact assemblies used for industrial diagrams.

9.5 Sketch a toggle switch symbol with one nonlocking and one locking contact.

9.6 Sketch the symbol for a normally closed safety interlock switch.

9.7 Sketch the symbols for the *PNP* and *NPN* transistors.

9.8 Illustrate three waveform shapes.

9.9 Make a sketch and show designation letter to identify: (a) a variable capacitor; (b) multiple capacitors on a diagram; (c) an electrolytic capacitor; (d) a tapped battery; (e) a chassis connection; (f) a plug and receptacle; (g) a choke flange on a rectangular waveguide; (h) a terminal board; (i) a semiconductor diode; (j) a silicon-controlled rectifier.

9.10 What symbols represent: (a) a ground connection; (b) a coaxial cable; (c) a relay coil; (d) a circular waveguide; (e) a single-shielded wire? Show the letter designations for each symbol.

9.11 Draw the following switch elements: (a) segment; (b) moving contact; (c) terminal; (d) locking contact.

9.12 Show two symbol arrangements of the following: (a) SPNC contacts; (b) SPST switch; (c) SP rotary switch with bridging contacts; (d) SP rotary switch with momentary contacts.

9.13 What symbols are used for: (a) meter; (b) a pilot light (give two alternates); (c) a pilot light with color identification?

9.14 Sketch the symbols for: (a) tapped; (b) adjustable; (c) magnetic core; (d) continuously variable; (e) mechanical linkage; (f) rotation; (g) shielded; (h) clockwise rotation.

9.15 Make sketches of the following symbols: (a) a tapped resistor; (b) a variable resistor; (c) a relay coil (show two alternates); (d) a power transformer primary, center-tapped secondary winding, and one filament winding; (e) an inductor with *dc* winding; (f) a three-phase transformer with a symbol for a "delta" connection.

9.16 Show the following semiconductor elements: (a) collector; (b) base; (c) *P*-type emitter; (d) *N*-type emitter; and the symbols for: (e) zener diode; (f) tunnel diode; (g) silicon-controlled switch; (h) field-effect transistor with two insulated gates; (i) varactor; (j) temperature measuring thermocouple. Identify by letter designations where applicable.

9.17 Show a symbol buildup for: (a) a telephone type relay with two form A and one form B contacts; (b) a three-pole, six position rotary switch with two bridging contacts and one nonbridging contact; (c) an alternate form of (b); (d) a plain waveguide connected to a choke flange; (e) a three-pole circuit breaker; (f) a three-section fixed capacitor, shielded type.

Draw the symbols in Exercises 9.18 through 9.23 on A-size sheets with lines spaced $2\frac{3}{4}$ inches apart and on $1\frac{3}{8}$-inch centers. Identify each symbol by name and give the letter designation above it.

9.18 (a) relay coil (give 2 alternates); (b) double-pole, single-throw switch; (c) two-conductor shielded cable; (d) single plug and receptacle; (e) ten-post terminal board; (f) crystal diode; (g) variable capacitor; (h) fixed resistor; (i) fuse (three alternates).

9.19 (a) chassis connection; (b) ground connection; (c) coaxial cable; (d) choke flange; (e) rectangular waveguide; (f) hybrid junction; (g) three-wire polarized receptacle; (h) three-connection jack.

9.20 (a) zener diode; (b) unijunction transistor; (c) *PNP*-type transistor; (d) silicon-controlled rectifier; (e) tunnel diode; (f) field-effect transistor.

9.21 (a) ammeter; (b) crystal; (c) a permanent magnet generator; (d) 1-phase motor; (e) fluorescent lamp, 4-wire; (f) ballast lamp; (g) incandescent pilot lamp with red jewel; (h) a counter.

9.22 (a) four-pole, single-throw switch; (b) two-circuit pushbutton; (c) polarized relay with two form A and one form B contacts; (d) three forms of relay coil symbols; (e) a two-section slide switch; (f) normally open interlock switch; (g) circular waveguide.

9.23 (a) one-way repeater; (b) amplifier; (c) two arrester symbols; (d) a series motor; (e) thermistor; (f) variable resistor with a tap connection; (g) thermocouple.

9.24 The symbols in Fig. 9-27 are drawn incorrectly. List: (a) the errors; (b) the device represented by the symbol; (c) the correct letter designation.

9.25 Using 4 by 4 grid paper, draw six typical complete symbols by combining the symbol elements shown in Fig. 9-28. Identify each symbol by name and letter designation.

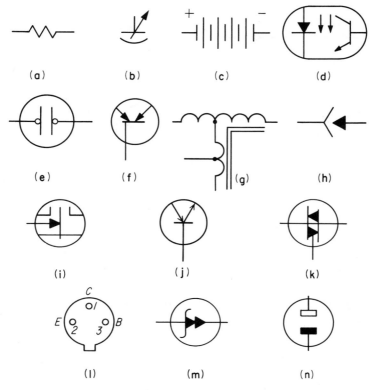

FIG. 9-27. Symbol Problems: Symbol Errors.

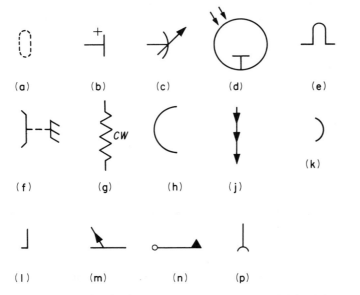

FIG. 9-28. Symbol Problems: Symbol Elements and Partial Symbols.

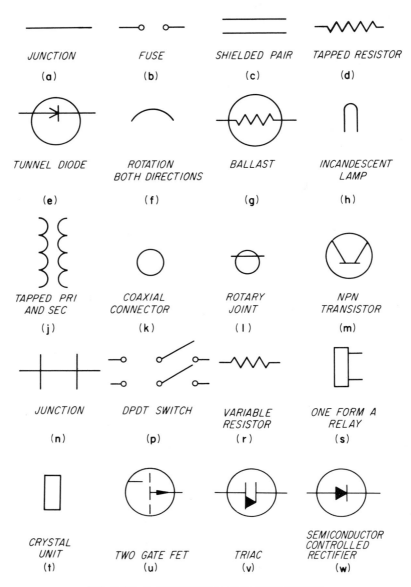

FIG. 9-29. Symbol Problems: Incomplete Symbols.

9.26 Complete the symbols shown in Fig. 9-29 on 8 by 8 grid paper, and list the items added to each symbol.

9.27 To show similarity in symbols, draw and label the following and their nearest counterparts: (a) a contact; (b) connector contacts; (c) a junction; (d) capacitor.

9.28 Draw the following symbols on 4 by 4 graph paper, using an enlarged scale and indicating angles, grid reference lines, and other details for each symbol: (a) an *NPN* transistor; (b) a photoemissive diode; (c) a ground connection; (d) a two-phase power transformer; (e) a varistor; (f) a 3-conductor jack; (g) a photovoltaic cell.

9.29 Select three symbols from the following and draw them on an enlarged scale; that is, transistors twice their usual size and other symbols in proportion. List the classification below each symbol (e.g., semiconductor); the type (e.g., *PNP*); and the letter designation: (a) a photon-coupled isolator; (b) twin variable capacitor with mechanical linkage; (c) a shaded-pole motor; (d) dc type neon lamp; (e) *PNP* transistor; (f) shorting-type bar switch; (g) three-contact polarized receptacle.

9.30 Draw the following symbols used in cable television system drawings: (a) underground routing; (b) 4-way splitter; (c) 2-output directional tap.

10

REFERENCE DESIGNATIONS

Considering the many components that appear on a typical schematic diagram, some means had to be provided to identify them and to correlate this identification with the parts lists, assembly drawings, and manuals.

Reference designations were devised for this purpose. These designations consist of a combination of letters and numbers that identifies both the symbols (and the components they represent) and their location on the circuit diagram. They are not, however, to be considered as an abbreviation for the names of the items. The letters used in the reference designation identify the class of the component, as shown in Table 10-1. These same combinations are used interchangeably to identify electronic, electrical, and mechanical components, as well as assemblies and subassemblies.

Reference designations are used not only on the schematic diagrams but also on the connection or wiring diagrams, parts lists, detail or assembly drawings, and technical manuals. They are the main link between the electrical circuitry and the mechanical details of electronic equipment, for without these reference designations it would be impossible to differentiate between the identical components that appear throughout the equipment or to locate the components on the mechanical drawings.

Reference designations may be classified according to two general divisions —those used for consumer or relatively simple equipment and those used for military and complex equipment such as computers. Because of the greater complexity of much of the military equipment, military reference designations require a somewhat different approach than those used for the simpler equipment. In many respects, however, identical practices are followed for both.

Table 10-1 Item Name Designation Letters (ANSI Standard Y32.2-1975)

adapter, connector	CP	crystal unit, piezoelectric; crystal unit, quartz	Y
amplifier	AR		
amplifier, magnetic (except rotating)	AR	current regulator (semiconductor device)	CR
amplifier, operational	AR	cutout, fuse	F
amplifier, rotating (regulating generator)	G	delay function	DL
		delay line	DL
amplifier, summing	AR	diode, breakdown	VR
antenna	E	diode, capacitive	CR
antenna, radar	E	diode, semiconductor	CR
arrester, lightning	E	diode, storage	CR
assembly, inseparable or nonrepairable	U	diode, tunnel	CR
		disconnecting device (connector, receptacle)	J
assembly, separable or repairable	A		
attenuator (fixed or variable)	AT	disconnecting device (connector, plug)	P
audible signalling device	DS		
autotransformer	T	disconnecting device (switch)	S
backward diode	D, CR	discontinuity (usually coaxial or waveguide transmission)	Z
ballast tube or lamp	RT		
barrier photocell	V	divider, electronic	A
battery	BT	dynamotor	MG
block, connecting	TB	electronic divider	A
blocking layer cell	V	electronic function generator	A
blower	B	electronic multiplier	A
brush, electrical contact	E	equalizer; network, equalizing	EQ
bus bar	W	facsimile set	A
cable, cable assembly (with connectors)	W	fan; centrifugal fan	B
		ferrite bead rings	E
capacitor bushing	C	field effect transistor	Q
capacitive diode	CR	filter	FL
capacitor	C	fuse	F
cavity, tuned	Z	fuseholder	X
cell, battery	BT	gap (horn, protective, or sphere)	E
cell, solar	BT	generator	G
choke coil	L	Hall element	E
chopper, electronic	G	handset	HS
circuit breaker	CB	hardware (common fasteners, etc.)	H
coil, radio frequency	L	head (with various modifiers)	PU
coil (all not classified as transformers)	L	headset, electrical	HT
		heater	HR
computer	A	horn, electrical	LS
connector, receptacle, electrical	J	hydraulic part	HP
contact, electrical	E	indicator (except meter or thermometer)	DS
contactor (manually, mechanically, or thermally operated)	S		
		inductor	L
contactor, magnetically operated	K	inseparable assembly	U
core, air; magnet; magnetic; storage	E	instrument	M
coupler, directional	DC	insulator	E
coupling (aperture, loop, or probe)	CP	integrated circuit package	U

275

Table 10-1 (Continued)

integrator	A	reactor	L
interlock, mechanical	MP	receiver, radio	RE
interlock, safety, electrical	S	receiver, telephone	HT
inverter, motor-generator	MG	receptacle (connector, stationary	J
jack	J	portion)	
junction (coaxial or waveguide)	CP	recorder, elapsed time	M
junction, hybrid (magic tee)	HY	recording unit	A
lamp	DS	rectifier (semiconductor device,	CR
lamp, fluorescent	DS	diode)	
lamp, glow	DS	rectifier, semiconductor controlled	Q
lamp, incandescent	DS	rectifier (complete power supply	PS
lamp, pilot	DS	assembly)	
lamp, resistance	RT	regulator, voltage	V
lamp, signal	DS	relay	K
lampholder	X	resistor	R
light emitting diode	DS	resistor, current regulating	RT
line, artificial	Z	resistor, terminating	AT
loop antenna	E	resistor, thermal	RT
loudspeaker	LS	resistor, voltage sensitive	RV
mechanical part	MP	resolver	B
meter	M	resonator (tuned cavity)	Z
microcircuit	U	rheostat	R
micromodule	U	rotary joint (microwave)	E
microphone	MK	semiconductor controlled switch	Q
mode suppressor	Z	semiconductor controlled rectifier	Q
mode transducer	MT	semiconductor device, diode	CR
modulator	A	sensor (transducer to electrical	A
motor	B	power)	
motor-generator	MG	servomechanism, positional	A
multiplier, electronic	A	shield, electrical	E
network, equalizing	HY	shield, optical	E
network, general (where specific	Z	shifter, phase	Z
class letters do not fit)		shunt, instrument	R
network, phase changing	Z	signal light	DS
oscillator, magnetostriction	Y	slip ring (ring, electrical contact)	SR
oscillograph	M	socket (see page 277)	X
oscilloscope	M	solenoid, electrical	L
pad	AT	speaker	LS
part, miscellaneous electrical	E	squib, electric	SQ
part, miscellaneous mechanical	MP	squib, explosive	SQ
(bearing, coupling, gear, shaft, etc.)		stabistor	CR
phase changing network	Z	strip, terminal	TB
photodiode	CR	subassembly, separable or repairable	A
phototransistor (isolator)	A, U	subdivision, equipment	N
phototube, photoelectric cell	V	switch	S
plug, electrical (connector, movable	P	switch, interlock	S
portion)		switch, semiconductor controlled	Q
potentiometer	R	taper, coaxial or waveguide	T
power supply	PS	teleprinter	A

Table 10-1   (Continued)

transformer	T	varistor, asymmetrical	D, CR
transistor	Q	varistor, symmetrical	RV
transmission line, strip-type	W	vibrator, interrupter	G
transmission path	W	voltage regulator (semiconductor	VR
transmitter, radio	TR	device)	
triode, thyristor	Q	waveguide	W
triac, gated switch	Q	waveguide flange (choke)	J
tuner, E-H	Z	waveguide flange (plain)	P
varactor	D, CR	winding	L

10.1 REFERENCE DESIGNATIONS FOR CONSUMER EQUIPMENT

In some respects, these designations have a simpler approach than military designations because the equipment that they identify is less complex. The practices
explained here are for electronic and electrical equipment. Practices for industrial control equipment, which follow a different set of identification standards,
are described in Chapter 14.

Reference Designation Format

The designations may be given on the schematic diagram as C31, C20A,
C20B, L14, R73, etc. The numerical designation follows the class letter or letters
on the same line, without a space or a hyphen. It is also the same size as the
letter or letters. The suffixes A and B in the examples above indicate that the
capacitors are part of capacitor C20, see Fig. 10-1(m).

Reference Designation Placement

Some of the typical designations for components are shown in Fig. 10-1.
Note that there is no specific placement for reference designations; they may be
placed above, below, or on either side of the graphic symbol. Other pertinent
information is also placed next to the symbol. Therefore, in drawing a schematic diagram sufficient space must be allocated around the graphic symbols to add
such information as capacitance value, terminal numbering, inductance value,
resistor value and tolerance or rating, and identification of capacitor terminals.
    In Fig. 10-1(o) are shown practices of symbol designation placement that
should be avoided unless limited by space.

Reference Designation Details

Some examples of reference designation practices are shown in Fig. 10-1.
    The capacitor designations are shown in Fig. 10-1(a) through (d) and (m).
Note that suffixes have been added to the designation in Fig. 10-1(m) to distinguish the two sections of capacitor C20. The voltage rating has also been
included for each section.

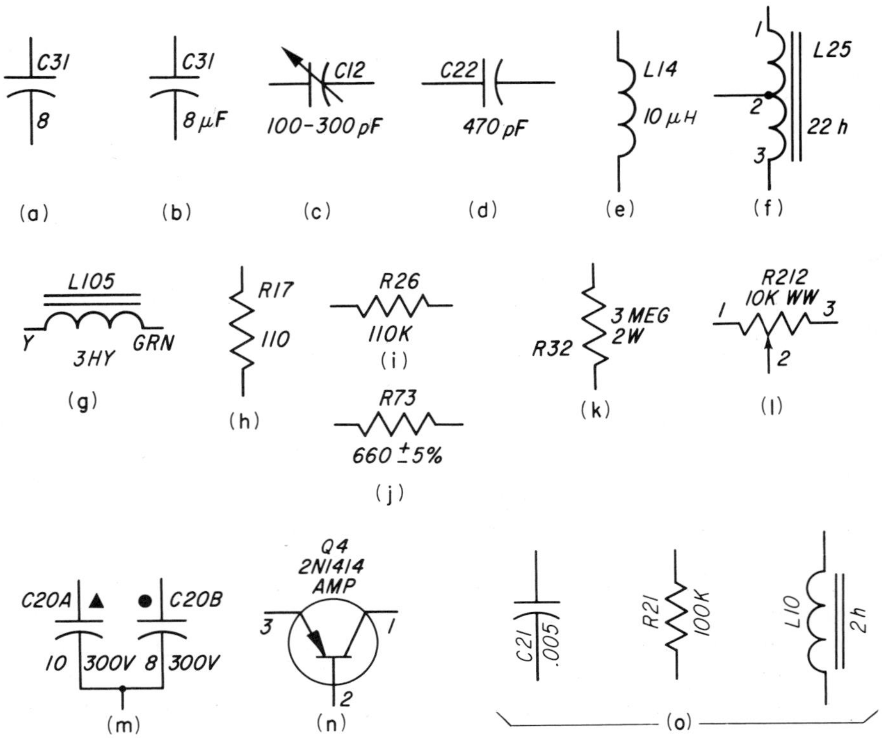

FIG. 10-1. Typical Component Reference Designations.

The inductance examples in Fig. 10-1(e) through (g) indicate not only the value of inductance but also the terminal connections, Fig. 10-1(f), and the color coding, Fig. 10-1(g).

The resistance symbols in Fig. 10-1(h) through (l) show such information as values of resistance, rating in watts, and terminal numbering. The type of resistance (wire-wound) is also given in Fig. 10-1(l).

The transistor graphic symbol Fig. 10-1(n) gives the following information: the transistor reference designation, the transistor type, and the circuit type in which the transistor is used—in that order. Transistor terminals are identifed in the basing diagram for the transistor by elements or numbers.

Reference Designation Table

When changes are made in the equipment, the schematic diagram is also changed by adding or deleting graphic symbols. Because of such changes, a reference table (Fig. 10-2) is often made a part of the diagram to eliminate an unnecessary search for a designation that might have been removed. This is particularly important with such component types as capacitors, inductors,

REFERENCE DESIGNATIONS			
HIGHEST USED			
C82	L23	Q10	R72
OMITTED			
C15	L10	Q8	R23
C17			R24
C31			R27

FIG. 10-2. Reference Designation Table.

resistors, and switches which represent a major proportion of the graphic symbols on the diagram.

The remaining symbols are not renumbered when a graphic symbol is deleted. The deleted reference designations are listed in the table along with the highest designations used. Therefore, it is very important to keep the tables up to date whenever changes are made on the schematic diagram.

The placement details for reference designations and other information given on page 277 apply to both civilian and military equipment.

Sequence of Numbering

It has become standard practice to number each class of components from left to right, beginning in the upper left-hand corner and following horizontally and vertically across the diagram. This makes it easier to locate the reference designations and, thus, the symbols they identify on the schematic diagram.

In complex diagrams that contain serveral major horizontal subdivisions (Fig. 10-3), each subdivision or "layer" is numbered progressively. The components of each class are numbered consecutively from 1 to 199, 200 to 299, and so on.

Component Values

To prevent possible error and to conserve space, the values of the three basic components—capacitance, inductance, and resistance—are designated on schematic diagram as follows:

1. *Capacitance.* 0–9999 pF, in picofarads; more than 10,000 in microfarads, with 1,000,000 pF equal to 1 μF or 1 microfarad. Thus, 100,000 pF would be given on the diagram as .1 μF, and 9900 pF would be given in picofarads.

2. *Inductance.* This value is expressed as either a henry (H); a millihenry (mH), which is equal to one thousandth of a henry; or a microhenry (μH),

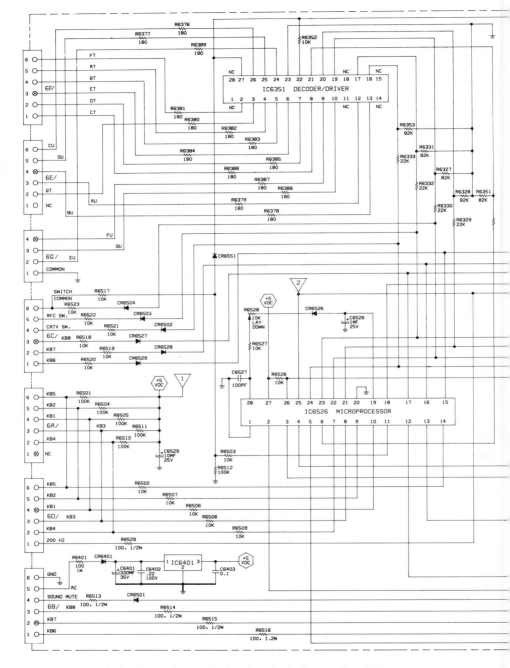

FIG. 10-3. Reference Designation Numbering in a Complex Diagram.

which is equal to one millionth of a henry. However, it is preferable to use the fewest possible ciphers. For example, 8 mH is preferred to .008 henry or 8000 μH. Note, too, that the plural is henrys.

3. *Resistance.* 1–99,999 in ohms; more than 100,000 ohms in megohms, with 1,000,000 ohms equal to 1 megohm, or 1 MEG. Values of 1000 through 1,000,000 ohms, ending in three zeros are specified in kilohms, with 1 kilohm, or 1K, equal to 1000 ohms. It is preferable to use 33K instead of 33,000 ohms and .5 MEG instead of 500,000 ohms. Note that the comma is omitted in values having four digits—2400 ohms, for instance. The omega or ohm value of resistors is understood and, therefore, is omitted from the resistor value designation.

Component Value Notes

The following notes or similar ones are usually used to avoid repetition of component value designations on the schematic diagrams:

UNLESS OTHERWISE SPECIFIED:
1. RESISTANCE VALUES ARE IN OHMS OR KILOHMS (ABBREVIATED K).
2. CAPACITANCE VALUES BELOW UNITY ARE IN MICRO-FARADS AND GREATER VALUES IN PICOFARADS.
or
3. CAPACITANCE VALUES ARE IN MICROFARADS.
or
4. CAPACITANCE VALUES ARE IN PICOFARADS.

Thus, the capacitance in Fig. 10-1(a) would be 8 pF (picofarads), according to Note 2, or 8 μF (microfarads), according to Note 3.

10.2 ELECTRICAL AND ELECTRONIC REFERENCE DESIGNATIONS FOR MILITARY AND COMPLEX EQUIPMENT

Because of the complexity of many military and other electronic equipments, which contain recurring identical subassemblies, assemblies, and units, a special reference designation system has been devised to identify all of the components that are part of such equipment. This system is described in ANSI Standard Y32.16-1975, *Reference Designations for Electrical and Electronics Parts and Equipments.* The use of this standard has been made mandatory by the Department of Defense for all schematic diagrams for military equipment.

Three methods of forming and applying reference designations are described in this standard: Unit Numbering, Location Numbering, and Location Coding. The basic method described here—Unit Numbering—serves as the basis for the other two, which physically locate the item in the unit, assembly, or subassembly either by sequential numbering of areas or by coordinate coding. The student is referred to the above standard for further information on these two methods.

Equipment Division Terms

The following definitions apply to the system divisions (see also Fig. 10-4):

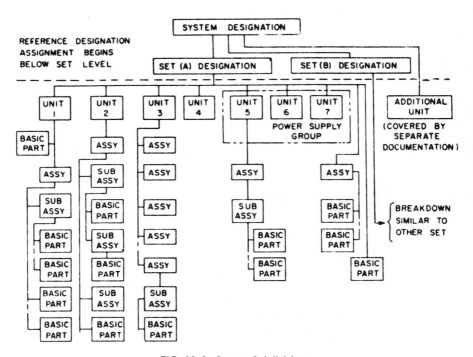

FIG. 10-4. System Subdivisions.

Basic Part—one piece or two or more pieces that are so joined together that they will be destroyed if the part is disassembled. Examples: resistor, relay, microelectronic device.

Subassembly—two or more parts that are joined together but which can be replaced separately. Examples: terminal board with mounted components; resistor board subassembly.

Assembly—two or more parts of subassemblies that are joined for a specific task. An assembly is generally a larger or more complex assemblage of parts. Examples: radio-frequency amplifier; power supply assembly.

Assembly, Microelectronic Device—an assembly of nonrepairable parts or circuits. Examples: integrated circuit package; microcircuit.

Unit—a combination of parts, subassemblies, and assemblies that are mounted together. Examples: radio receiver; electric motor.

Group—a combination of units, subassemblies, or assemblies that form a subdivision of a set or system but which are not capable of a complete operational function. Example: antenna group.

Set—a unit or group of units, which, together with required assemblies, subassemblies, and required basic parts, serves an operational function. Examples: radar set; radio transmitting set. (A typical set is shown in Fig. 10-5.)

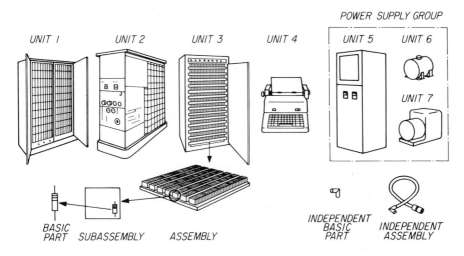

FIG. 10-5. Typical Set Breakdown.

System—a combination of two or more sets which may not be functionally interrelated themselves, but together with assemblies, subassemblies, and basic parts, perform an operational function. Examples: DEW—early warning radar system; ground controlled approach (GCA) system.

Functional Designation—words, abbreviations, or letter combinations describing the item's function as slew or yaw. It is not a reference designation.

Reference Designation Application

Figure 10-5 shows the breakdown of a typical set by unit, assembly, subassembly, and basic part.

Reference designations are assigned only to basic parts, subassemblies, assemblies, and units.

Unit Identification

Each unit within a set is assigned a consecutive number. The number is omitted, however, if there is only one unit. If mounted within a common cabinet or a common rack, or joined with other units, these single units should be classified as assemblies.

Basic Part and Subassembly Identification

The reference designation of a basic part that is not an integral part of a subassembly is made up of the unit number, a letter or letters to identify the class of part (see Table 10-1), and a number to identify that specific part. These numbers are assigned consecutively within each class and start with one. Subassemblies are identified in a similar manner. For example:

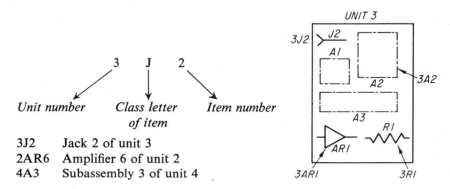

3J2   Jack 2 of unit 3
2AR6   Amplifier 6 of unit 2
4A3   Subassembly 3 of unit 4

The reference designation of a part of a subassembly is made up of the unit number; the letter "A" to identify it as a subassembly; the number of the specific subassembly; a letter or letters to identify the class of part (see Table 10-1); and the number of the specific part in that class. These numbers are assigned consecutively, beginning with the number "one." Subassemblies of subassemblies are identified in the same way.

Figure 10-6 illustrates how this method is applied in a typical unit. The numbers in parentheses show how the progressive subdivisions of this unit are identified with the item or part number. This same relationship may also be seen in the example:

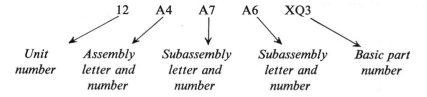

Therefore, the designation 12A4A7A6XQ3 is a socket (X) for transistor 3, of subassembly 6, of subassembly 7, of assembly 4 in unit 12.

Subassemblies and basic parts that are not integral with units are identified in a similar manner, but the unit number is omitted. For example, in a cable assembly W15 with a connector plug on each end, the connectors are identified as *W15P1* and *W15P2* as illustrated in Fig. 10-7.

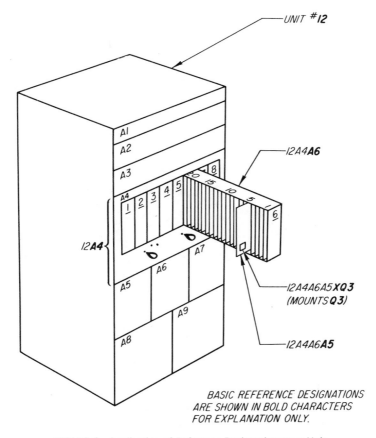

UNIT #*12*

12A4*A6*

12*A4*

12A4A6A5*XQ3*
(MOUNTS *Q3*)

12A4A6*A5*

BASIC REFERENCE DESIGNATIONS
ARE SHOWN IN BOLD CHARACTERS
FOR EXPLANATION ONLY.

FIG. 10-6. Application of Reference Designations to a Unit.

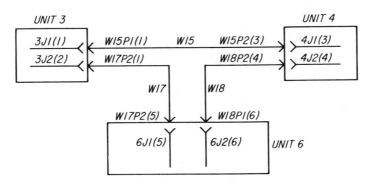

FIG. 10-7. Typical Connector and Inter-Unit Cable Designations.

285

Partial Reference Designations

Partial reference designations may be used on circuit diagrams and drawings if the units and subassemblies are readily identified. The partial designations are limited to the class letter or letters and the subassembly or part, such as AR1C1, A1AR1, and A1 in Fig. 10-8.

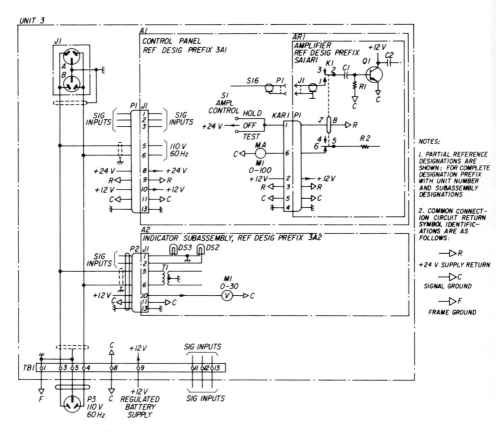

FIG. 10-8. Typical Reference Designations on a Schematic Diagram.

When parts of more than one unit or subassembly appear on the same drawing, positive identification should be provided by including enough of the reference designations. For example, one should use A5C4 for capacitor 4 of subassembly 5. A note should be included on the drawing to indicate that:

PARTIAL REFERENCE DESIGNATIONS ARE SHOWN; PREFIX WITH UNIT NUMBER OF SUBASSEMBLY LETTER OR LETTERS FOR COMPLETE DESIGNATION.

Reference Designation Assignment

The method of forming reference designations and of assigning them on schematic diagrams, described on page 277, also applies to reference designations for military schematic diagrams. This applies to each item—unit, assembly, etc.

Special Requirements

When identical units form a set, they should be identified with different unit numbers. All the identical subassemblies and parts in such units should have the same reference designation but different unit numbers. For example, resistor 12 in unit 5 is designated 5R12, and the corresponding resistor in unit 7 is 7R12.

Similarly, different designations should used to identify any identical assemblies that form a unit. All subassemblies and parts in identical assemblies should have the same designations but different assembly numbers. For example, assembly 4 of unit 5(5A4) and assembly 6 of unit 7 (7A6) have identical designations. A resistor in 5A5 is designated 5A4R12, and the corresponding resistor in 7A6 is designated 7A6R12.

The letter "E" is not used for terminal identification if such parts as terminal boards, transformers, and sockets already have assigned terminal numbers.

Connectors. The movable connector on a mating pair should be designated "*P*" (see Fig. 10-9, connector *P3* on *A7*).

The stationary connector of a mating pair should be designated "*J*" or "*X*," note connector *J2* on *A7* in Fig. 10-9.

Other connector designations depend upon the conditions of use and can be seen in Fig. 10-9. This figure illustrates connector designations for plug-in items. Note that mating receptacles for connectors on cables are identified as "*J*."

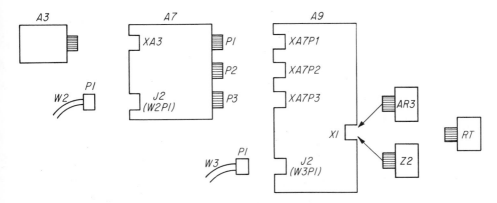

FIG. 10-9. Connector Designations for Plug-In Items.

The reference designation for mating connector and adapter should be shown in parentheses next to the connector if their mating connector is not shown on the schematic diagram drawing.

Relays. Relay elements are identified by suffixes added to the basic relay symbol "K," a dash and then the terminal designations separated by a comma, as for example K6-3, 4.

Another identification method, when relay elements are separated on a diagram, is to add a suffix letter for each element to the reference designation of the relay, starting with A. This method is also used for separated switch elements.

Only one of these methods should be used for relay identification on a diagram.

Sockets. Sockets or fuse holders that are always used with a single assigned part, subassembly, or unit (such as a transistor, fuse, or a printed-circuit board with or without connectors) should be identified by the letter "X," the class letter or letters, and a number to identify the associated part. For example, the socket for transistor Q4 would be identified XQ4.

Potted, embedded, or hermetically sealed subassemblies, modular assemblies, printed-circuit boards, and integrated circuit packages that are replaceable as a single item should be identified as parts. The parts within these subassemblies should be given reference designations only when required.

Repeated-use circuits. The same basic reference designation should be assigned to corresponding parts within each circuit. Such circuit groups should be identified from one to another by assigning a separate "N" designation to each group within a given assembly, as for example N1C1, N2C1, N3C1 (see Fig. 10-10). They identify capacitor C1 common in all three groups. Complete reference designations should be shown for all separated items of each repeated use circuit.

Reference Designation Numbering

To maintain their consecutive numbering, parts or subassemblies are not renumbered after reference designations have been assigned, even though additions and deletions are sometimes required. Deletions should be given in the handbook parts list as, for example, "C22, C23 deleted," or "4R4 through 4R8 deleted." Additional parts take the next consecutive number designation in the class to which they are added.

Reference Designation Marking

The designations should be marked legibly and permanently next to each subassembly and part, on the chassis, the partition, the back of the front panel, and so forth. They should not, however, be marked on replaceable parts or subassemblies.

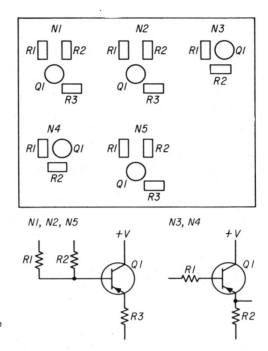

FIG. 10-10.   Repeated-Use
Circuit Practices.

Assemblies or parts that are not integral with the units should be marked on the assembly or part.

Identical subassemblies (see page 286) should bear partial reference designations, so that these subassemblies may be used in multiple application. If space limitations prevent showing the location of these parts and subassemblies on a diagram, marking should be placed on the unit or subassembly where it will be visible when they are viewed. Otherwise, the diagram should be a part of the record in the engineering drawings, handbooks, etc.

Designations for parts that are enclosed in separate and removable shields may be marked on the shields if part replacement does not result in the destruction of the shields or compartments, and if the shields are not interchangeable with other similar parts in the unit.

The designation of a fixed connector should be marked near the connector on the wiring side, and the reference designation or functional designation of the mating connector should be marked on the plug-in side. In cases where marking is not possible, a diagram showing the part and socket location should be placed on the chassis or structure, where it will be visible.

When abbreviated designations are marked on the unit, a note should appear on the unit such as,

REF DESIG PREFIX — — — —

Insert the proper prefix in the blank space.

SUMMARY

A series of reference designations is developed to identify the many similar components represented by graphic symbols on a schematic diagram. Such designations are also useful on connection diagrams, parts lists, and assembly drawings. They help to differentiate between identical components appearing on the equipment and to locate them on mechanical drawings.

QUESTIONS

10.1 What is the purpose of reference designations?

10.2 Discuss the various places where reference designations appear.

10.3 Give ten examples of reference designations used in consumer equipment, including examples of multiple unit components, such as capacitors.

10.4 How are components numbered on schematic diagrams of both the simple and complex types?

10.5 What are the various designation values of capacitance? Inductance? Resistance?

10.6 What are the preferred methods for specifying capacitance? Inductance? Resistance? Cite examples.

10.7 Give three examples of component value notes as they would appear on a schematic diagram.

10.8 What is a subassembly? Give two examples.

10.9 List item designation letters for: (a) antenna; (b) filter; (c) choke coil; (d) cable assembly; (e) integrated circuit package; (f) semiconductor diode; (g) relay; (h) resistor; (i) terminal; (j) winding.

EXERCISES

10.1 Sketch several examples of reference designation placement, relative to the symbols. Show the type of information that is generally given.

10.2 Make a sketch of a reference designation table as it would appear on a schematic diagram.

10.3 Sketch a breakdown block diagram for a typical equipment. Start with a set and work down to parts or components.

10.4 Draw an example of a subassembly identification by reference designation.

11

SCHEMATIC DIAGRAMS

A schematic diagram is merely a condensation of the electrical circuit data that is represented by the graphic symbols discussed in Chapter 9 and identified by the reference designations outlined in Chapter 10. In industrial applications, such a diagram is known as elementary diagram.

A beginner sometimes confuses a schematic or elementary diagram with a wiring or connection diagram, since the two terms are used loosely. The difference is quite simple. The schematic diagram shows the various components by accepted symbols disposed in an orderly manner. The shortest routes are used to show the interconnections of these symbols. The wiring diagram, on the other hand, shows the various components in their relative positions. Each connection is drawn separately and follows its path through the equipment. The components are generally represented by physically recognizable symbols, rather than graphic symbols.

The schematic diagram is usually the first step in an electronic project because it presents and lists the components that are to be included in the equipment. It later becomes the master drawing for the identification and rating of the various components on the mechanical drawings, for the design layout, and for ordering the components.

Depending upon his experience and knowledge of circuit drawing, the draftsman may be given considerable latitude in drawing the schematic diagram. He is expected to know the latest practices in symbols, reference designations, notes, and other relevant information because the project engineer may not be familiar with them. As the draftsman gains experience he will find that the knowledge he acquired previously on similar projects and also his ability to locate catalogs, standards, and other data will be useful to him in drawing such a diagram.

The process of laying out a schematic diagram based on the engineering sketch consists of: (1) collecting data about the components that are depicted on the diagram; (2) organizing the general layout of the diagram, based upon circuit functions or equipment subdivisions; (3) identifying the components by reference designations and other data; (4) adding adequate notes to the drawing; and (5) having the diagram conform in size and details to the drawings that will be produced for mechanical assemblies and details.

The diagram shown in Fig. 11-1 is comparable to a section of the complex diagram in Fig. 11-2, which shows a typical layout that contains complex circuit information in a systematic arrangement.

Although the schematic diagram may be prepared from a freehand sketch, such as shown in Fig. 11-1, the original layout may require considerable modi-

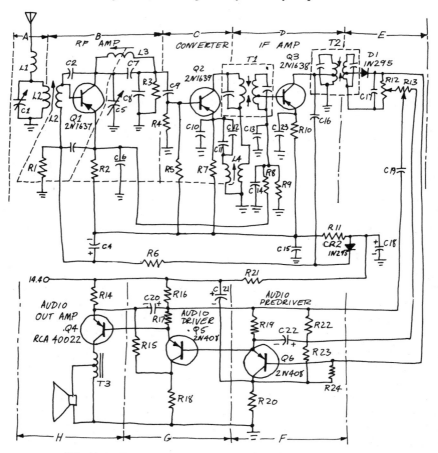

FIG. 11-1. Typical Diagram Sketch Showing Diagram Subdivisions.

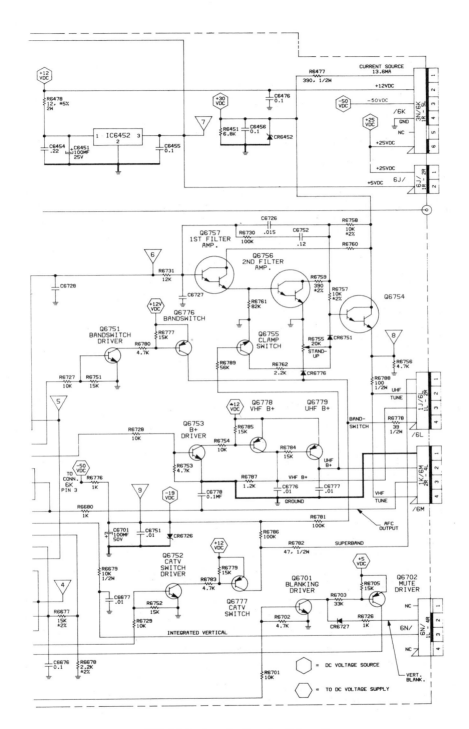

FIG. 11-2. Part of a Complex Schematic.

fication before a desirable circuit arrangement is secured. Whenever possible, component symbols should be located in accordance with circuit functions.

There are two basic fields involved in schematic diagram drawing—the electronic communication field, which applies to both consumer and military equipment, and the industrial field, where the major application of electronics is in control devices. The elementary or schematic diagram and the connection diagrams used in the industrial field will be covered in Chapter 14.

Some of the schematic diagram practices shown are covered in Y14.15-1966 (R1973), *Electrical and Electronics Diagrams*, and *Supplements*—Y14.15a-1970 and Y14.15b-1973.

Many published diagrams do not follow the ANSI standards of graphic symbols or diagram practices, either because the draftsman may not be up to date on the latest changes, or because it may be assumed that the reader will not be familiar with the latest practices and, therefore, will not notice the variations. Such diagrams, however, may be used as practice examples and redrawn according to the procedures outlined in this chapter.

11.1 BASIC COMPOSITION OF DIAGRAMS

The freehand schematic diagram shown in Fig. 11-1 should be carefully analyzed to understand its composition and the basic processes involved in its construction. It is subdivided into sections or groups of symbols that have a definite relationship to each other. The circuit in this figure represents a receiver, consisting of input, Fig. 11-1(A); radio-frequency amplifier stage (B); converter stage (C); intermediate-frequency amplifier (D); second detector (E); audio predriver (F); audio driver (G); and audio output amplifier (H). Thus, by breaking the diagram into several functional parts, it is possible to organize even a complex diagram and still maintain a high degree of orderliness and readability.

All schematic diagrams have several common elements: graphic symbols, symbol arrangement, stage arrangement, signal and power flow, component identification, and operating controls.

The circuit shown in Fig. 11-1 embodies these features. Notice that the signal flow is from left to right in the upper part of the diagram—a procedure that is followed whenever possible. The symbols are also grouped around each transistor in a definite pattern.

Symbol Arrangement

Schematic diagrams are generally drawn to show the components in stages, with each stage centered around a transistor, as illustrated in Fig. 11-3. The symbols are drawn functionally around each stage, depending upon the signal or current flow. This arrangement has become standard practice with the technical personnel who use schematic diagrams.

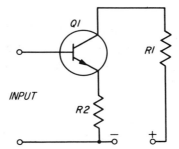

FIG. 11-3. Basic Transistor Circuit.

Signal and Power Flow

As previously mentioned, the normal signal flow on schematic diagrams is from left to right and from top to bottom, when the diagram has several layers or levels of stage symbols. When the output of one stage feeds through the coupling circuit into the input of the next stage, the circuit arrangement is known as a cascaded circuit.

Power supply connections to individual stages are generally arranged from bus lines drawn below the stage symbols. The higher voltage level is drawn at the top bus line, with the others progressively decreasing to zero or minus voltage. Voltage terminations from a stage are indicated with an arrow that shows the voltage level at the termination, instead of by carrying the circuit to a common bus line.

Component Identification

A basic one-stage transistor circuit (Fig. 11-4) shows typical transistor component identification details. These include: reference designations, such as *Q3*, *R1*, and *C2*; component values such as *100K*, *.01*; component voltage

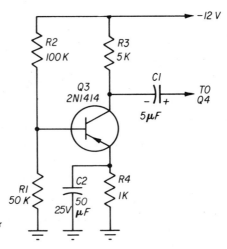

FIG. 11-4. Typical Transistor Component Identification.

rating such as 25 V; transistor designation such as 2N1414; and the destination designation, "*TO Q4*," to indicate a separated circuit on a complex diagram.

Preliminary diagrams may not require complete component information because they may be subject to many changes. After the design of equipment is complete or "frozen," however, the diagram is considered final and should carry all of the required information.

Operating Controls

Although schematic diagrams usually do not show components in their physical relationships, they may include mechanically operable components such as motors, switches, actuators, relays, capacitors, cams, etc. In such a case, it may be necessary to indicate motion, linkage, or position of the component-actuated device. An example of this is the mechanical linkage between coils *L2*, *L3*, and *L4* in Fig. 11-1.

11.2 INDIVIDUAL STAGE LAYOUT

By grouping the various graphic symbols that are related to each other within a rectangular area or a block, the schematic diagram can be subdivided into a series of blocks. These individual blocks comprise transistor stages, coupling circuits, tuning circuits, filters, power supply stages, and other auxiliary groups.

Symbol Requirements

Diagrams may be subdivided into two types, according to their use—production line and technical manuals, such as instruction books and engineering reports. The ultimate use of the schematic diagram determines to a large extent the size of symbols and the overall size of the diagram itself, needed for legibility.

Clarity is, of course, also a basic requirement in a schematic diagram, especially those prepared for production line work where readability is of prime importance. Consequently, graphic symbols should not be drawn small merely to obtain a smaller diagram.

An envelope diameter of $\frac{5}{8}$ inch will be sufficient for transistor symbols. The clarity gained from using this symbol size will more than offset the possible need for a larger diagram size.

At the present time, there is no definite standard for graphic symbol sizes, and, therefore, there is no need to draw a crowded, hard-to-read diagram to conserve space.

A diagram using the larger transistor symbol sizes may be reduced one-half photographically for technical manuals and other uses and still be legible.

Transistor Stage Layout

Modular layout principles can also be applied to transistor circuit layout. Figure 11-5 shows a basic transistor layout. The numbers indicate grid spacings in eighths of an inch.

The circuit in Fig. 11-5(a) is a typical single-stage transistor layout. It uses the horizontal-voltage type of layout to conserve vertical space. The arrowhead on the emitter pointing toward the base indicates a *PNP*-type transistor. The emitter is forward-biased in accordance with the polarity indicated, while the collector is reverse or negative-biased. The stage layout, Fig. 11-5(b), has an *NPN*-type transistor as indicated by the reversed emitter arrowhead and the polarities of the feeder or bus lines.

The draftsman must remember to indicate correct polarities of the power supply voltages for the *PNP*-type transistor, Fig. 11-5(a), or the *NPN*-type, Fig. 11-5(b). It is also customary to indicate the polarity on the emitter bypass capacitor C_E which is an important factor if polarized capacitors are used.

It should also be noted that the emitter arrowhead in the *PNP*-type transistor, Fig. 11-5(a), points in the conventional direction of current flow from the positive source (V_{EE}) to the collector, which is connected to the negative $(-V_{CC})$ through the load resistor R_L. The emitter arrowhead and the supply voltage connections in the *NPN*-type transistor, Fig. 11-5(b), are reversed.

The vertical feed layout in Fig. 11-5(c) requires less horizontal and more vertical space for each stage. The horizontal bus connections in Fig. 11-5(c) refer to the collector voltage line (V_{CC}), collector lead, emitter lead, and the emitter voltage line (V_{EE}), respectively.

The bus lines can be spaced four or more grid spaces below the stage component ground symbols, and whenever possible the transistor symbols are drawn on the same horizontal level.

Coupling Circuits

There are three basic methods of transistor stage coupling in use—resistance, direct, and transformer.

The resistance-capacitance type of coupling, Fig. 11-6(a), shows the use of the emitter bias resistor and bypass capacitor. The direct-coupling type, Fig. 11-6(b), requires a minimum of components, and is thus the cheapest to build. The transformer coupling type, Fig. 11-6(c), matches the input and output impedances of the transistor to attain maximum power gain.

Typical Transistor Circuits

Some typical transistor circuits are illustrated in Fig. 11-7. Note that the transformer secondary windings in Fig. 11-7(b) have been separated to allow straight-line connections to the transistor bases.

Other typical circuits for various transistors are shown in Fig. 11-8. The

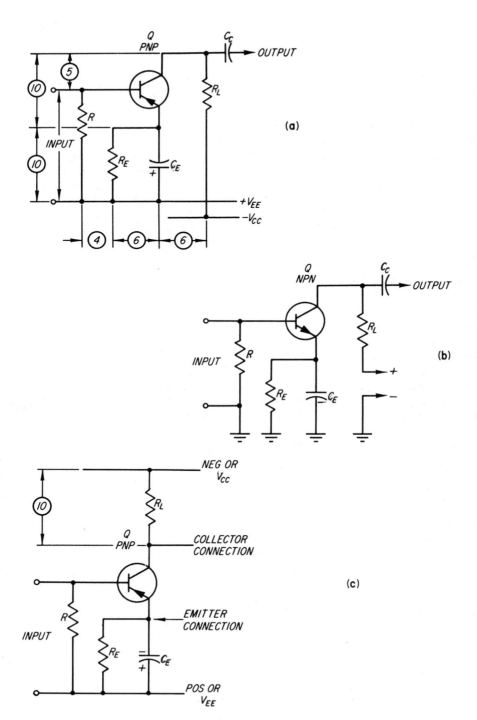

FIG. 11-5. Transistor Block Layout.

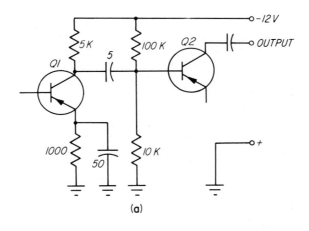

(a)

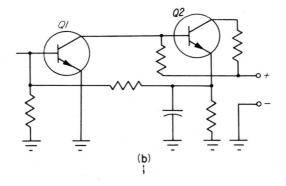

(b)

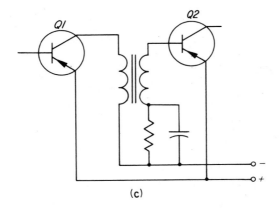

(c)

FIG. 11-6. Transistor Coupling Circuits.

299

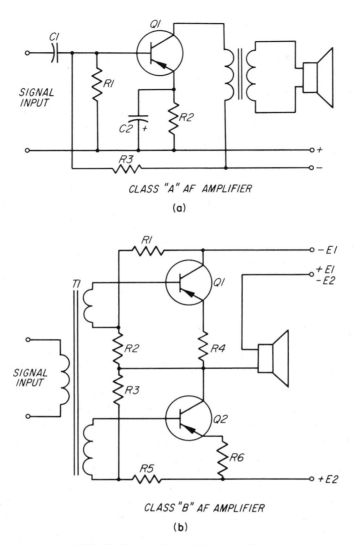

CLASS "A" AF AMPLIFIER

(a)

CLASS "B" AF AMPLIFIER

(b)

FIG. 11-7. Some Typical Transistor Circuits.

time-delay circuit shown in Fig. 11-8(a) contains a relay and uses the unijunction transistor for control. A circuit for a field-effect transistor is shown in Fig. 11-8(b) while in Fig. 11-8(c), a voltage-reference or zener diode is used to maintain a constant output voltage. The transistors are drawn offset to provide direct connections without a crossover.

Semiconductor Base Diagrams

The diagram in Fig. 11-9(a) shows the base of the transistor as viewed from the bottom. The circles, which identify transistor element connections, are consecutively numbered in clockwise rotation, starting with the lead adja-

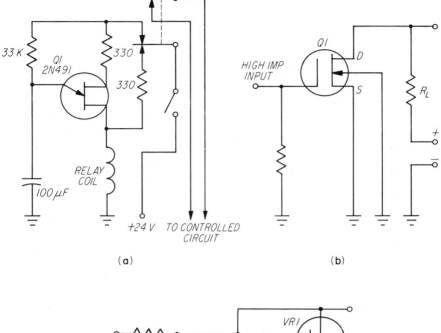

(a)   (b)

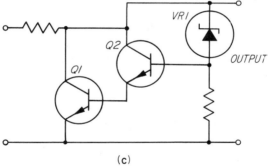

(c)

FIG. 11-8. Transistor Circuits. [(a) and (b)—Courtesy Semiconductor Department, General Electric Company.]

cent to the index tab or other reference point. Note that the elements are also identified by abbreviations that are placed next to each circle.

Although these base diagrams are arranged for wiring diagram circuitry, they also identify each transistor element by a number. This provides the information needed to number the transistor elements on the schematic diagram as shown in Fig. 11-9(b).

See Fig. 9-17 for identification of integrated circuit connections on the schematic diagrams.

Dual-in-line packages and flat packs should have their terminals identified to correspond with their technical manual identification.

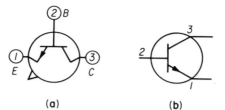

FIG. 11-9. Semiconductor
Base Diagrams.

(a)   (b)

11.3 SWITCH CIRCUITRY

Switches present some circuitry problems. This is especially true in complex circuit diagrams where the switches may be used to select a number of widely separated circuits. Switch contact configurations vary considerably, and the switches themselves may be of a multisection, rotary, or other mechanically operated type.

Multiple Section Switches

Rotary type switches, which may be in the form of a single deck, a wafer, or an assembly of several wafers, are used a great deal in electronic equipment for circuit switching, voltage control, and meter scale changing. In general, these switches have a common shaft that rotates the rotary sections of the switch deck assembly simultaneously. A mechanical index mechanism determines the exact location of the switch rotor at each position of the stationary contacts. This mechanism sets the switch rotor or rotors at 30-degree intervals, which is the most common indexing, or at any other desired indexing separation. The stationary switch decks carry the fixed contacts, and each deck rotor carries the movable contacts to interconnect with the fixed contacts.

Section Separation

Since each deck or wafer, or each side of the wafer, is electrically independent from the other, it is possible to separate each section from the others on the circuit diagram. This practice is also followed with relay contact sections, which will be discussed later in this chapter. However, the draftsman must remember to show such separated sections in the same relative positions and to identify each deck and also its front and rear surfaces.

Typical Switches

A typical switch symbol is shown in Fig. 11-10. The designation *"Front"* applies to the switch deck surface nearest to the switch control knob; *"Rear"* applies to the opposite side. Decks or wafers may be identified by letters beginning with *"A"* for the deck nearest the control knob and followed with *"B,"*

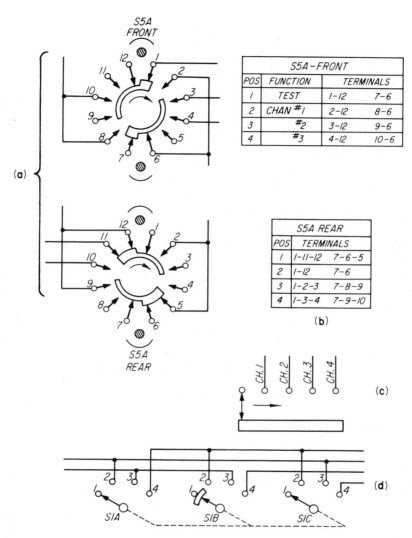

S5A – FRONT			
POS	FUNCTION	TERMINALS	
1	TEST	1–12	7–6
2	CHAN #1	2–12	8–6
3	#2	3–12	9–6
4	#3	4–12	10–6

S5A REAR	
POS	TERMINALS
1	1–11–12 7–6–5
2	1–12 7–6
3	1–2–3 7–8–9
4	1–3–4 7–9–10

(b)

FIG. 11-10. Switch Layouts and Tabulations.

"C," and so forth. Thus, the first switch deck would be identified as "*S1A Front*" or "*S1A Rear*," with contacts identified by "*1F*" or "*1R*" and so on, for the front and rear surfaces, respectively. The switch contact terminals are spaced to correspond with the spacing of the detent mechanism. A switch outline drawing that shows deck shape is used to show the orientation of switch terminals to the switch mounting or studs rather than using contact numbering or some other method.

Most switches do not have the terminals identified by numbers, so that identification numbers may be assigned arbitrarily. The numbering in multiple deck switches should correspond to a common reference point. To avoid errors, terminal numbering should be continuous. That is, all terminals should be numbered whether connections are made to them or not.

The direction of switch rotation should also be indicated within the symbol, preferably near the rotor contacts. To establish the correct circuit connection sequence, the movable members or rotors of each switch wafer should be drawn in their exact mechanical contour. The contacts should also be shown in their respective lengths. The long common contact members, such as *6* and *12* in the front section and *1* and *7* in the rear section in Fig. 11-10(a), make continuous contact throughout the switch rotation, while the remaining short contacts make progressive connections as the switch rotor turns.

Two basic switch rotor contours are used—the break-before-make type, shown extending to contact #1 in *S5A* "*Front,*" and make-before-break type, extending to contacts #5 and #6 in *S5A* "*Rear.*" Note that, in the latter case, the contact member makes connection to contact #7 as well as to #5 and #6.

Tabular Switch Data

In many instances, a separate table, such as in Fig. 11-10(b), is included in the border area of the diagram. This table designates switch connections as they occur during the switch rotation. This is helpful in transferring the switch data to such mechanical details as engraving a switch control panel. The connected circuits for each switch position are listed in the table as "*1* to *12,*" "*7* to *6,*" and so forth.

Simple Switches

The rotary switches just shown are the most complex symbols that the electronic draftsman will have to draw on a schematic diagram. There are also many simpler switch controls in use such as those shown Fig. 11-10(c) and (d). In Fig. 11-10(c), the common arm of the switch is represented by a straight bar; the direction of rotation is indicated by an arrow; and the action of the switch may be rotary even though the symbol represents linear motion. The multisection rotary switch shown in Fig. 11-10(d) does not indicate the direction of rotation because it is normally considered clockwise. The dashes that extend from the switch arms for mechanical interconnection or "ganging" indicate switch motion direction. The center arm is of the make-before-break type.

Switch Layout Details

Switch layout, along with relays and other complex symbols, presents certain problems in keeping crossovers to a minimum and allowing ample space for switch identification. Since many connection lines converge within the limited area around the switch symbol or symbols, it is desirable to make several trial layouts to ascertain the best possible arrangement. Again, if the switch sections are separated on the diagram, the final trial layout serves as a circuit reference for diagram checking later. The decision of whether to segregate the switch sections or to separate them throughout the diagram is best made after considering the number of long connection lines that will be needed if the switch details are consolidated in one area. Long connection lines hinder the readability of the diagram. Although the beginner is apt to be discouraged by switch complexities even after several trial layouts, he will be able to overcome the complexities more easily as he gains experience in circuitry. In any circumstance, the simplest switch symbol arrangement should always be used.

The mechanical linkage between switch wafers should be omitted on complex diagrams because it is understood that the wafers are operated by a common shaft.

Terminal Identification

Multiterminal switches that do not have markings of their own are identified by arbitrarily assigning numbers or letters to the symbol on the diagram. A location diagram showing such identified terminals then becomes part of a note on the diagram, as shown in Fig. 11-11. This practice also helps in circuit testing and preparation of the wiring or connection diagrams described in Chapter 12.

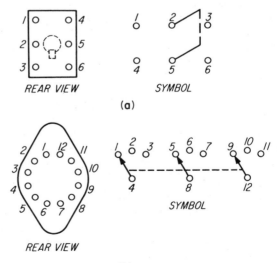

FIG. 11-11. Switch Terminal Identification.

11.4 LOGIC DIAGRAMS

The basic graphic symbols for electronic and mechanical components were presented in some detail in Chapter 9. In addition, complex switching circuits such as digital computers use special symbols. These so-called logic symbols are formed by combining two or more basic graphic symbols to save time and to reduce circuit complexity.

The graphic symbols for logic diagrams are covered in *Graphic Symbols for Logic Diagrams* (*Two State Devices*), *Y32.14-1973*, issued by the American National Standards Institute.

Digital Computers

The information in digital computers is processed by space sequences of signals, such as "one" or "zero," which are commonly referred to as "bits." The one and zero signals may be a positive level for one and a negative level for zero; a relay may be energized for one and deenergized for zero; magnetic marks may be made for one and none for zero, or the one and zero conditions may represent "add" or "not add" or "true" or "not true."

These computers use a series of basic circuits that are interconnected. One design uses a number of standard, plug-in chassis which contain a number of circuits. The input and output points of the circuits are brought out to the connector pins on the chassis. Thus, any desired digital circuit may be obtained by interconnecting the various input and output points.

Several basic logic elements are used to represent the processing data: the AND gate, the NAND gate, the OR gate, the NOR gate, flip-flop, binary counter, time delay unit, signal converter, power amplifier, transducer, or such storage memory devices as magnetic drum, tape, and semiconductor type.

Logic Terminology

Some of these terms are:

AND GATE	a multiple input gate that produces a (logical) "one" when all inputs are at a (logical) "one"
NAND GATE	a multiple input gate that produces a (logical) "one" when at least one input is at a (logical) "zero"
OR GATE	a multiple input gate that produces a (logical) "one" when at least one input is a (logical) "one"
NOR GATE	a multiple input gate that produces a (logical) "one" when all inputs are at a (logical) "zero"
FLIP-FLOP	a two-state (the "one" and "zero" state) stable circuit with two input terminals, which can be flipped or flopped from one state to the other by an input signal

BINARY COUNTER	a series of flip-flops interconnected so the total number of input pulses is represented by "ones" and "zeros"
BIT	a unit of information
READOUT	visual showing of computer process information in the form of tape, punched cards, etc.
STORAGE or MEMORY	a device for storing information mechanically, magnetically, electronically, etc.
SHIFT	movement of computer characters one or more places to the right or left
SHIFT REGISTER	short or long term storage for series or parallel operation.

Logic Symbols

A logic symbol is a graphic representation in diagrammatic form of a logic function. The logic diagram may use uniform shape symbols in the form of rectangles, or it may use distinctive shape symbols for its most commonly used functions and rectangles for the others. In either case, an internal abbreviated label identifies the function when in a rectangular form. Military logic diagrams require the use of distinctive shape symbols.

Some of the logic symbols in common use are shown in Fig. 11-12. The general symbol for an AND function is shown in Fig. 11-12(a), with the uniform shape on the left and the distinctive shape on the right. The distinctive shape symbol for a NAND function in (b) illustrates the use of a small circle at the output to indicate logic negation. The two types of symbols for an OR function in which the output assumes the one-state if one or more inputs assume the one-state are shown in (c). The distinctive shape symbol for a NOR function in (d) again uses a small circle at the output to indicate logic negation.

The symbol in Fig. 11-12(e) is for an EXCLUSIVE OR function in which the output with two inputs assumes the one-state if one, and only one, input assumes the one-state. The ELECTRIC INVERTER symbol in (f) represents a function in which the output assumes the one-state if, and only if, the input assumes the one-state.

The flip-flop, Fig. 11-12(g), which stores a single bit of information, is another device used in computers. The symbol should be drawn with its length equal to twice its width.

The shift register, Fig. 11-12(i), is a binary register that may be shifted to the right or left, one stage at a time. The symbol proportions are $1:1$ for the register and $1:\frac{3}{4}$ for the flip-flops.

A single shot (SS) or one-shot function is represented by the symbol in Fig. 11-12(j), which is rectangular in shape. A similar symbol (k) is used for a Schmitt trigger. This trigger is a device that is actuated when the input signal exceeds a definite threshold voltage.

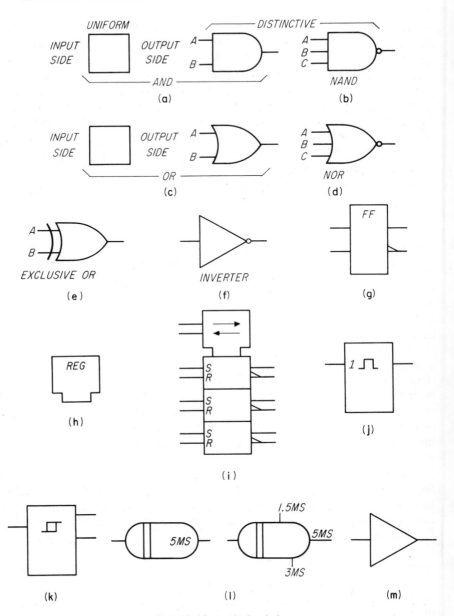

FIG. 11-12. Logic Symbols.

The symbol for a time-delay, Fig. 11-12(1), has two vertical lines to indicate the input side. The time-delay duration is included within the symbol or adjacent to the tap output.

An amplifier, Fig. 11-12(m), may be either linear or nonlinear or current or voltage type, and with one or more stages. The amplifier output assumes the one-state if, and only when, the input assumes the one-state. The symbol for an inverting AMPLIFIER is made by adding a small circle at the output.

Function Identification

The following combinations of letters are used within the uniform logic symbols to identify functions:

BO	Blocking Oscillator	S	Set
C	Clear or reset	SR	Shift Register
CF	Cathode Follower	SS	Single Shot
EF	Emitter Follower	ST	Schmitt Trigger
FF	Flip-Flop	T	Toggle or Trigger
(N)	Number of bits		
RG	Register		
RG(N)	Register, N stages		

Logic Diagram Details

Some of the circuits and their logic symbols are shown in Fig. 11-13. Note that the numerals represent the input and output pin numbers.

"Tagging" lines are included within the logic symbol (Fig. 11-14) to locate

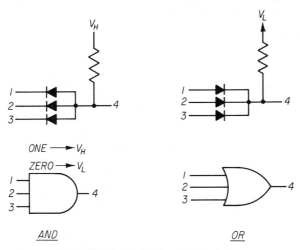

FIG. 11-13. Logic Circuit Equivalents.

OR FUNCTION

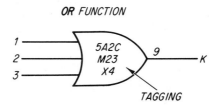

FIG. 11-14. Identification of Logic Circuit Location.

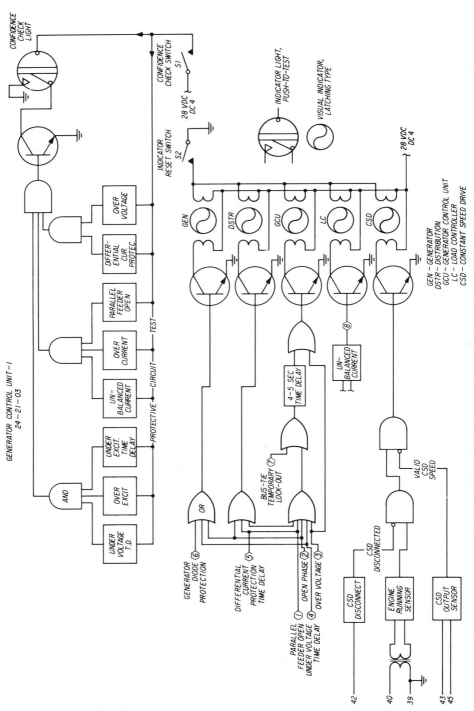

FIG. 11-15. Typical Detailed Positive Logic Diagram Using Distinctive Shapes.

the logic function on the drawing or within the equipment. The three lines identify:

5	drawing sheet number	M23	module code
A2C	circuit uniqueness	X4	abbreviated reference designation for socket 4 (see page 286)

In addition, input and output pin numbers are placed around the symbol periphery to identify the connections.

A typical positive logic diagram with distinctive shapes is shown in Fig. 11-15.

Waveform symbols may be placed next to the signal lines to indicate voltage levels, pulses, or pulse trains (PT).

Some of the practices that should be followed in individual function representation are shown in Fig. 11-16, e.g., an AND function with multiple inputs.

Templates, such as the one shown in Fig. 11-17, are available for drawing logic symbols.

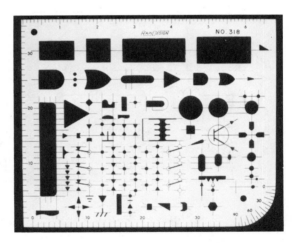

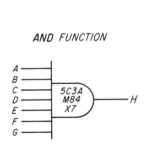

AND FUNCTION

FIG. 11-16. Logic Function Representation.

FIG. 11-17. Logic Symbol Template.

11.5 FILTER AND TUNED CIRCUITS

These are used in applications requiring frequency discrimination or tuning by various means.

Filter Circuits

Filters generally consist of combinations of capacitors and inductors that are either air-core or magnetic-core, depending upon the frequency that is desired to be passed or suppressed. They fall into three main types—low-pass (*LP*), high-pass (*HP*), and band-pass (*BP*); see Fig. 11-18.

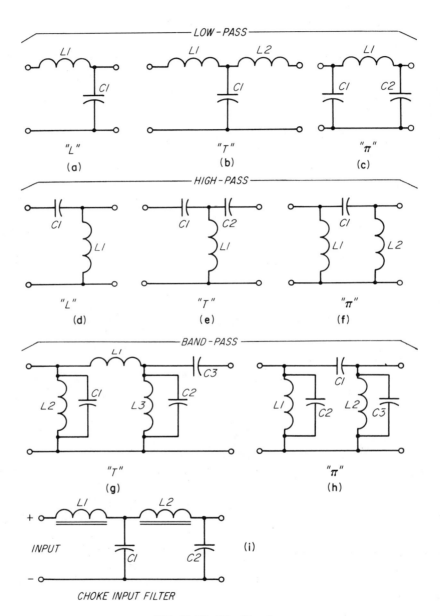

FIG. 11-18. Filter Circuits.

A low-pass filter allows passage of all frequencies below a specified (cutoff) frequency, and the high-pass filter does just the opposite. It rejects all transmission above the specified cutoff frequency. The band-pass filter allows passage of a certain band of frequencies but cuts off passage of frequencies below and above that band.

As shown in Fig. 11-18, filters are identified by their shape, i.e., the figure formed by their symbols. Thus, the "L" is formed by an inductance and a capacitor, Fig. 11-18(a); the "T" consists of two inductances and a capacitor between them, Fig. 11-18(b); the π indicates an inductance and a capacitor at each end, Fig. 11-18(c). The "L" and "T" types of the low-pass variety are used in power-supply circuits to filter out the ac component. The circuit shown in Fig. 11-18(i), consisting of an input choke followed by a π-type filter, is a combination used in power-supply circuits.

Some coupling circuits are actually portions of a filter network. Complex filters may be built as separate units. These are drawn on the schematic diagram in either a block form or as a separate diagram (see page 317).

Tuned Circuits

In addition to filter circuits, there are tuning circuits that have an inductance connected in parallel with a tuning capacitor, as shown in Fig. 11-19(a), or in series, Fig. 11-19(b). Tuning circuits are used to select a resonant frequency by varying either the inductance or the capacity. Capacitance variation is more frequently used, since it is easier to accomplish mechanically.

An adjustable iron-core inductor, Fig. 11-19(c), has a movable core for

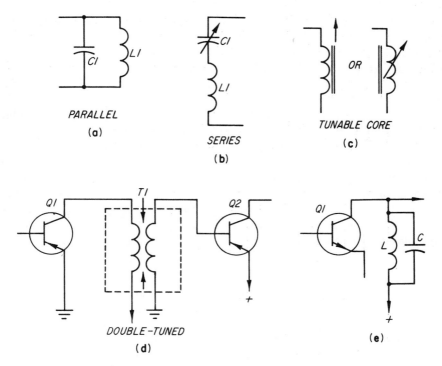

FIG. 11-19. Tuned Circuits.

the same purpose, but because of mechanical problems, it does not lend itself to continuous operation.

A double-tuned, iron-core, interstage transformer between stages, Fig. 11-19(d), has the winding symbol located so that the top lead is in line with the base of the transistor. Transistor envelopes are drawn on the same horizontal level, and the preceding transistor lead is brought down accordingly.

A single-coil, capacity-tuned or "tank" circuit is shown in Fig. 11-19(e). The "L" and "C" symbols may be drawn in a horizontal arrangement, or it may be preferable to conserve horizontal space by locating them vertically as shown.

11.6   DIAMOND AND ANGULAR LAYOUTS

Because of the nature of some circuits, a diamond layout of connection lines and symbols may be preferred to add to the readability.

Bridge Circuits

Among the more common circuits using this type are the Wheatstone bridge, Fig. 11-20(a), and the rectifier circuit bridges, Fig. 11-20(b). The Wheatstone bridge has an unknown resistance X and a center zero galvanometer across two arms of the bridge. By adjusting the value of $R3$ until the galvanometer reads zero, the null or zero balance point is reached. The diamond-shaped circuit can be recognized instantly on the diagram, as compared to the conventional arrangement shown in Fig. 11-20(c).

Other examples of diamond-shaped bridge circuits are the full-wave rectifier and voltage doubler circuits shown in Fig. 11-20(b).

The phase-shifting network, Fig. 11-20(d), is a variation of the bridge circuit or split diamond. The crossovers should be drawn at a 45-degree angle, and the component symbols should be drawn at right angles to the slanted lines.

The practice of drawing multiphase transformer circuits in a particular shape, depending upon the circuitry, has been replaced by the newer method in which all the transformer winding symbols are drawn parallel and the connection method indicated by a special symbol (see page 235). The winding termination markings on such transformers, which indicate polarity, should be carefully noted on the diagram.

Angular Lines

On occasion, angular lines are used to illustrate the functions of some of the circuits or to simplify circuit connections. Some typical examples are shown in Fig. 11-21. A dual potentiometer control is illustrated in Fig. 11-21(a), and connections between shift registers in a logic schematic diagram are shown in Fig. 11-21(b).

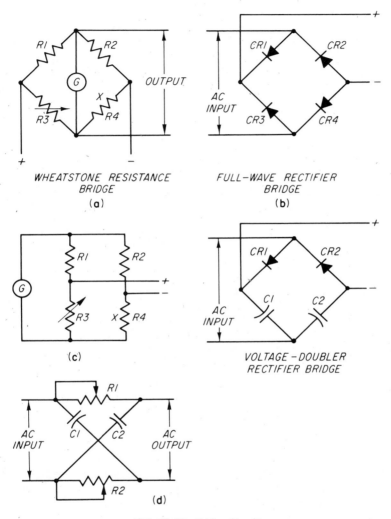

FIG. 11-20. Bridge Circuits.

The use of angular connection lines should be kept to a minimum. They should be restricted to cases such as those illustrated or to similar applications, rather than used indiscriminately.

11.7 SIMPLIFIED LAYOUTS

The logic symbols previously described are only one method to simplify diagrams. Other means are employed to avoid showing all the symbols and connection lines on a complicated circuit diagram. This is done by using insert

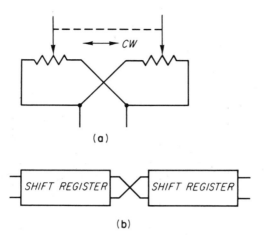

(a)

SHIFT REGISTER   SHIFT REGISTER

FIG. 11-21. Angular Circuit Lines.

(b)

diagrams such as those in Fig. 11-22 to replace detailed circuits, partial circuits, subdivided symbols, and repeating patterns.

Insert Diagrams

In many instances, electronic equipment consists of a number of sub-assembly units with a plug-in or a terminal board arrangement for external connections. A number of such units may be mounted on a common mechanical chassis assembly, or the subassembly may be a printed-circuit board with a terminating connector that connects to the rest of the circuit.

These groups of components are separated from the remainder of the circuit diagram by drawing them together in an enclosure made of alternating long and short dashes or one long and two short lines.

Examples of such subassemblies are filter networks, relays, plug-in amplifiers, integrated circuits, printed-circuit boards, and logic cards. They may terminate in a plug, terminal boards, terminal pins, or even wire leads, depending upon which is easiest to remove. If no replacement is expected, they may be permanently wired in place. They may also be assembled within a mechanical enclosure as a package or encapsulated to protect the components from moisture and mechanical displacement.

To facilitate servicing, the outgoing terminals or connector pins are numbered on the diagram when the plug-in method is used. The use of such subassembly connection devices is also indicated on the diagram by drawing the appropriate symbols and numbering and labelling them as required.

Several diagrammatic arrangements of a plug-in type subassembly for an interstage resistance coupled circuit are shown in Fig. 11-22. Note that considerable space is available around each component in Fig. 11-22(b) for its designation and value details. The circuit in Fig. 11-22(a), however, is easier to understand electrically because the component symbols are drawn in a

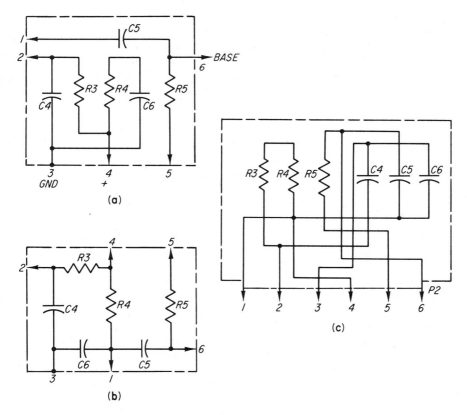

FIG. 11-22. Insert Diagrams.

logical, progressive order. A sequential terminal arrangement with the same symbols is illustrated in Fig. 11-22(c). The numerous crossovers necessary in this layout complicate the circuit unnecessarily and make it difficult to follow the diagram.

The numbered arrowed terminations in all of these circuits represent the male pins in a connector. Pin *3* is connected to the electrical shield in Fig. 11-22(a) by means of a dot on the enclosing shield, which is presumed to be grounded. In Fig. 11-22(c), the short dash lines between the pin terminals indicate that the pins are part of the plug assembly *P2*.

Sectionalized Circuits

Some of the circuits on a schematic diagram may readily be separated from the rest of the diagram without affecting its readability. An example of this is the power-supply circuit in Fig. 11-23. Another example is a control

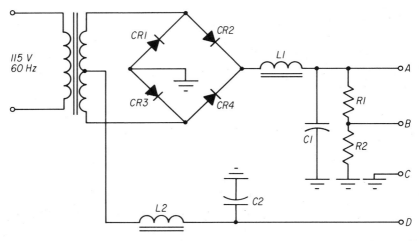

FIG. 11-23. Power Supply Circuit.

circuit, which is connected to the remainder of the diagram by two or three connection lines.

It is a common practice to locate such circuits at the lower left or lower right of the diagram. Outgoing connection lines may be extended to the remainder of the circuit, or they may terminate by voltage designations or by capital letters. These terminations are picked up on the body of the diagram with the same voltage notations or letter terminations.

This practice also makes it possible to place the sectionalized circuit or circuits wherever space is available on the body of the diagram.

Elimination of Connection Lines

In many instances, diagrams can be simplified by eliminating long connection lines that extend across all or part of the diagram. However, care must be taken to adequately identify circuits at both ends of the circuit interrupted by a break.

Connection lines may be eliminated under the following circumstances:

1. When the ground or chassis symbols can be used for individual component returns.

2. Where connection lines can be terminated and identified by a suitable designation. In this case, the circuit would be continued elsewhere on the diagram or on another diagram if it is a complex circuit that has several individual diagrams.

The connection lines in a diagram may also be simplified by interrupting such supply lines as transistor $+$ and $-$ voltages and ground returns. The

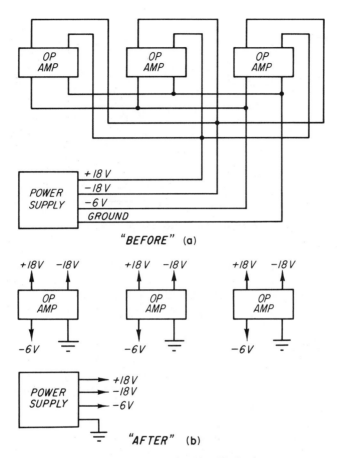

FIG. 11-24. Connection Line Elimination.

"*BEFORE*" and "*AFTER*" supply connections of three operational amplifiers and their power supply are shown in Fig. 11-24. The practice shown in Fig. 11-24(a) may be slightly confusing if there are many stages involved. This confusion may be alleviated to some extent, however, by interconnecting the supply lines of several closely located stages and bringing them out as shown in Fig. 11-24(b).

When a group of long connection lines extends across the diagram for control purposes, for instance, one line may be drawn in full across the diagram. The remaining lines are interrupted and identified in the same manner at each end, that is, control #1, control #2, etc.

The connection lines may also be combined into a single cable, as shown in Fig. 11-25, with bends in each line to indicate the direction of the conductors that join the cable. To maintain the continuity of the circuit, each end of such

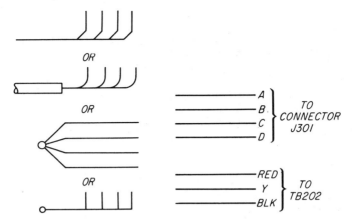

FIG. 11-25. Connection Line Grouping.

FIG. 11-26. Destination of Interrupted Connection Lines.

groupings should be identified by numbers, colors, letters, or other similar means.

Interrupted lines may be grouped and bracketed to indicate their destination as shown in Fig. 11-26.

Ground Symbols

It is rather unusual to find a common ground connection line used on a diagram because, as a rule, the individual ground returns terminate in a ground or chassis symbol. For the sake of appearance, the ground symbols are drawn on the same level as in Fig. 11-27(a). If necessary, the component symbols

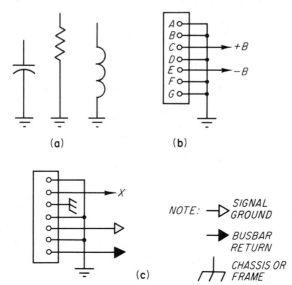

FIG. 11-27. Ground Symbols.

are staggered vertically to allow sufficient space for identification, but the ground symbols remain on the same horizontal level.

If there are a number of ground returns on a terminal strip or a connector and there is not enough space for individual symbols, they may utilize a common connection line, see Fig. 11-27(b). When such ground returns are combined, appropriate symbols should be used to indicate the circuits that are not returning to a true electrical ground. These circuits should then remain separated on the diagram. The special symbols for such circuit returns as signal ground and a ground bus bar are shown in Fig. 11-27(c).

Identification of Component Parts

Parts of terminal boards or connectors that are separated on the schematic diagram may be identified as shown in Fig. 11-28(a). When separate terminals of connectors or terminal boards are grouped on the diagram, they may be identified by their reference designation, which is followed by the terminal identification number or letter; see Fig. 11-28(b).

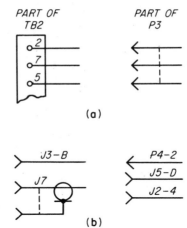

FIG. 11-28. Identification of Component Parts.

Junctions

This is an appropriate time to consider junctions of connection lines on schematic diagrams. Although it is sometimes felt that the junction dot may be omitted at junction points, especially if the connections are offset as in Fig. 9-20(d), there are many instances when connection lines cross each other on the diagram. When the junction dot system is used, there can be no question whether or not there is a connection at the crossover. Conversely, confusion may arise if junction dots are not used. Consequently, this book follows the junction dot system because it requires very little effort to execute and prevents possible error on the part of the reader, the production worker, or the technician using the finished diagrams. A dot $\frac{1}{16}$ inch in diameter is large enough for most diagrams.

11.8 RELAY CIRCUITS

In some instances, it is desirable to simplify the circuit by subdividing certain symbols such as switches and relays. Switch practices have already been described.

A good example is a multicontact relay. Subdividing its contact assemblies throughout the circuit diagram helps to simplify the diagram and eliminate long connection lines, which are difficult to follow. Confining many relay connections to one area contributes to a multiplicity of crossovers, which in turn may result in confusion and possible errors.

The typical multicontact relay circuit shown in Fig. 11-29(a) could be a part of a complex circuit diagram. Diagram circuitry can be simplified by locating the relay contact assemblies near the controlled circuits, Fig. 11-29(b). If possible, the relay contact assemblies should be shown in vertical position and in the normal or deenergized position. In addition to the regular relay identification, each relay contact assembly may also be identified by a capital

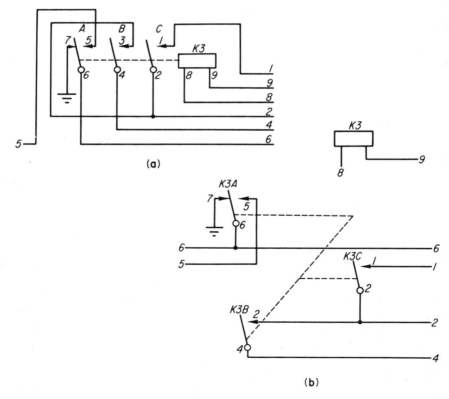

FIG. 11-29. Typical Relay Circuit.

letter (*K3A*, *K3B*, and so on), a procedure similar to that followed in identifying capacitor sections and other components. In this instance, the relay coil is identified on the diagram by *K3*. If convenient, long dash lines can be used to interconnect the various contact assemblies and to indicate their mechanical linkage. However, such lines should not be so long that they are confusing.

In drawing a relay diagram, a preliminary sketch should be made of the complete relay circuit with all of its contact assemblies and circuit terminations identified by letters as described above. This sketch may be used as a reference drawing to fit the separate parts of the relay on the diagram.

11.9 ESTIMATING SPACE REQUIREMENTS

Before the draftsman begins to draw a schematic diagram, he must solve two problems. First, he must determine the size of the finished diagram before he can select the drawing sheet size; second, he must draw and locate the graphic symbols according to established procedures.

Diagram drawings are "dimensionless." That is, there are no definite dimensions for the draftsman to follow except those that he determines at the outset, according to the following described procedures.

Although graphic symbols were described at length in Chapter 9, some basic rules are needed to establish their location and spacing on the diagram. Component symbols must be located in a definite relationship to such major components as transistors and control devices, depending upon the circuit function of the symbol. Space must be provided around each symbol on the diagram for lettering of the reference designations and for component identification. Sufficient space is also needed for interconnection lines between major diagram subdivisions; both crowding and wasted space must be avoided in drawing these lines. The draftsman must do considerable advance planning before he starts to draw the finished diagram.

Grid-Paper Layouts

One approach to the problem of size is to use ruled grid paper for both the preliminary layouts and the finished diagram. This grid paper is usually available in sheets or rolls, with squares of 8 by 8 or 10 by 10 to the inch and with inch divisions printed with a heavier line. Because the ruling is printed in blue ink, it will not show on diagram prints.

Symbols and other diagram components may be assigned arbitrary spaces on the basis of this grid spacing; i.e., a symbol that is eight grid spaces long would be one inch long. Thus, a common basis is established for symbol dimensions and spacing, and consequently the size of the finished diagram is also established.

There are a number of time-saving advantages in the use of grid paper for trial layouts and finished diagrams. After making one or more trial layouts, it is possible to segregate major circuit subdivisions and thus arrive at the best possible space utilization for each. Such trial layouts may be quickly drawn freehand on the grid paper, giving adequate space for symbol placement and identification, and then fitted to the best space advantage within the diagram area.

Estimating Space by Symbol Number

Another method of estimating the size of the drawing sheet needed for a given schematic diagram is on the basis of the number of symbols it contains. This relationship is illustrated in Table 11-1, which shows the approximate number of graphic symbols that will fit on the various drawing sheet sizes.

Table 11-1   Drawing Size Compared with Number of Graphic Symbols

Drawing Size	Drawing Area (sq. in.)	Reduced by	New Area	Number of Symbols Allowing $2\frac{1}{2}$ sq. in. per Symbol
B 11 × 17	187	25%	140	55
C 17 × 22	374	20	300	120
D 22 × 34	748	10	650	250
E 34 × 44	1496	10	1340	525
F 28 × 40	1120	10	1008	400
H 28 × 50	1400	10	1260	500
J 34 × 50	1700	10	1530	600
K 40 × 50	2000	10	1800	700

Table 11-1 is based upon transistor envelope diameters of $\frac{5}{8}$ inch, resistor symbols of $\frac{5}{32}$ inch high, and capacitor symbols of $\frac{5}{16}$ inch wide, and other symbols in proportion. The approximate size for symbols may be selected on the plastic symbol templates that are generally used for symbol drawing.

The table also allows for data normally included on the schematic diagram, such as reference designations, component values, circuit values, circuit subdivisions, drawing notes, and so forth.

Subdividing a complex diagram into a number of individual units that are interconnected will tend to reduce the number of symbols that will fit within a given drawing area.

The number of interconnection lines drawn between the various segments of the diagram will also reduce the possible number of symbols. This may be alleviated by procedures outlined on page 318. The freehand sketch or sketches that are furnished as part of a complex diagram layout should provide some indication of the number of interconnection lines that will be included on the diagram. Diagrams containing closely spaced connection lines are hard to read, and when such lines are traced across the diagram, it is possible to make errors. Therefore, connection lines on military schematic diagrams have to be spaced at least $\frac{3}{32}$ inch apart after reduction, and thus allowance must be made for this requirement on any diagram that will be reduced.

11.10 MISCELLANEOUS SCHEMATIC DIAGRAM PRACTICES

Some drawing practices that were mentioned earlier are illustrated in Fig. 11-30.

The methods of indicating equipment marking—words, letters, etc.— enclosed within a rectangle are shown in Fig. 11-30(a).

Where equipment subdivisions are separated on the schematic diagram by line enclosures, subdivisions may be identified by their functional title, as in Fig. 11-30(b), with larger lettering than the remainder of the diagram and with underlining for emphasis.

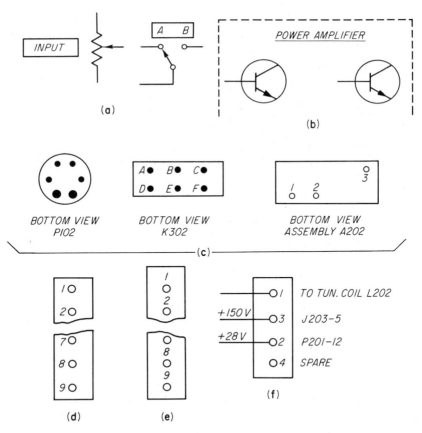

FIG. 11-30. Miscellaneous Practices.

Representative insert diagrams of such plug-in items as connectors, relays, or mechanical assemblies are shown in Fig. 11-30(c). Notice that the male or pin contacts are indicated by a filled circle in these illustrations and that the female or socket contacts are shown with a circle. Notation is also made of the reference designation for each component and the view shown. Other information included might be the name of the manufacturer, the catalog number, and the lead identification by color and number.

It might be difficult to locate specific symbols on a complex diagram that contains many hundreds of symbols, especially if there are numerous capacitors and resistors. A solution would be to list the designations in a tabulated form, employing the zoned type of drawing sheet for the diagram and designating the zone location for each component.

However, when deletions, additions, or other changes are made in these two symbol types, care must be exercised to make the corresponding revisions in their listings in the tabulations.

Letters or numbers that identify terminal boards or other connector devices should be located either in line with or slightly off center from the terminal points, Fig. 11-30(d), not centrally between the terminals, Fig. 11-30(e), which is confusing.

Whenever possible, all terminations on a schematic diagram should be located on its outer periphery and not within the body of the diagram.

Such terminations should be identified with their destination points. The identification should be complete, giving the terminal board designation and its terminal number or letter; the connector designation and its pin number; the general unit designation, and so on, see Fig. 11-30(f). Such identification should be centered on the terminal as shown. To avoid confusion, any additional matter such as voltage value, polarity etc., should be listed away from the destination point identification as on terminals 2 and 3.

11.11 REFERENCE DESIGNATIONS

A system of reference designations, or letter-number combinations, has been adopted to identify the various components and devices as represented by their graphic symbols in both civilian and military equipment. These reference designations were described in detail in Chapter 10. Some examples of symbol lettering, including the reference designations, were shown in Fig. 10-1.

11.12 DIAGRAM LAYOUT

As mentioned before, the lack of dimensional guide lines is one of the chief difficulties that the draftsman will encounter in circuitry work. Consequently, he is forced to "design" or "lay out" the diagram on the basis of the information he receives in the form of a freehand sketch or a revised print of an existing diagram. He may also have to form a new, more complex diagram from a compilation of circuits taken from several existing diagrams. When such diagrams are made as the mechanical design of equipment progresses, they may require considerable additions, deletions, and alterations, or possibly even a complete new approach before a satisfactory finished diagram is drawn.

However, if the alterations on an existing diagram are minor, the draftsman's task will be quite simple because he can use the previous diagram as a guide line for the layout. Then, it is only necessary to allocate sufficient additional space for the new symbols and their identification.

Component Information

Before proceeding with the diagram, the draftsman should get complete and correct information about each component used on the new diagram or revision. This information may be obtained from several sources: the mechanical and electrical component drawings, the parts lists, the manufacturers' catalogs, the drafting room manuals and standards, technical bulletins, the project engineer, or other similar sources.

All of this data should be carefully noted on the freehand circuit sketch, a marked-up print of the existing circuit, or any other source of circuitry that serves as the basis for the diagram to be drawn. When space limitations prohibit such a procedure, a list of components may be prepared, giving all pertinent information and relating it to the working sketch by number, letter designation, or some other means.

Layout Considerations

After complete information has been obtained, the next step is to make one or more rough preliminary layouts to determine the space requirements of the diagram, the relative arrangement of symbols or major groups of symbols, and so forth. The designer of mechanical assemblies has to follow a similar approach to determine if his drawing sheet will be large enough.

Another factor to consider in diagram layouts is the width of the drawing paper, which is limited to 34 or 40 inches. This becomes a problem in complex diagrams, which contain a number of horizontal layers of segregated component symbols. The use of the diagram is also a consideration. For example, a long diagram is more acceptable for instruction books than a wide diagram that requires several horizontal folds. In this case, the "bottom fed" type of diagram shown in Fig. 11-5(a) is more practical because it requires less vertical space for each transistor stage.

Diagram Layout

It is advisable to make preliminary layouts on cross-section or grid-ruled paper with eight spaces to the inch. Although the first, trial layout may not necessarily be the best one, it will serve as a guide for the next, better proportioned layout. The draftsman should strive for a well-spaced, uncluttered layout, even if he has to use the next larger size drawing sheet. If the diagram is made during the preliminary stages of equipment development, he should always keep in mind that additional component symbols may be added later.

The complete circuit should be divided into functional segments, with each

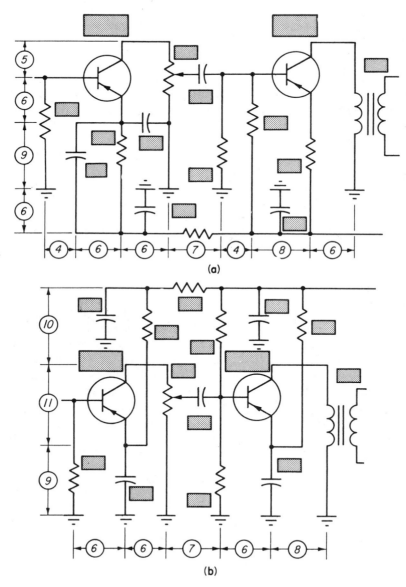

FIG. 11-31. Preliminary Layout.

segment allotted a certain amount of diagram space. This circuit breakdown is quite helpful in making trial layouts, which can be rearranged or improved upon later.

The two-stage, resistance-coupled amplifier circuit shown in Fig. 11-31 illustrates the allotment of space requirements. Because the two stages are similar, they establish the basic overall space requirements, to which the inter-stage coupling components must be added.

Of the two alternate layouts shown in this sketch, that of Fig. 11-31(a) has a cleaner appearance and less crowding between symbols and consequently is commonly used for electronic communication equipment diagrams. The shaded blocks represent the areas that are left for reference designations, component values, and other identifications. Space also should be allowed near the transistor envelopes to list transistor pin numbers. The circuit in Fig. 11-31(a) is called "bottom fed" because the feeder lines are located below the basic stage components.

Estimating Space Requirements

After a freehand trial circuit arrangement has been drawn, the grid spaces allocated for the various component symbols can be marked on the sketch between the vertical and horizontal lines that extend from the center of the symbols or connection lines (see Fig. 11-31). The grid spaces given below represent average figures for the majority of schematic diagrams using $\frac{1}{8}$-inch grid spacing. Thus, average spacings on the diagram are:

Component	Grid spaces
Capacitors	4–8
Inductors	8
Resistors	8

Other dimensions	
Diagram items	*Grid spaces*
Transistor envelope diameter	5
Resistor symbol length	4
Capacitor symbol width	2–3
Lettering height	1 (or $\frac{1}{8}$ inch)
Connection line spacing	2
Spacing between groups of connection lines	3–4

Although horizontal spacings between symbols have been indicated (Fig. 11-31), they are merely representative and depend upon the circuit components and arrangement. Larger or more complex component symbols, such as multiposition switches, relays, and others may need more space. Terminations such as terminal boards, connectors, and so on, require additional space for identification and lettering.

Symbol Locations

Symbols should be located on grid lines with sufficient space left for lettering. It may be necessary to stagger the symbols on adjacent vertical lines to obtain space for lettering. To avoid crowding of symbols or lettering, the location of symbols should be altered as required.

Potentiometer symbols that connect to a transistor element should be drawn preferably below the center of the symbol to obtain greater lettering area.

Symbol Lettering

It is sometimes difficult to locate space on a diagram for such lettering as component value, tolerance, type, reference designation, polarity, terminal identification, rotation, etc. Several methods of presenting such information were illustrated in Fig. 10-1. The lettering is usually done after the circuitry details have been completed and, in some cases, may require ingenuity to eliminate the appearance of crowding or ambiguity. Although horizontally aligned symbols give the diagram a better appearance, it may be necessary in some instances to offset or "stagger" them to obtain sufficient lettering space.

While vertical lettering may be used to gain space, Fig. 10-1(o), it detracts from the readability of the diagram and therefore is not recommended.

Diagram Layout Procedures

The layout of a schematic diagram, complex or relatively simple, can be expedited by following a certain routine. This not only saves time, it also results in a better finished product. The beginner in circuitry drawing will benefit by following the procedures outlined here:

1. Obtain the freehand sketch or sketches or other circuit information that are to serve as the basis for the schematic diagram.

2. Check this material for full information about component identification; check the subdivision of the diagram, if any, and the identification of such subdivisions; and check for possible revisions if the equipment is in the developmental stage, and allow extra drawing space.

3. Make freehand trial layouts of the diagram sections on graph paper and establish the approximate dimensions of each section, allowing symbol spaces according to the practices outlined earlier in this chapter.

4. Make freehand trial layouts of such diagram appendages as auxiliary circuits and details of complex component symbols.

5. Make a rough trial layout of tables and notes to establish the approximate space they require.

6. Make a trial assembly of the diagram section layouts and all other material to arrive at the rough complete diagram layout, allowing space for connection lines between the diagram sections and auxiliary diagram information. The shape of this rough diagram assembly should fit on a standard drawing sheet or on standard roll-size paper if it is a large diagram. The required diagram area should usually not impinge upon the revision-title block area of the sheet.

7. Modify the rough diagram layout as necessary.

8. Attach the rough trial layouts used in the complete rough diagram assembly to a sheet equal to drawing size selected.

9. Roughly sketch in the connection lines between various diagram sections and add connectors, terminal blocks, and other termination symbols.

10. Draw the finished diagram, using the rough layout assembly as a guide. The finished diagram may be drawn on 8 by 8-spaced graph paper, cut to standard drawing sheet size, or the graph paper may be slipped below the standard drawing sheet to act as a spacing guide.

Following these procedures eliminates the necessity for drawing a complete preliminary diagram after completing the first nine steps, and then drawing the finished diagram, and saves time and effort as well. However, it may help the beginner acquire proficiency and confidence if he draws such a preliminary diagram before the finished diagram. After he has gained a little experience this should be unnecessary.

The draftsman's work is simplified if the original diagram sketch, furnished by the originator, has such a circuit format that little can be gained by trying to rearrange it on the finished diagram. In such cases, it is only necessary to establish the overall dimensions of each circuit subdivision, based upon the dimensional practices and spacings given earlier in this section.

Improvements may be incorporated in the finished diagram as the drawing progresses. Slight modifications may be necessary to provide space for an extra connection line or additional space for lettering.

All of the symbols are usually drawn first, starting with such basic symbols as transistors or electron tubes. All like symbols are drawn at one time, especially if they are on the same horizontal or vertical level. If a change in symbol position is expected, it may be advisable to draw the connection lines to the approximate symbol position and fill in the symbol later. This eliminates redrawing the symbol from one position to another.

Short connection lines, such as those that extend from the transistor envelope to the immediate circuit components of the transistor, are drawn next. Ground symbols may all be drawn at the same time, bringing vertical feeder lines down to the approximate position of the horizontal feeder connection lines, which are drawn next.

Junction points and crossovers should be checked carefully before lettering is added. Although there have been some attempts to omit dots at junction points, using them is more logical and results in less errors by the draftsman or the eventual user of the diagram. The American National Standards Institute has adopted the use of junction dots in their *Electrical and Electronics Diagrams*, ANSI Y14.15-1966 (R1973). The use of dots also avoids the necessity of off-setting junction connections, which would otherwise be required.

Lettering of component values and reference designations is added next. Occasionally, this may require moving some of the symbols to provide additional space.

Such diagram appendages as insert diagrams, flat pack, transistor, integrated circuit, dual-in-line package, connector, switch, and relay base layouts, and other similar material should be drawn next in the border of the diagram. These are followed by notes and other explanatory data.

After the diagram has been completed, it should be checked for the following:

1. Indication of all mechanical linkages, enclosures, and shielding
2. Adequate identification of all terminations
3. Complete and clear identification and designation of components. The practices outlined in Chapter 10 for military equipment reference designations should be followed for military schematic diagrams
4. Proper identification of separated symbols and circuits, and auxiliary circuits
5. All extra identification symbol markings, such as polarity, color, terminal number, transformer interconnections, switch wafer, waveform, voltage, tolerance, rotation, etc.
6. Inclusion of all complete tabular data
7. Inclusion of special note references to symbols and other identification and, when required, the use of standard abbreviations
8. A minimum of crossovers and bends
9. Complete drawing title, identification of equipment depicted on the diagram, and other relevant data
10. Consecutive numbering of components in left-to-right and top-to-bottom sequence, unless dictated otherwise by military requirements
11. Correct and proper identification of flat pack, transistor, integrated circuit, dual-in-line package, auxiliary, insert, and connector layout diagrams
12. Adequate reference notes for all reference notations on the diagram
13. Selection of reference designation letters in compliance with military or industry standards
14. Adequate cross references for related equipment or parts lists
15. Complete titles or other identification for major subdivisions on the diagram.

The recommended 8 by 8-grid spacing and $\frac{1}{8}$ inch lettering makes the diagram suitable for a 2:1 reduction for technical manual purposes. If no reduction is contemplated, a 10 by 10-grid spacing and $\frac{1}{10}$ inch lettering may be used to reduce diagram size, along with smaller symbols and smaller line-to-line spacing. However, readability is sacrificed in complex diagrams when the diagram contents are crowded to achieve a questionable reduction.

Typical Transistor Diagram

A typical layout is shown in Fig. 11-32. Certain procedures help to obtain a clean-cut diagram and to simplify the connection line work.

For example, crossovers are reduced by aligning the terminals on board *TB101* with their circuit connections instead of arranging the terminals in alphabetical or numerical order. Placing the voltage feeder or bus lines above the main part of the diagram or below the ground returns also helps to reduce the crossovers.

An abbreviated form of terminal identification gives the destination from the individual terminals to other units or components; i.e., terminal "*C*" is identified as being connected to a minus source. A terminals may also identify the destination point at the opposite end, as for example "*TB209/2*" identifies the terminal as being connected to terminal 2 on terminal board *TB209*.

Main circuit subdivisions such as *PREAMPLIFIER* and *DRIVER* are labelled above the main body of the diagram and are sometimes underlined for emphasis. Certain lettering, such as *INPUT* in Fig. 11-32, is boxed to indicate a similar marking on the equipment itself.

Typical Complex Layout

An example of a large, complex layout is shown in block form in Fig. 11-33. This drawing was too large for standard drawing sheets, so it was drawn on a roll graph paper, with 8 by 8 spacing, 32 inches wide by 80 inches long. The blank areas between the rows of blocks are occupied by interconnection lines, which extend from the blocks across the drawing. Space has been reserved for notes, transistor base diagrams, and other items on the right-hand side of the diagram.

Each unit was first laid out in a rough trial layout to establish its overall dimensions. Then, the number of interconnection lines was determined and their relative positions altered as required to obtain a minimum of crossovers and bends. By spacing these lines $\frac{3}{16}$ inch apart, they meet the military space requirements of $\frac{3}{32}$ inch when the original drawing is reduced 50 percent. Thus, the photographed copy is only 16 by 40 inches, a size that can be readily folded for use in manuals.

The circuit for the "*Fine Tuning Indicator*" unit, in the upper center of Fig. 11-33, is shown in Fig. 11-34. As noted before, symbols on the same horizontal plane are drawn first, starting with the transistor symbols. Special attention should be given to the location of transistor symbols because it establishes the layout of the rest of the diagram.

As the beginner acquires more and more experience, drawing complex

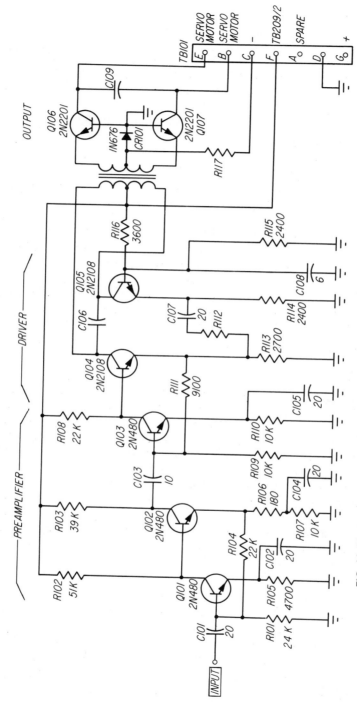

FIG. 11-32. Typical Transistor Circuit. (Courtesy Semiconductor Department, General Electric Company.)

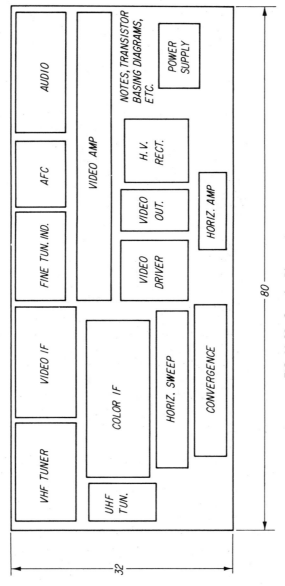

FIG. 11-33. Complex Diagram Layout.

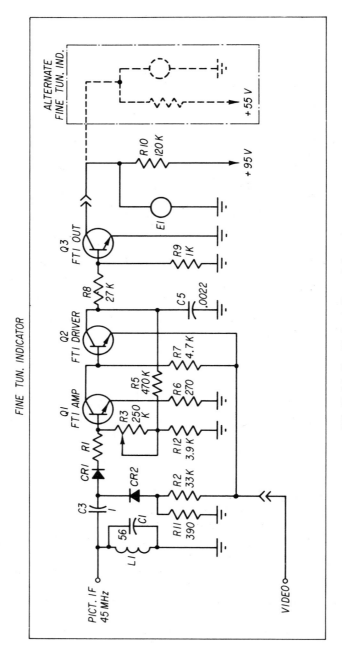

FIG. 11-34. Fine-Tuning Indicator Schematic.

diagrams becomes less problematical, and he will be able to draw some of the diagrams directly from the initial sketch supplied by the originator.

J-K Flip-Flops

The logic and schematic diagrams for one flip-flop of a dual J-K master-slave flip-flop are shown in Fig. 11-35. The J-K inputs are controlled by the clock

FUNCTIONAL LOGIC DIAGRAM (EACH FLIP-FLOP)

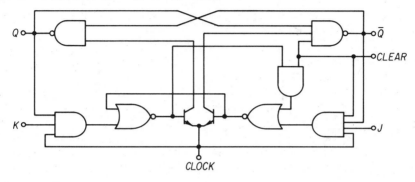

SCHEMATIC (EACH FLIP-FLOP)

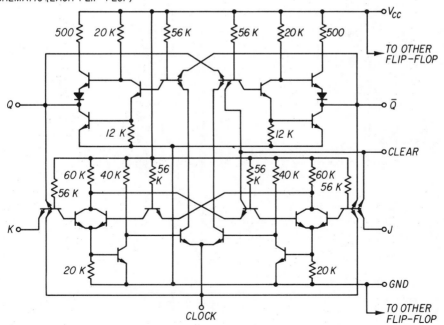

FIG. 11-35. Logic and Schematic Diagrams for J-K Flip-Flops. (Courtesy Texas Instruments, Inc.)

pulse, which also regulates the state of the coupling transistors connecting the master and slave sections.

These diagrams are representative of some of the more complex integrated circuits.

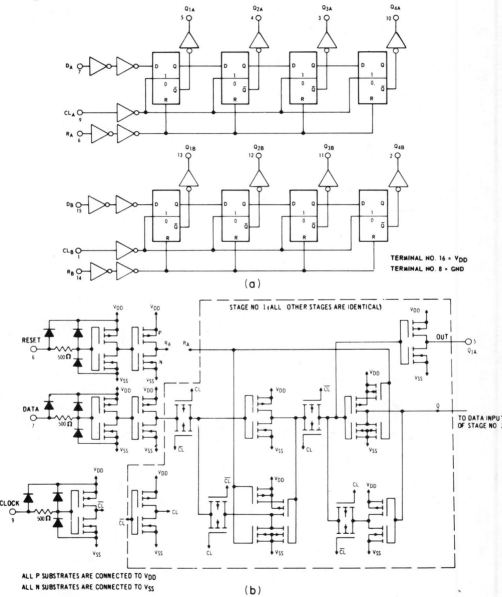

FIG. 11-36. Static Register. (a) Logic Diagram. (b) Schematic Diagram for One Stage.

Integrated Circuits

Figure 11-36 shows the logic diagram and a partial schematic of a static register. In (a) is the complete logic diagram of two independent, 4-stage serial-input/parallel-output registers. All register stages are master-slave flip-flops. In (b) is the schematic diagram for stage #1 only. The remaining seven stages are identical and can be represented by insert blocks that show only the input and output terminals for each register.

Drawing the complete schematic diagram would result in undue complexity. Thus, generally, only a logic diagram is drawn.

Checking the Finished Diagram

It is advisable to carefully check the completed diagram in all respects before turning it over to the regular checker. The details of checking schematic diagrams are given in Chapter 16.

SUMMARY

Before he can draw a schematic diagram properly, the electronic draftsman must be familiar with graphic symbols and reference designations, as well as with the standardized practices involved in laying one out. Transistor or other basic units are the focal points to which other component symbols are related. Space should be left around each graphic symbol for reference designations and other identification material. Logic diagrams require a special treatment of their own. Other special layouts include filter and tuned circuits and switch and relay circuitry. Connection lines can be eliminated to simplify the drawings. However, a certain routine should be followed in drawing a schematic diagram, whether it is simple or complex.

QUESTIONS

Sketches should be included with answers when necessary.

11.1 How does a schematic diagram differ from a wiring diagram?

11.2 Describe the steps involved in preparing a schematic diagram.

11.3 List the major components that generally serve as major subdivision points in a schematic diagram.

11.4 Describe the various component identification details, and give an example.

11.5 Describe some of the basic requirements that must be followed in graphic symbol layout.

11.6 What components are likely to require an indication of mechanical motion on the diagram? Give an example.

11.7 Describe three types of filter circuits. Identify each by name.

11.8 What are the three basic transistor interstage circuits? Describe each briefly and make a sketch, identifying each type.

11.9 What is a logic symbol? Sketch three examples, and identify them by name.

11.10 Define AND, NOR, NAND gates; FLIP-FLOP; READOUT; MEMORY; and SHIFT REGISTER.

11.11 Illustrate transistor lead identification with a sketch.

11.12 How are integrated and hybrid circuits shown on a complex diagram?

11.13 Describe a simplified layout and sketch an example.

11.14 What is a sectionalized circuit?

11.15 What are some of the practices followed in drawing complex schematic diagrams?

11.16 List the various items that should be checked after a schematic diagram is completed.

11.17 Draw a full-wave rectifier circuit in both the angular and the parallel layouts.

11.18 What size of drawing sheet would be needed for a schematic diagram containing 200 graphic symbols?

11.19 Draw an example of an insert diagram.

11.20 Make "before" and "after" comparison sketches illustrating how to eliminate connection lines.

11.21 List some of the steps involved in laying out a diagram. Describe each step briefly.

11.22 In what order are the various parts of a typical schematic diagram drawn?

EXERCISES

Develop the preliminary freehand layouts on cross-section paper and finished diagrams on plain drawing vellum, on vellum with cross-section paper placed underneath, or on vellum with disappearing grid lines. Use schematic diagram templates whenever possible. Indicate all junctions by dots. Reference designations, abbreviations, mechanical linkages, shielding, polarity signs, voltage values, and lettering should be included where necessary. See Chapters 9 and 10 for information on graphic symbols and reference designations and other sources of information. Consult transistor manuals for correct element connections. Other layout arrangements may be made of the circuits shown, but completed circuits should be checked according to the steps outlined in Chapter 16. Redraw commercial diagrams following the practices outlined in this book.

11.1 Sketch a typical transistor circuit, and identify the transistor elements.

11.2 Make a sketch showing the direction of current flow in a *PNP*-type transistor stage.

11.3 Sketch a rotary-wafer switch with both surfaces used for connections. Identify the various parts.

11.4 Select three transistor circuits from a transistor manual and sketch them on 8 by 8 grid paper. Identify input and output connections, power supply line, and the component symbols by reference designations.

11.5 Draw a circuit of a double-*T* low-pass filter, consisting of four inductances and two capacitors, and include input and output terminals in the form of two terminal blocks.

11.6 Draw a complex filter network consisting of a two-section, *T*-type band-pass filter, followed by one section of a *π*-type band-pass filter. Show input and output terminals.

11.7 Illustrate two *PNP*-type transistors connected with the following interstage types of coupling: (a) transformer; (b) resistance-capacitance; (c) direct-coupled.

11.8 Draw a schematic diagram consisting of sections A, B, and C of Fig. 11-1. Select a suitable scale so the drawing will be $8\frac{1}{2} \times 11$ or an *A*-size drawing sheet.

11.9 Using Fig. 11-1 for reference, make a diagram showing each stage as a block and include antenna and loudspeaker symbols for input and output. Show the wiring from the power input terminal to the individual stages.

11.10 Draw a power transformer supply circuit consisting of a two-bladed plug connected to the primary of the power transformer through a series fuse and a double-pole, single-throw toggle switch. Two female outlet receptacles, a voltmeter, and a neon indicator light with a series resistor are also connected across the primary winding. Use an $\frac{1}{8}$-inch grid and assign reference designations to all components.

11.11 A partial diagram in outline form of an FM tuner using two FET transistors is shown in Fig. 11-37. Draw this circuit on an 8 by 8 grid in complete symbolic form. Assign reference designations to all components. In the figure, *L* represents inductance symbols; *C*, capacitor symbols; *R*, resistor symbols; and *G*, ground symbols. Draw mechanical coupling symbol between *C*(VAR 1) and *C*(VAR 2). Both inductances (*L*) should have a center-tapped connection.

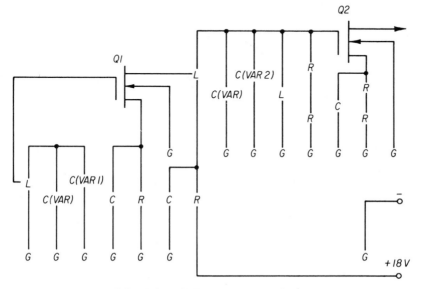

FIG. 11-37. FM Tuner Schematic Diagram.

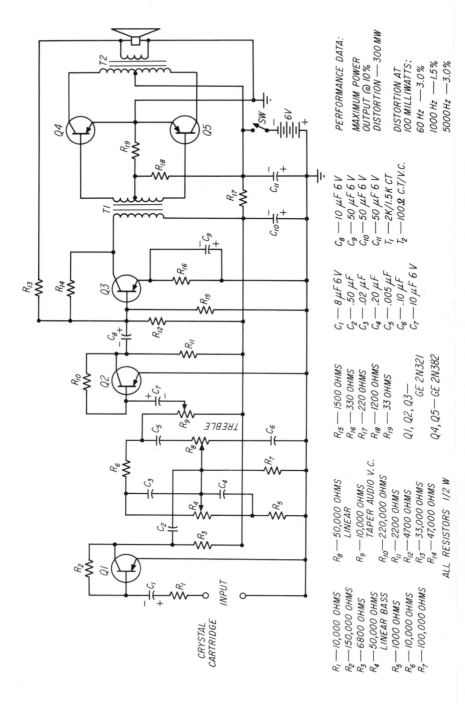

FIG. 11-38. 6-Volt Phono Amplifier. (Courtesy Semiconductor Department, General Electric Company.)

CRYSTAL
CARTRIDGE
INPUT

TREBLE

R_1 — 10,000 OHMS
R_2 — 150,000 OHMS
R_3 — 6800 OHMS
R_4 — 50,000 OHMS
　　　LINEAR BASS
R_5 — 1000 OHMS
R_6 — 10,000 OHMS
R_7 — 100,000 OHMS

R_8 — 50,000 OHMS
　　　LINEAR
R_9 — 10,000 OHMS
　　　TAPER AUDIO V.C.
R_{10} — 220,000 OHMS
R_{11} — 2200 OHMS
R_{12} — 4700 OHMS
R_{13} — 33,000 OHMS
R_{14} — 47,000 OHMS

ALL RESISTORS 1/2 W

R_{15} — 1500 OHMS
R_{16} — 330 OHMS
R_{17} — 220 OHMS
R_{18} — 1200 OHMS
R_{19} — 33 OHMS

Q_1, Q_2, Q_3 —
　　　GE 2N321

Q_4, Q_5 — GE 2N382

C_1 — 8 μF 6 V
C_2 — .50 μF
C_3 — .02 μF
C_4 — .20 μF
C_5 — .005 μF
C_6 — .10 μF
C_7 — 10 μF 6 V

C_8 — 10 μF 6 V
C_9 — 50 μF 6 V
C_{10} — 50 μF 6 V
C_{11} — 50 μF 6 V
T_1 — 2K/1.5K CT
T_2 — 100 Ω C.T/V.C.

PERFORMANCE DATA:

MAXIMUM POWER
OUTPUT @ 10%
DISTORTION — 300 MW

DISTORTION AT
100 MILLIWATTS:
　60 Hz — 3.0%
　1000 Hz — 1.5%
　5000 Hz — 3.0%

11.12 Draw the circuit of a 6-volt phono amplifier in Fig. 11-38 and place the component values next to their respective graphic symbols. Insert the symbol for crystal cartridge in place of the input terminals shown. Return all grounded connections to individual grounds rather than the common ground shown. Identify all transistors by their type and number and their designations. Include the general notes and the complete title shown in the title block.

11.13 The saturated flip-flop circuit in Fig. 11-39 uses *PNP*-type transistors and *R-C* coupling components. Draw this circuit on $\frac{1}{8}$-inch grid, giving complete transistor symbols and assigning reference designations to all components. Include an 8-terminal board for all connections on the right-hand side and assign suitably identified spares.

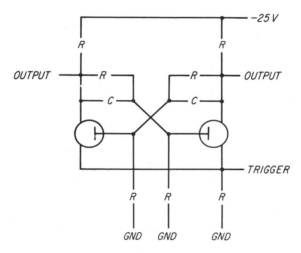

FIG. 11-39. Basic Saturated Flip-Flop Using PNP Transistors. (Courtesy Semiconductor Department, General Electric Company.)

11.14 Select a suitable grid for a square-wave generator schematic, Fig. 11-40, correcting symbols as necessary and assigning reference designations to all com-

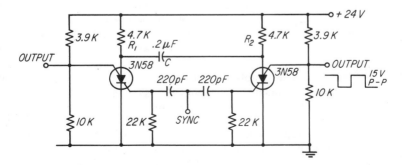

FIG. 11-40. Square-Wave Generator Circuit. (Courtesy Semiconductor Department, General Electric Company.)

ponents. Terminate all connections except output on right-hand side of the diagram in a 4-pin receptacle of the *MS*-type.

11.15 Draw a schematic diagram of the 10-watt power amplifier circuit in Fig. 11-41. Include a phono jack at the input in place of the terminals shown and two pin connectors for battery input and speaker connections. Correct symbols and reference designations to conform to practices previously described.

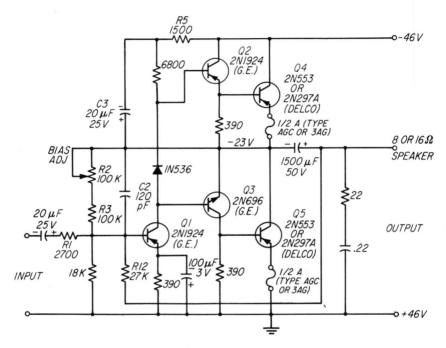

FIG. 11-41. 10-Watt Power Amplifier. (Courtesy Semiconductor Department, General Electric Company.)

11.16 Draw the circuit shown in Fig. 11-42 on $\frac{1}{8}$-inch grid and include appropriate reference designations for all components. Identify each stage by title above the diagram, and draw the power supply in the lower left-hand corner of the diagram. Correct symbols to correspond with existing standards. Include such general notes as may be required. Show a coaxial connector for the 300-ohm feed-line input.

11.17 A multiple resistor assembly contains three single-pole multiposition rotary switches with nonbridging contacts and a single-pole, three-position bridging-type contact switch for selecting connections to each multiposition switch. Thirty-six resistors, each of standard value (Table 4-1) and a ± 10 percent tolerance, are wired so any resistor can be selected from the twelve on each multiposition switch. The first switch has resistors starting with 10-ohm value, then 12, etc.; the second starting with 100, then 120, etc.; the third starting with 100,000, then 120,000, etc. The three-position switch has the common arm connected to one output terminal, and each of the three contacts are connected to

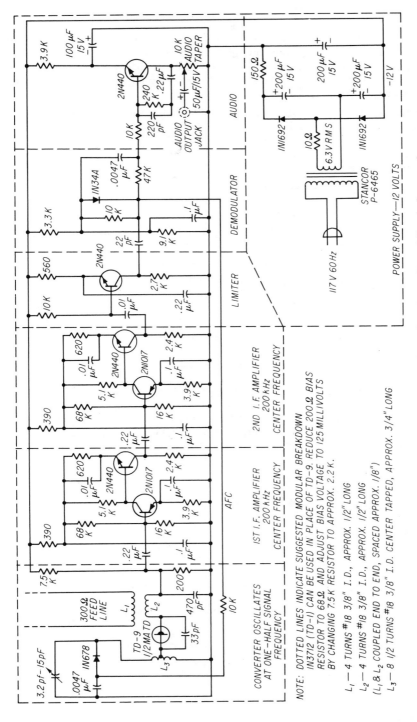

FIG. 11-42. FM Tuner. (Courtesy Semiconductor Department, General Electric Company.)

345

multiposition switch arms. One end of each resistor connects to the second output terminal. The three multiposition switches are to be identified as low, medium, and high. Draw a complete schematic diagram showing the rotary switch symbols and their linkage, and identifying all components by reference designations and resistors by values.

11.18 Draw a schematic diagram of a 27-mHz crystal oscillator, Fig. 11-43, using standard graphic symbols as follows: (1) capacitors 1, 6, 7, 11; (2) resistors 2, 4, 5; (3) 27-mHz crystal 3; (4) *NPN* transistor 10; (5) transformer primary 8 and secondary 9. Assign reference designations and show a coaxial connector for load terminals.

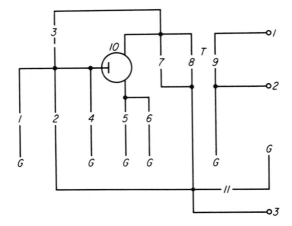

FIG. 11-43. 27-mHz Crystal Oscillator.

11.19 Draw a detailed schematic of the logic diagram shown in Fig. 11-44(a). The AND circuits #1 and #2 in (a) contain diodes in the input circuits with the

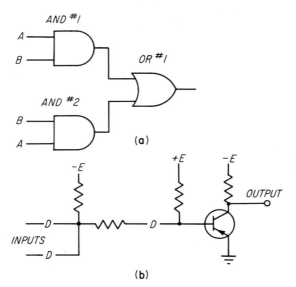

FIG. 11-44. Block Diagram of a Logic Circuit.

anodes connected to the inputs as in (b); the diode connected to the transistor has the anode connected to the transistor base. The transistor is of the *PNP* type. The schematic for the OR circuit #1 is the same as for the AND circuits.

11.20 A printed-circuit board, used in a 30-watt audio amplifier, is shown in Fig. 11-45, together with associated components. From this data, develop the schematic diagram for the circuit and identify the components by their reference designations.

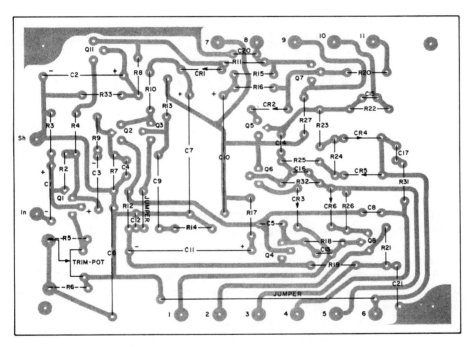

FIG. 11-45. Bottom View of a Printed Circuit Board with Components Shown as Electrical Symbols.

12

CONNECTION OR
WIRING DIAGRAMS

Although the schematic or elementary diagram provides the basic circuit connection information, it is not complete enough to cover the wiring of numerous identical electronic assemblies on a production basis or to use in instruction or servicing manuals.

Its counterpart is the connection or wiring diagram, which shows the terminal-to-terminal wiring within the equipment, module, control panel, or other electronic assembly.

The complexity of presenting such wiring information in an acceptable, easy-to-understand form has resulted in the adoption of several government and industry standards that outline the practices to be followed.

Among these standards are: MIL-STD-12D of the Department of Defense; ANSI Standard Y32.2-1975, *Graphic Symbols for Electrical and Electronics Diagrams*; and ANSI Standard Y14.15-1966 (R1973), *Electrical and Electronics Diagrams*, and *Supplements*—Y14.15a-1970 and Y14.15b-1973.

References to the component wiring or wiring diagram refer to the internal connections of a unit, equipment, chassis, etc. A diagram that shows the external wiring between separate equipments is known as an interconnection diagram and is much simpler in its makeup.

12.1 CONDUCTORS

There are two principal methods of connections in electrical circuits—by copper conductors in the form of braid, insulated wire, or cable, and by metallic waveguide conductors made of aluminum, brass, copper, silver, or other metals.

For convenience in soldering, the majority of copper conductors are tinned, whether they are of a solid or stranded construction.

Solid Wire Sizes

Round copper wire sizes are identified in accordance with American Wire Gage (AWG) standards. Sizes #10 through #36, shown in Table 12-1, are most important to the electronic draftsman. Wire sizes are specified by diameter in inches and by the cross-sectional area in circular mils.

Some useful points to remember can be gleaned from Table 12-1. For instance, #10 wire is approximately 10,400 circular mils in cross section, and this number is divided in half every three wire sizes. Also, there are 32 feet of #10 wire per pound (bare), and this length doubles every three sizes. The resistance of #10 wire is .031 ohm per pound, and this value quadruples every three sizes. Thus, #16 gage wire is approximately 2600 circular mils in cross section, has 128 feet per pound, and resistance of .500 ohm per pound.

Table 12-1   Round Solid Copper Wire Table

AWG Size	Nom. Diam. (in.)	Area (Cir. Mils)	Bare (Ft/lb)	Bare (Ohms/lb)	Recommended Current Amp. (max.)
10	.1019	10,380	31.8	.031	30
11	.0907	8234	40.1	.050	
12	.0808	6530	50.6	.080	18
13	.0721	5178	63.8	.128	
14	.0641	4107	80.4	.203	12
15	.0571	3257	101.0	.322	
16	.0508	2583	128.0	.513	7
17	.0453	2048	161.0	.815	
18	.0403	1624	203.0	1.296	4.5
19	.0359	1288	256.0	2.060	
20	.0320	1022	323.0	3.280	3
21	.0285	810	408.0	5.210	
22	.0253	642	512.0	8.290	1.8
23	.0226	509	648.0	13.200	
24	.0201	404	818.0	20.900	1.1
25	.0179	320	1031.0	33.300	
26	.0159	254	1300.0	52.900	.7
27	.0142	201	1639.0	84.200	
28	.0126	159	2067.0	133.900	.4
29	.0113	126	2607.0	213.000	
30	.0100	100	3287.0	338.000	
31	.0089	79	4145.0	538.000	
32	.0080	63	5227.0	856.000	
33	.0071	50	6591.0	1361.000	
34	.0063	39	8311.0	2165.000	
35	.0056	31	10,480.0	3441.000	
36	.0050	25	13,210.0	5473.000	

Knowing the cross-sectional area of a wire in circular mils is useful in determining the allowable current to be carried by a given conductor, based upon a specified number of circular mils per ampere. The last column in Table 12-1 lists the recommended maximum current for the various wire sizes, based upon approximately 350 circular mils per ampere. This information is useful in computing the size of conductors in harness assemblies.

Stranded Conductors

The "hook-up" wire used in most chassis wiring is generally size #22 and of stranded construction; i.e., it consists of a number of small wires or strands twisted together. This stranding provides the conductors with flexibility, which is especially important in larger chassis where wiring entails a number of conductors in a harness assembly.

The equivalent solid conductor size is determined by the cross-sectional area of each strand, multiplied by the number of strands. Thus, ten strands of #32 wire, with a cross-sectional area of 63 circular mils each, have a total of 630 circular mils and are the equivalent of #22 wire (see Table 12-1).

Stranded wire is identified by the number of strands, followed by a diagonal bar, and then the AWG wire number. Thus 7/30 represents seven strands of #30 wire.

Conductor Insulation

Solid or stranded conductors have to be covered with an insulating material to keep them apart from adjacent wires, to keep them away from grounded metallic structures, and to maintain a high circuit resistance between such wiring.

Many insulation materials are employed for such conductor covering. The type used depends upon the voltage, temperature, humidity, and other environmental conditions. Many plastic materials such as polyvinyl chloride and Teflon in various wall thicknesses are widely used. In many instances, an additional covering of lacquered glass braid or a jacket of nylon is specified. Individual conductors are shielded by means of a braided tinned copper shield over the plastic insulation.

Note that the use of polyvinyl chloride insulation for conductors or other purposes is prohibited in military airborne equipment.

Bus Wire

Solid tinned copper wire, popularly known as bus wire, is frequently used for short point-to-point connections, and ranges in size from #22 to #12. Hard-drawn copper is often used to give rigidity to such connections.

When required, the insulation of bus conductors may be varnished sleeving, made of impregnated woven cotton or rayon, or plastic tubing of polyvinyl chloride, fiberglass, or Teflon. This tubing insulation is popularly known as "spaghetti" or sleeving, and it comes in a variety of colors, sizes, and thicknesses.

Copper Braid

Connections that require considerable flexibility, such as the grounding of equipment mounted on isolating shock mounts, or RF transmitter circuits, use flat tinned-copper braid. This braid is available in a variety of widths to carry the expected current.

Conductors and Cables

A great variety of insulated conductors and cables are used in the electronic industry. Some of them are shown in Fig. 12-1.

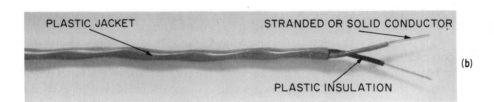

(a)

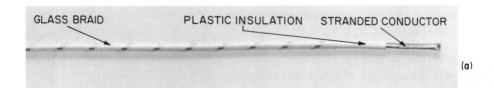

(b)

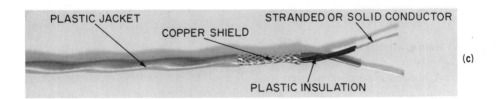

(c)

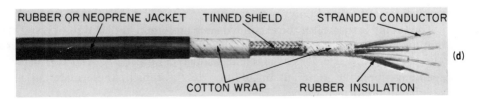

(d)

FIG. 12-1. Typical Conductors and Cables. (Courtesy Alpha Wire Corporation.)

The single conductor, shown in Fig. 12-1(a), is of the color-coded type found in chassis wiring and harness assemblies. Its plastic insulation may be one of a variety of solid body colors, or its body color may be white with one or two colored tracers added for individual conductor identification. Such conductor color marking is essential in harness assemblies because it eliminates the need for identification tags on the ends of individual conductors.

Conductors used for connecting alternating current are generally twisted together in pairs, Fig. 12-1(b), which prevents the possibility of ac hum pickup by other wiring.

Conductors likely to radiate electromagnetic energy that might affect other conductors generally have an external shield, which consists of flexible copper braid woven over the insulation, Fig. 12-1(c), with or without a protective jacket. Such shields are connected to chassis, if grounded, or to other points at ground potential.

Usually, a cable, Fig. 12-1(d), is an assembly of two or more conductors, although this is not true in the case of RF coaxial cables (Fig. 4-12), unless the outer shield is considered as the second conductor.

Conventional cables may have numerous variations. For instance, they may be made up of various numbers of conductors, individually shielded or plain, in solid body colors or with tracer identification, of different wire sizes, and with or without an external jacket or shield.

Coaxial cables are used in radio-frequency circuits as the transmission lines between the equipment and antenna and in equipment interwiring.

The center conductor in a coaxial cable is centered within the outer shield, either by means of special insulating washers or an insulating material such as polyethylene or Teflon that completely fills the space. The size of the center conductor is determined by the current it is to carry. The impedance of the cable is determined by the spacing between the conductor and the outer shield.

Heat-Shrinkable Tubing

A special type of expanded heat-shrinkable, nonmelting insulating tubing with a "memory" has been developed to protect exposed wire connections at connectors, terminal boards, and spliced joints, or to encapsulate semiconductors, capacitors, and other components. It is available in many materials such as polyolefin, PVC, and polyvinylidene fluoride.

The tubing is formed in the expanded state and upon application of heat, by a hot air blower or other heat source, shrinks 50 percent or more in diameter to its preexpanded size. It is available in diameters ranging from $\frac{3}{64}$ to 4 inches depending upon the material. In addition to the tubing, these materials are available in preformed shapes.

Some examples of application are shown in Fig. 12-2.

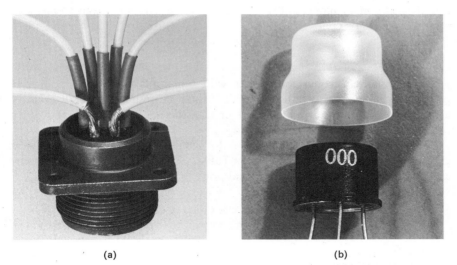

(a)   (b)

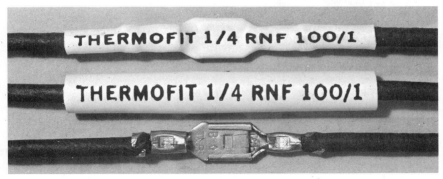

(c)

FIG. 12-2. Heat Shrinkable Tubing Applications. (a) Connector Leads. (b) Transistor Preform. (c) Splice.

12.2 CONDUCTOR IDENTIFICATION

A system of individual conductor identification is needed in complex electronic wiring assemblies to simplify circuit testing and tracing of individual circuits. Such a system is especially important in harness assemblies because individual conductors can no longer be traced by eye; thus, there must be some system to indicate the individual conductor connections to the components.

Although conductor coding systems vary, they may include such items as: conductor color or colors; AWG wire size, solid or stranded; conductor insulation designation by name, abbreviation, or drawing number; and circuit destination by color or code.

Conductor Color Codes

In complex electronic equipments with numerous connection points, the preferred system uses conductors with insulation or a solid color braid, or a braid with a white (or transparent or translucent) background and one or two stripes for additional differentiation.

In less complex equipments, solid color conductors are used to identify the chassis wiring by function, as shown in Table 12-2. Thus, transistor base circuits are identified by a green conductor and power supply by a red conductor.

Table 12-2   Functional Color Coding of Chassis Wiring

Function	Color	Code	Abbreviation
Grounds, grounded elements	Black	0	BK
Heaters or filaments	Brown	1	BR
Power supply, +	Red	2	R
Transistor emitters*	Yellow	4	Y
Transistor bases	Green	5	G
Anodes (plates) and transistor collectors*	Blue	6	BL
Power supply, minus	Violet	7	V
	(Purple)		PR
AC power lines	Gray	8	GY
	(Slate)		S

*Also applies to diodes, semiconductor elements, and other elements with operations similar to transistors.

The interconnection wiring between complete equipment units uses white insulation for all conductors, with the circuit identification printed in black either on the insulation or on sleeves, tags, or pressure-sensitive tapes attached to the conductors.

Color Abbreviations

Conductor colors are generally abbreviated as shown in Table 12-3. The first column, which is preferred, represents abbreviations that quickly identify the colors they represent. The abbreviations in the second column are reduced to a single letter, which is convenient if space is limited but requires an explanatory table as part of the diagram notes to clarify the abbreviations. Some of these abbreviation color letters are taken from the first letter of the color designation and others from the final letter.

Multiple color combinations are written with the body color first and followed by the abbreviations for the first and second stripes with a hyphen or diagonal separating them, i.e., W-R or W/R.

Thus, a white wire with a red and an orange stripe would be designated as

Table 12-3   Conductor Color Abbreviations

Color	Code	Preferred Abbreviation*	Secondary Abbreviation†	MIL-STD-12 D
Black	0	BK	K	BLK
Brown	1	BR	N	BRN
Red	2	R	R	RED
Orange	3	O	E	ORN
Yellow	4	Y	Y	YEL
Green	5	G	G	GRN
Blue	6	BL	B	BLU
Violet	7	V	V	VIO
(Purple)		(PR)		
Gray	8	GY		GRA
(Slate)		(S)	S	
White	9	W	W	WHT

*Adopted from ANSI Standard, *Electrical and Electronics Diagrams.* Y14.15–1966 (R1973).
†From EIA Standard RS-359.

W-R-O or WHT-RED-ORN. To avoid confusion, the same color abbreviation method should be followed throughout the set of drawings.

Conductor Insulation Abbreviations

In addition to identifying conductors by color abbreviations, it may also be necessary to use abbreviated designations to identify their insulation coverings.

For example, a #20 AWG wire with a single nylon covering, single cotton covering, and cotton braid would be identified on the drawing by "20NCCB." The conductor may also be identified by the wire size and a code letter for the insulation. For example, in 20G, the number 20 identifies the wire size, and "G" is a code letter for the particular type of covering. Coded insulation coverings have an identification table on the drawing.

When most or all of the conductors are of the same type, a drawing note may be used to identify the conductor size and type. For example:

ALL CONNECTION WIRES ARE 7/30 STRANDED, WITH CODE L INSULATION UNLESS OTHERWISE SPECIFIED.

Conductor Marking

As mentioned before, white background conductors are used for interconnection wiring between equipment units. The designations for such conductors may be in the form of letters, numbers, or symbols that are made directly on the insulation by hot stamping with a special machine. These markings are generally made with black characters to give the greatest contrast.

Pressure-sensitive tapes or adhesive labels with printed characters, short marked sleeves, or tags are also suitable for this purpose. Such *markers* are attached or wrapped at the conductor terminations or along the conductors as required for identification. These methods are needed to mark conductors or cables that cannot be marked otherwise, for example, conductors with glass insulation, shielded wires, or such multiconductor cables as harness assemblies.

The marking means and their locations should be shown on the harness and other conductor assembly drawings.

Wiring Techniques

Considerable time and effort have to be devoted to the design and production of interconnection wiring in electronic equipment. Conductors must be selected and specified on the basis of their current carrying capacity, along with the method of attaching them to termination points, either by soldering or wire-wrapping. Wiring is accomplished by the point-to-point method or by cable or wiring harnesses.

Point-to-point wiring. In this method, the individual wires are run from one connection point to another by the most direct route. Components, such as resistors or capacitors, are frequently wired directly between such points. To prevent low-frequency radiation, ac leads are generally twisted together rather than run directly.

Cables or harnesses. Although point-to-point wiring is sufficient for most commercial equipment applications, military electronic equipment specifications generally call for cable or harness wiring. This is described in detail in Chapter 15. Cabling assures uniformity of individual connections in complex equipments and simplifies trouble shooting. It is possible to identify individual conductors through wire color-coding and color systems in circuit interconnections.

Soldering. Soldering techniques are of the individual joint soldering and wave soldering types.

Both solid and stranded wires can be used for interconnecting wires. This wiring process requires an operator to locate the points to be connected, select the correct interconnection wires previously cut to length, and solder the wires to the connector pins or other termination points.

Wave soldering is more applicable to printed board interconnections and is accomplished by dipping the connection side of the board into a "wave" of molten solder for a few seconds.

Termi-Point®* connections. This is a mechanical system of making electrical connections from a stranded or solid wire to terminal posts on a connector or printed-circuit board.

The stripped wire end is laid against the face of the terminal post and a

---

*®Registered trademark of AMP, INC.

special spring clip is slipped over the wire and the post to form a gas-tight joint. Either a special hand tool or automatic equipment is used to strip the wire end and apply the clips.

Solderless wrap connections. Orginally used to wire central telephone equipment, the solderless wrap connection method has been used in such applications as the fire-control system in nuclear submarines. It has an advantage over the conventional hand-soldering technique because it can be adapted to numerically controlled machine wiring, which is self-checking, thus eliminating the chances of human error.

Solderless wrap requirements. The preferred terminal shape is square or rectangular in cross section, with at least two sharp corners. Other terminal shapes are used, but most configurations have something that makes them less desirable than rectangular or square terminals. Presently, panels can be wired on a .100 inch-square grid system with automatic "Wire-Wrap" machines. The wires to be connected by hand tools must have their ends stripped for a distance determined by the wire gage and by the number of turns to be wrapped around the terminal. A special tool is used for wire-wrapping, Fig. 12-3(a). The bare wire-end is inserted in the tool, which is then positioned over the terminal, and upon pressing the trigger, the wire is automatically wrapped around the terminal, Fig. 12-3(b). A high-pressure, gas-tight, metal-to-metal contact is made at each of the four corners of the terminal; the number of turns is deter-

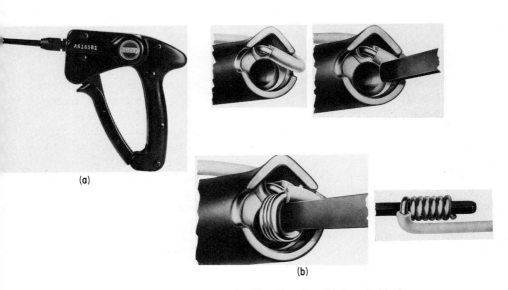

(a)

(b)

FIG. 12-3. (a) A Special Tool for Wire-Wrapping. (b) Steps in Making a Solderless Wrap Connection.

mined by the diameter of the wire and the mechanical requirements of the individual operation.

Terminals may be from .025 to .045 inch square, or they may take the form of rectangulars of a similar cross-sectional area. Various tools handle wire gages from #18 down to #32 AWG. An automatic "Wire-Wrap" machine for production wiring of electronic equipment, which uses #26 and #30 AWG wire, is shown in Fig. 12-4.

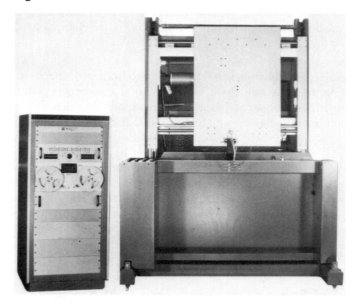

FIG. 12-4. Automatic "Wire-Wrap" Machine.

Multiwire®* interconnection system. In addition to the wire-wrapped panels and multilayer printed-circuit boards, another newer system has emerged, namely the multiwire interconnection system.

Basically, the first step in this process is to prepare an etched format for power and ground planes, Fig. 12-5. Next, an adhesive material is applied to the board to hold the wiring in place.

The wiring machine, Fig. 12-6, operating on numerical control, lays the wire pattern on the board, point to point, using #34 copper wire with polyimide insulation. Each wire starts and ends at the hole locations. Upon completion of the wiring pattern, the wires are pressed into the adhesive and cured. After encapsulation in epoxy, the component holes are drilled, and copper is deposited in the holes by the additive plating process, Fig. 12-7. Thus each wire is bonded to the wall of its hole.

*® Registered trademark of Kollmorgen Corporation.

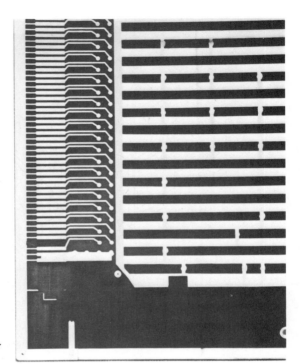

FIG. 12-5. Etched Format for Power and Ground Planes.

FIG. 12-6. Multiwire Wiring Machine.

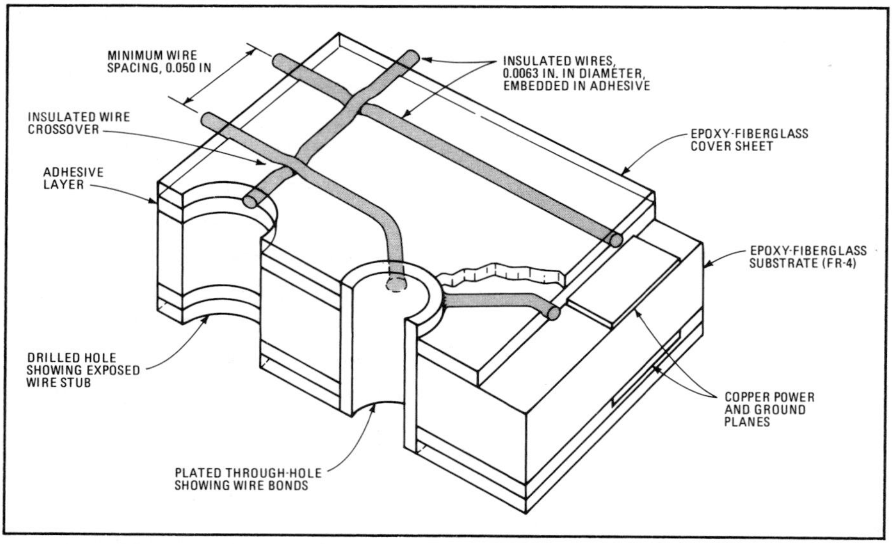

FIG. 12-7. Cross Section of a Drilled Hole in Multiwire Interconnection System.

The assembly of printed-circuit components then follows as in conventional board assembly.

Artwork: Narrow lines and close spacing should be avoided in artwork supplied by the customer. It can be drawn on .007-inch Mylar or other stable material and be two times scale if it is produced manually. A 100-mil grid is preferred. Pad diameter should be at least $\frac{1}{32}$ inch wider than the hole diameter.

Multiwire system of interconnections is particularly suited for connecting the chip carriers. They require a density of 80 conductors per square inch for 6.3-mil wire or 100 conductors for the 4-mil wire.

Chip carriers may have terminals spaced as close as 20 mils apart, which would require the use of a number of layers in a multilayer board assembly to accommodate all of the terminals. On the other hand, only two layers of multiwire wiring will accommodate all of the chip carrier terminations.

12.3 COMPONENT DETAILS

There are three basic parts to a connection diagram. They are (1) the components, which are represented by suitable geometric or other symbols and identified with the same reference designations used on the schematic diagram; (2) the symbol terminations or other terminating devices; and (3) the connection lines, which represent the wiring conductors. Simple symbol presentation is recommended to save drafting time and to enhance the diagram's readability.

Component Representation

Components are represented on the connection diagram in numerous ways, depending upon their physical complexity.

Many of the simpler components can be represented by their principal physical shape. That is, round or nearly round components are represented by a circle, and all other shapes are represented by a rectangle. It is important to adequately identify termination points on the component symbol. Several methods of indicating terminal positions relative to the mounting surface or some other reference point are shown in Fig. 12-8(a).

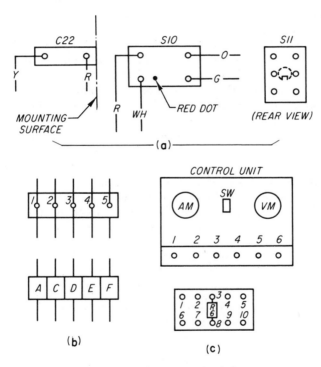

FIG. 12-8. Component Symbols.

Such components as terminal boards or strips are represented by small circles within a rectangle, with the circles used to indicate individual terminals, or by a rectangle that is subdivided into smaller rectangles, with the terminal circles omitted, see Fig. 12-8(b). Note that the terminal circles or small rectangles are arranged to suit circuit convenience, rather than following a numerical or alphabetical sequence. Terminal circles should be at least $\frac{1}{8}$ inch in diameter to allow for possible diagram reduction.

Component symbols are not necessarily limited to the individual components. They may also represent assemblies with terminations, as in Fig. 12-8(c).

Among the more involved component symbols on the connection diagrams are those used for transistors and integrated circuits.

Some of the more common semiconductor symbols for use on connection diagrams are shown in Fig. 12-9. In (a) are depicted transistor symbols with the components as viewed from the bottom. The designations "*E*," "*B*," and "*C*" are for emitter, base, and collector, respectively. Case shape is indicated and reference points are shown. Designations "*1*", "*2*," "*3*," may be used to identify the terminations in place of "*E*," "*B*," and "*C*" and so forth. This numbering should agree with the numbering shown in transistor manuals.

Figure 12-9(b) shows integrated circuit assemblies housed in TO-5 cans. Since it is impractical to show the many graphic symbols within each symbol, their external connections are identified by sequential number as related to a reference point shown.

The four lead TO-72 symbol in Fig. 12-9(c) represents either a single gate FET with the leads: #*1*–drain; #*2*–source; #*3*–insulated gate; #*4*–substrate; or it could be a dual gate FET with the leads: #*1*–drain; #*2*–gate #*2*; #*3*–gate #*1*; and #*4*–substrate. Lead identification should be provided by a separate note on the diagram.

In Fig. 12-9(d) are shown integrated circuit assemblies of the dual-in-line packaging (*DIP*) type, a flat pack, and a power transistor. The assemblies in (b), (c), and (d) are shown as viewed from the bottom or lead side, with the index and other markings serving as reference points.

Identification of individual leads can be either on the outer periphery as in (a), (b), and (c) or internally as in (d), depending upon available space on the component symbol and on the connection diagram.

Symbol Details

Some of the symbol details are shown in Fig. 12-10. Small circles, at least $\frac{1}{16}$ inch in diameter, are used to represent terminal connections, rather than their actual shapes. A five terminal male connector is shown in Fig. 12-10(a). Its terminal designations must be the same on the diagram as on the actual component, and they should also agree with the schematic diagram and other drawings.

Potentiometer connections, illustrated in Fig. 12-10(b), indicate the view shown. The terminal numbering should agree with the numbering assigned to the same symbol on the schematic diagram.

The capacitor terminal connections in Fig. 12-10(c) are the designations appearing on the capacitor.

To permit unobstructed approach of connection lines, the component reference designations and terminal markings should preferably be placed within the component symbol, or the identification numbers or letters may be offset as in Fig. 12-8(b).

If connectors with multiple terminals are drawn according to their actual

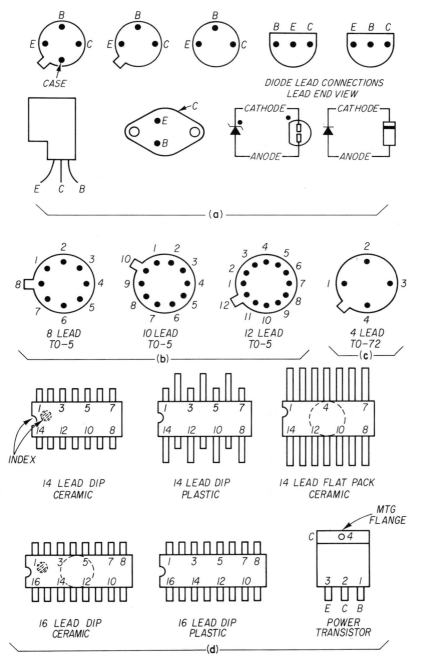

FIG. 12-9. Semiconductor Details.

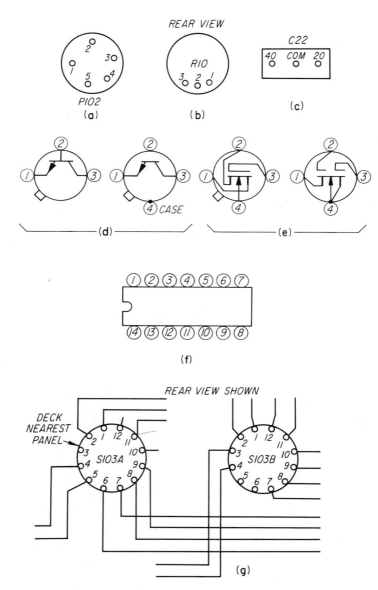

FIG. 12-10. Symbol Details.

physical appearance, it may be necessary to enlarge the overall symbol to allow passage of the connection lines to the terminals.

Component symbols such as meters, circuit breakers, fuses, and protective relays should have the rating, meter range, and other information placed next to the symbol or tabulated elsewhere on the diagram.

An alternate method of representing semiconductor devices, such as transistors, on the connection diagrams is shown in Fig. 12-10(d), (e), and (f). The graphic symbol of the device is drawn within the component circle, while the terminal connections are small numbered circles on its periphery. Their numbering, in relation to a reference point, should correspond to the terminal numbering appearing in transistor manuals. The terminals or leads are shown in the same relationship as they would appear as viewed from the bottom of the device.

Divided Symbols

It will be necessary to separate the decks of such complex component symbols as multiple-deck rotary switches to present the wiring information adequately. The decks should be separated only enough for connection line wiring, see Fig. 12-10(g). Common, easily recognized reference points, such as mounting studs, should be indicated, and the terminal designations should agree with similar symbol markings on the schematic diagram.

Conductor Identification

There are various coding schemes in use for conductor identification. A few examples are given in Fig. 12-11.

The identification of conductor size shown in Fig. 12-11(a) indicates 7 strands of #30 wire with solid green covering. If only the color is indicated, it may be inserted within the connection line as in Fig. 12-11(b). Individual conductors extending between components may be identified by numbers as in Fig. 12-11(c). Coaxial conductors have the RF cable identification, Fig. 12-11(d), as well as the coaxial cable recognition symbol.

Conductor Destination

Conductor destination may be indicated on complex diagrams as shown in Fig. 12-11(e). The destination *TB101-2* refers to terminal board *101*, terminal #2. Another example, shown in Fig. 12-11(f), identifies the destination of the

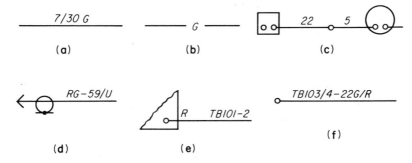

FIG. 12-11. Conductor Identification and Destination.

conductor as terminal board *TB103*, terminal #*4*, while the conductor itself is identified as #*22*, solid wire, solid-green color, with type *R* insulation.

These destination designations are useful in a complicated circuit to test the circuit, to check continuity and voltage, and to trace connections.

Lettering

Connection diagrams require considerably more line lettering than schematic diagrams. In addition to the lettering of symbol reference designations, considerable lettering is also required at the terminations for proper identification. In many instances, connection lines are also identified by color, insulation, etc. as described in Section 12.2. When necessary, because of space limitations, lettering on vertical lines may be used, reading from the right-hand side of the drawing.

Connection Lines

In general, connection lines are drawn in a uniform width throughout the connection diagram. However, heavier lines may be used to emphasize bus bars, ground returns, or power circuits; or conversely, control circuits may be lighter than the rest of the diagram.

Light dash lines are used to indicate the mechanical linkage between switch sections, and long and short dash lines are used to indicate the grouping of components or assemblies. Dash lines of normal weight are used to show alternate wiring or connections, or wiring concealed behind panels or components. Wiring that enters through a grommet is also frequently shown by dashes extending beyond the grommet to indicate interposition of the chassis or panel between the observer and wiring.

12.4   EQUIPMENT VIEWS

In connection diagrams it is customary to arrange component symbols in relative positions as seen by a person who views the equipment during the wiring process.

The view represented on the connection diagram should show the wiring or terminal side of the equipment or chassis. If components are mounted on the inside bottom of the chassis and also on adjacent sides as in Fig. 12-12, the view is arranged with the sides that have a common edge or the corner next to the bottom or principal view, as if the sides were opened flat.

Since most equipment wiring views are more complex and involve additional features—such as wired panels attached to the chassis, connections extending from the bottom or principal wiring view to the top of the chassis, or connections extending to the outside surfaces of chassis sides—it is frequently necessary to provide additional views. These auxiliary views are all drawn relative to the principal view of the wiring side of the equipment. The wiring view in Fig. 12-13

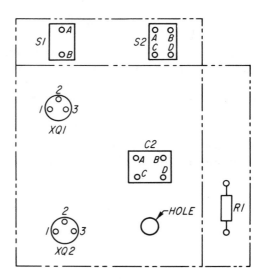

FIG. 12-12. Connection Diagram Views.

BOTTOM VIEW OF CHASSIS

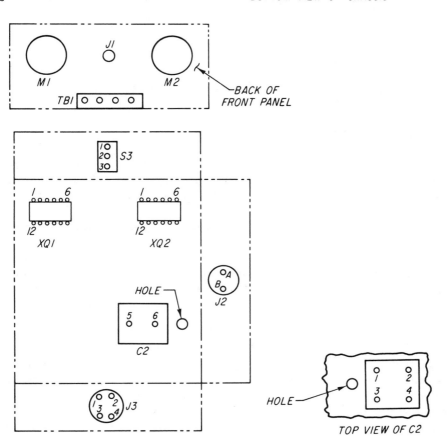

FIG. 12-13. Connection Diagram Auxiliary Views.

is composed of the bottom chassis view, together with auxiliary views of the back of the front panel and the top view of capacitor *C2*.

The draftsman must be careful to keep the auxiliary views in proper orientation to the main view. The relationships of auxiliary views should be identified if they cannot be clearly shown.

Terminals or terminal boards that are mounted on more than one surface are oriented by revolving the planes into view as if they were developed in conventional orthographic projection (see Fig. 12-14).

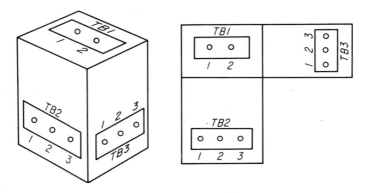

FIG. 12-14. Development of Multiplane Terminal Board View.

12.5 CABLE AND WIRING DRAWINGS

Like their mechanical counterparts, electronic products are assembled by following a standardized procedure. The components are assembled into small sub-assemblies, attached in turn to the main mechanical assembly, and then wired to form a complete unit or equipment.

Part of the wiring information may be covered in such details as the harness assemblies that compose the main wiring in many complex equipments, or it may be covered in the pre-wired component boards or the printed circuit-board assemblies. However, none of these supply the complete wiring information required for military electronic equipment, nor do the schematic diagrams.

Cable Drawings

One of the simpler drawings, involving both wiring and mechanical assembly, is the cable assembly drawing. The example shown in Fig. 12-15(a) has the cable assembly terminating in connectors, known in military applications as the "MS" type, consisting of a molded body that contains a number of floating metal inserts, of either the male (plug) or female (socket) type. Each insert is identified at its position in the molded body by a letter or number. The enclosing

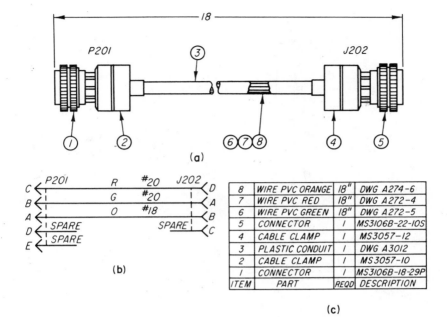

8	WIRE PVC ORANGE	18"	DWG A274-6
7	WIRE PVC RED	18"	DWG A272-4
6	WIRE PVC GREEN	18"	DWG A272-5
5	CONNECTOR	I	MS3106B-22-10S
4	CABLE CLAMP	I	MS3057-12
3	PLASTIC CONDUIT	I	DWG A3012
2	CABLE CLAMP	I	MS3057-10
I	CONNECTOR	I	MS3106B-18-29P
ITEM	PART	REQD	DESCRIPTION

(c)

FIG. 12-15. Drawing of a Simple Cable.

shell of aluminum or some other metal has a threaded wire clamp attached to anchor the outgoing wires which relieves the strain on the soldered connections at the inserts. An adjustable saddle is used on the wire clamp to handle various cable diameters.

The opposite end of the metal shell has a coupling ring, used to attach the connector shell to its mating connector.

The interconnecting wires are held together by plastic tubing that is slipped over them during the assembly, or they are laced together by a lacing cord that is used in harness assemblies.

To properly connect the various conductors to connectors *P201* and *J202*, a simple diagram, Fig. 12-15(b), is made a part of the drawing to identify the conductor colors and their insert terminations at each connector.

Instead of being limited to one assembly, such a drawing may serve as a general drawing for a number of such assemblies, with an individual parts list, Fig. 12-15(c), prepared for each.

Pictorial Point-to-Point Wiring

In some applications, it is desirable to present wiring details in greater detail than the conventional connection diagrams. These pictorial point-to-point diagrams include diagrams prepared for the instruction manuals for various types of do-it-yourself electronic kits, and for the critical wiring details of

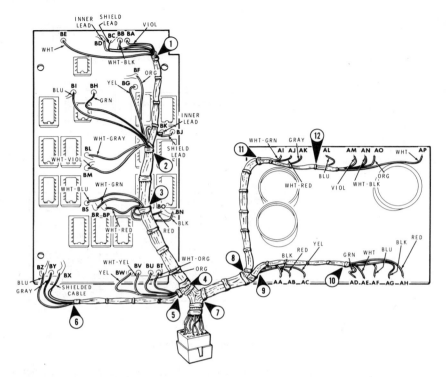

FIG. 12-16. Cabling Connections to a Printed Circuit Board. (Reprinted by permission of Heath Company.)

conventional connection diagrams. An example of part of a wired bottom view of a chassis with a printed circuit board is shown in Fig. 12-16.

The various components and their wiring are shown in their relative positions and are usually identified by numbered balloons or "call-outs." Except for possible reference dimensions or critical separation details such drawings are dimensionless. Conductors are identified by color code and seldom carry any additional stripes or other pictorial identification.

Notes are used extensively for additional wiring information and may contain any of the following:

1. Indication of cabled or harness assemblies as part of the wiring
2. Soldering specifications
3. Indication of pigtailed components, i.e., components with their own leads
4. Instructions for making insulated soldered joints
5. Specifications for making critical wiring assemblies

To conserve drafting time, many of the nonessential details of the component outlines of switches, tube sockets, and resistor and capacitor markings are omitted. This not only saves time, it also provides greater drawing readability.

A single bottom wiring view of the chassis may not be sufficient to give all of the wiring details, so that additional views may be required to show components or wiring that is hidden by other components.

The interconnections between the various component symbols, terminal boards, and connectors on the connection diagram also present a problem. To lay out the connection diagram pattern systematically, it is best to start with one or more rough sketches of the proposed interconnections and to separate a complex diagram into a series of smaller, more manageable assemblies. Such a rough sketch (Fig. 12-17) will eliminate the need to rework the final diagram drawing because of insufficient space for connection lines and interference with

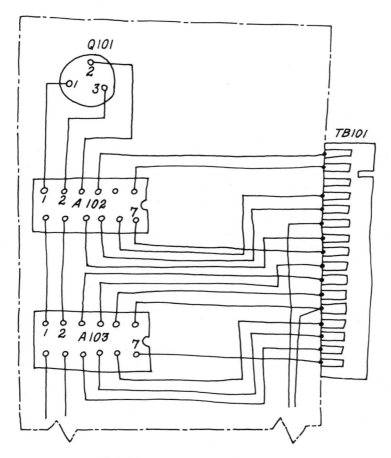

FIG. 12-17. Rough Sketch is Helpful.

component symbols. The sketch establishes to a major extent the space needed for these lines because it is based upon their number and interspacing.

Although components are usually placed on the diagram in their relative positions, the sketch will also help to relocate them if necessary to simplify the connection pattern.

The parts list should include not only the mechanical and electronic components and wiring material, but also such incidental parts as terminal lugs, sleeving, etc.

12.6 CONNECTION DIAGRAM LAYOUT

A well laid out connection diagram depends largely upon the planning and preparation spent in trying to achieve a clear, simple layout of lines and symbols, to represent the conductors and the components. This is of the utmost importance in complex diagrams where extensive intricacy in the connection line arrangement is likely to be a problem to the user. It is up to the electronic draftsman to use the trial and error method until he achieves an orderly arrangement.

Preliminary Details

The drawing of connection diagrams presents the same basic difficulty that is sometimes experienced in drawing schematics, namely, the lack of dimensions to estimate the drawing size.

The layout procedures outlined later in this chapter will help to establish the diagram dimensions that should be selected to keep within a standard drawing size sheet, if possible, or for the larger sizes, to reduce the number of horizontal folds required with a long, narrow diagram. The point-to-point or continuous-line diagram type that shows all of the connection lines requires a greater diagram area than the cable or interrupted-line type. However, it is preferred for production line use because it has superior readability.

Some of the following questions should be answered at the outset to eliminate possible reworking later:

1. If the diagram is complex and a harness is to be used, what connection wires will be eventually incorporated into a wiring harness?

2. What connection wires are critical, and therefore, what connection lines should be separated from the remainder of the wiring and so indicated on the diagram?

3. What connection wires should be shielded and so noted on the diagram?

4. What connection wires are high-voltage or carry pulse or radio frequency circuits?

5. What connection wires carry low-frequency, alternating current, such as power supply leads? Should they be paired and twisted?

6. What connection wires should be of the coaxial cable type and separated from the remainder of the wiring?

Before the layout work is started, a print of the schematic diagram should be marked up on the basis of this or similar information. The print will then serve as a reference should any question arise as to the exact procedures followed on the connection diagram.

Drawing Procedures

To present a neat appearance, connection lines should be spaced in definite increments. This procedure can be readily followed by using grid paper, ruled 10 by 10, 8 by 8, or other equal spacing per inch. Such paper is described in Chapter 17. The grid paper may be slipped beneath the conventional drawing paper to act as a guide in drawing connection lines on the diagram.

The beginner will do well to use grid paper for this rough layout because it helps to maintain the approximate spacing between the connection lines.

The first step is to draw the component symbols in the form of circles or rectangles, depending upon the component shape, and to locate them according to their relative positions in the equipment.

One point to be remembered in drawing connection diagrams is that connection lines run from one terminal to another, or to a mechanical or electrical tie point, not to junctions at the most convenient spot, as on the schematic diagrams.

On rare occasions, a splice or a joint may be made outside of a tie point, but it should be indicated by a note on the diagram. For example, several adjacent terminal points may be connected together by a bus bar without having the connection wire connect from one tie point to the next. The joints are made by connections from each terminal to the main bus bar.

After the component symbols have been drawn, the connection lines are drawn between the nearest component terminals, according to the circuit on the schematic diagram. These connections should be drawn lightly along the grid lines because as the diagram begins to take shape connection improvements can be made. The major problem in drawing a connection diagram is to keep line crossovers and jogs to a minimum.

Lines connected to related circuits should be grouped together to assist visual tracing of such circuits. Component terminals may be spread out, or the component symbol may be moved on the preliminary sketch to reduce jogging and crossovers (Fig. 12-18). In Fig. 12-18(a), no attempt has been made to keep the separation between connection lines to a uniform spacing. Instead, they have been drawn to fit within available space. The same circuit is shown in Fig. 12-18(b) with such changes as greater spacing between the terminals of *TB102* to accommodate the dual connections and relocating *PS105* to eliminate jogs in its connection lines. The lines extending to the right and left from

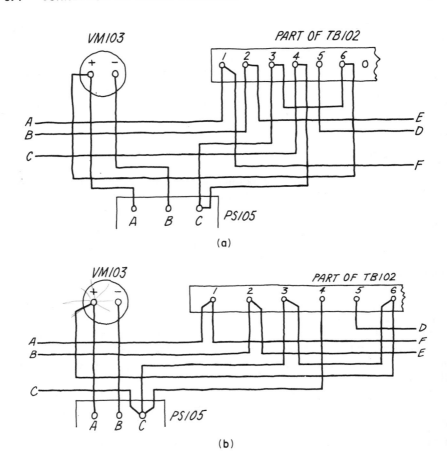

FIG. 12-18. Connection Line Layouts.

terminal *1* on *TB102* have been drawn on the same level. The same is true of terminals *2* and *3*.

The freehand sketches will determine the number of connection lines that extend between component symbols. Then, the required spacing between the symbols can be established on the basis of the number of lines and their spacing for the final drawing. Thus, a rough line pattern is determined before the finished diagram is started.

The sketches for sectionalized diagrams will have to be assembled to determine the overall diagram dimensions and to make necessary adjustments to connection line wiring.

As the work progresses, it will be necessary to check the circuit connection lines against the schematic diagram. Perhaps the easiest method is to follow the procedures described in Chapter 16 for checking wiring diagrams, using the schematic diagram as the master (see page 485).

Connection Line Spacing

These lines should be spaced $\frac{1}{4}$ inch apart or even more if the diagram is to be reduced photographically so that there will be at least a $\frac{1}{8}$-inch separation after reduction.

Long lines extending across the diagram should be separated into groups of three, with a double spacing between. Such separations enhance the diagram because the eye can follow the connection lines more easily than with groupings of ten or more lines. This group separation is also a military requirement.

Adequate space should be allowed around component symbols and their termination points for required lettering and other identification.

If symbols are grouped into individual subassemblies with boundaries indicated by dot and dash lines, adequate separation should be maintained between the lines and connection lines to avoid confusion.

Connection Line Details

Using space to the best advantage is more of a problem for the electronic draftsman in circuitry work than in the mechanical aspects of electronic drafting. This is especially true in connection diagrams where a multitude of connection lines must be directed to various sections and termination points on the diagram with a minimum of crossovers, jogs, and corners.

The square corner is probably the most common of the three types of corners in use, followed by the inclined line at 45 degrees and the rounded type. The latter types are used to indicate the direction of the conductors merging into a cable. The square corner is essential when each connection line is drawn out. It is also the simplest and quickest to draw and less likely to create confusion if crossovers and jogs are used.

Connection lines that cross should be drawn at right angles to each other. "Saddles" or loops are no longer used at the crossover points.

Some examples of connection line arrangements are shown in Fig. 12-19. The best space utilization is shown in the symmetrical layout Fig. 12-19(a), which provides twelve connection line circuits, with each circuit changing its direction 90 degrees rather than with crossovers or jogs. Symmetry is still maintained in Fig. 12-19(b), where four of the circuits continue straight through while the other eight change their direction. An example of the cornering of connection lines to get them around a component symbol or to approach a terminal board is shown in Fig. 12-19(c). When necessary, transposition of connection lines is best accomplished at the corner, Fig. 12-19(d), to maintain a uniform pattern that is easy for the eye to follow.

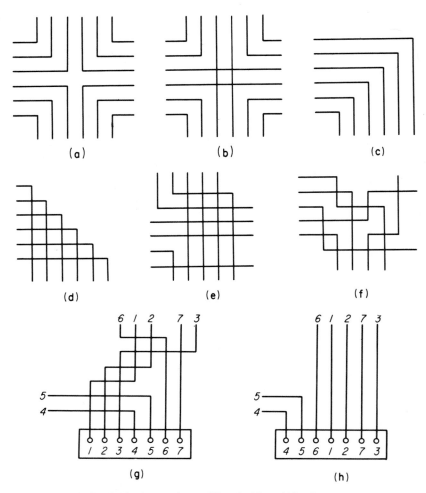

FIG. 12-19. Connection and Terminal Board Line Patterns.

The example in Fig. 12-19(a) represents an ideal condition without crossovers. However, in many cases, such crossovers are unavoidable, and jogs may even be introduced in the lines as shown in Fig. 12-19(e) and (f).

In simplifying connection line arrangement, both ends of each line must be considered because eliminating crossovers at one end may necessitate introducing them at the other or at some point in between.

To a large extent, the simplicity or complexity of the circuit pattern will be determined by the terminations of connection lines at terminal boards or component terminals. Terminals arranged in a sequential order, Fig. 12-19(g), may result in a complex, hard-to-follow line pattern that more than offsets the slight advantage of having the terminals sequentially arranged.

The rearrangement of the terminal sequences, Fig. 12-19(h), has a twofold advantage—the connection lines are easier to follow along their paths, and the space required at the terminal board area is reduced. Sufficient space should be left near the terminal board connections for any lettering that may be required, such as connection wire-color identification and circuit identification.

Since the terminal sequence on a terminal board establishes the connection line pattern, it is advisable to draw the connection lines across the diagram first and then to extend them to the terminal board or other termination source, keeping the line pattern as simple as possible and with a minimum of crossovers and jogs. The terminal numbering is added next in random sequence. Nonsequential terminal numbering is rarely detrimental to the diagram or circuit tracing.

Miscellaneous Details

Some of the details that apply to the various connection diagram types are shown in Fig. 12-20. A method of denoting prewired components on such an assembly as a component board is illustrated in Fig. 12-20(a) by dotted line

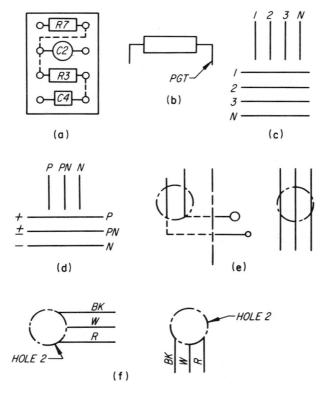

FIG. 12-20. Miscellaneous Diagram Details.

connections. Prewiring may also be indicated by solid connection lines if they are drawn within the component symbol outline. Components with permanently attached leads, such as fixed resistors or capacitors, have the leads indicated by the abbreviation "*PGT*," as in Fig. 12-20(b). Other wiring that is not a part of cable assembly may be designated "DIR" or direct when connections are made directly from one terminal to another, or "SUR" or surface when the connection lines extend from one terminal to another in a convenient manner.

Alternating-current component symbols should have the phasing indicated according to the physical construction of the component. If the construction is not a restricting factor, the phasing sequence should be arranged as in Fig. 12-20(c), with phases indicated vertically or horizontally by numbers. The direct-current polarity indication is arranged as shown in Fig. 12-20(d), with neutral indicated by "*PN*" or $\pm$.

Connection lines that extend through holes are shown in a variety of ways as illustrated in Fig. 12-20(e). When more than one hole appears on the diagram they are numbered from one up, to provide an easy reference. If the connection lines are interrupted on passage through the holes, it is necessary to identify the individual conductors by color abbreviations or some other means, and to identify the holes with a number as shown in Fig. 12-20(f). Such an identification may be necessary when the same hole appears in separate views on the diagram.

After the wiring paths have been established, it may be found that such obstructions as shields and partitions are in their way. In this case, the mechanical detail drawings of the equipment or assemblies should be modified to incorporate the holes or slots required for cable paths or individual conductors. Grommets, eyelets, or other mechanical protective means should be added on such drawings.

Connection diagram information may be supplemented by local or general notes listing conductor color and size, insulation, type of wire, shielding, soldering details, such conductor terminations as terminal lugs, protective insulation, cabling, and any other details that require an explanation.

12.7 CONNECTION DIAGRAM TYPES

In general, there are three basic connection or wiring diagram types: continuous line (the most familiar), interrupted path, and tabular or lineless.

The type selected depends upon such factors as the complexity of the equipment, the degree of detail required, the ultimate use of the diagram, and the diagram's size limitations.

The continuous line or point-to-point diagram, which is the predecessor of the other diagram types, is used for drawing less complex circuits. As its name implies, every connection line is shown between the terminations on the component symbols or terminal boards. Since the connections are the easiest to trace on this diagram type, it is less likely to result in errors when being read or used as a wiring guide, Fig. 12-21(a).

The highway or cable diagram, Fig. 12-21(b), is a variation of this diagram

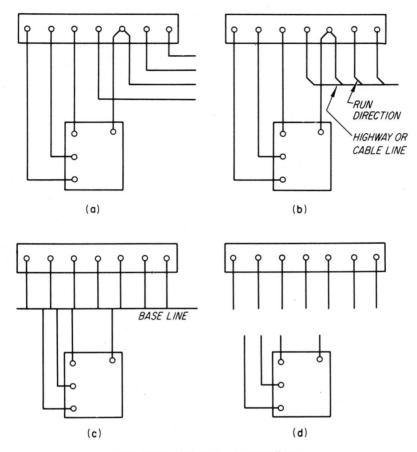

FIG. 12-21. Connection Diagram Types.

type. The individual connection lines are merged into single lines, known as highway or cable lines, which simplify the diagram and reduce the space required for the connection lines.

The interrupted line diagrams are of the base-line or feed-line types, Fig. 12-21(c) and (d). In the base-line type, the short connection lines from component symbols terminate at a base line, while in the feed-line type the connection lines do not terminate in a base line.

The "FROM" and "TO" connections in the tabular or lineless diagram type are listed in a tabular form instead of being represented by connection lines.

Point-to-Point Connection Diagrams

As mentioned before, these diagrams either show every connection line or combine them into single-line cables, interposed between component symbols.

The point-to-point connection diagram represents the original concept of illustrating the actual wiring between components, wire-by-wire, and serves as a master for preparing the cable or harness drawings. Therefore, it is important to locate the individual connection lines between the components in the same relative positions they would have if they were placed in the harness assembly.

When the connection lines are combined into single-line cables, the cables should be shown split into branches and positioned between the components if a harness drawing may possibly be developed from the diagram.

An example of a point-to-point diagram is shown in Fig. 12-22. The schematic diagram of a power supply circuit is illustrated in Fig. 12-22(a), and the corresponding connection diagram is shown in Fig. 12-22(b). The component symbols and connection lines in Fig. 12-22(b) are located according to their relative positions in the equipment. The component symbols have a minimum of detail and are drawn in proportionate size to facilitate identification. The terminations on individual component symbols should be shown and identified in their relative positions to prevent possible errors.

The component symbols should be identified by the reference designations that appear on the schematic diagram. The diagram may also supply such additional connection line identification as color and wire size, and such drawing notes as:

UNLESS OTHERWISE SPECIFIED, ALL WIRES ARE #22. DO NOT USE POLYVINYL CHLORIDE INSULATION AS PER SPECIFICATION MIL-E-5400.

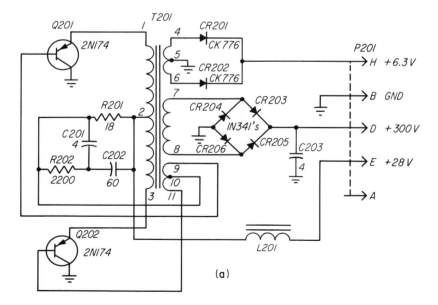

FIG. 12-22. Power Supply Circuit. (a) Schematic Diagram.

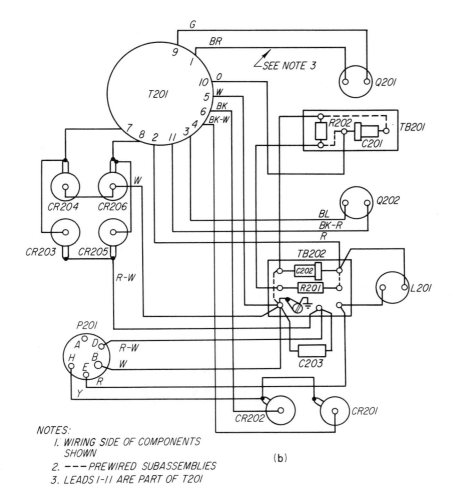

FIG. 12-22. (Continued) (b) Point-to-Point Connection Diagram.

UNLESS OTHERWISE SPECIFIED, ALL WIRES ARE A PART OF CABLE
ASSEMBLY B-1932.
SOLDERING SHALL BE IN ACCORDANCE WITH SPECIFICATION D372.
ALL WIRES ARE A PART OF T204.

Other local or general notes may also be used to describe processes or
components such as wire stripping, lacing, and terminals.

A junction box, containing mating connectors or terminal boards, may be
required when equipment assemblies or subassemblies are interconnected. There
are several steps in drawing this connection diagram that will help to reduce the
complexity of interconnection lines. As shown in Fig. 12-23, the terminations,

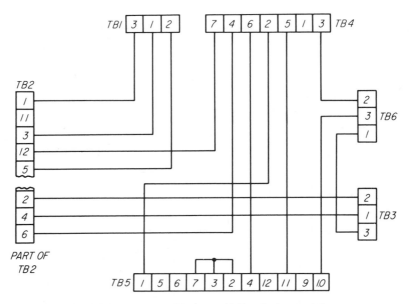

FIG. 12-23. Junction Box Diagram with Terminals out of Sequence.

which may be terminal boards or multipin connectors, are represented by graphic symbols. They are arranged to obtain the shortest jumper lines and the minimum of crossovers and jogs rather than by a numerical sequence of connection points.

Another helpful step is to split the terminal boards or connectors to eliminate possible jogs in the jumper lines (see terminals *2* and *4* of *TB2*).

Connections between the various terminal boards are made by starting at their corners and assigning terminal numbers in whichever sequence results in the shortest jumper lines. This helps to keep the center area clear for long connection lines.

Spare terminals are drawn for possible wiring additions later and should be located with "free" space in front of them for such additions. This procedure helps to eliminate extensive drawing revisions at a later date. Thus, terminal #*1* of *TB4* would be better located next to terminal #*7* where there is more space for wiring addition.

Once again, preliminary sketches help in working out the various details before the final diagram is started. The student can see the difference between the arrangement in Fig. 12-23 and a sequential arrangement by redrawing the diagram with the terminal board terminals in numerical order and comparing the two diagrams.

Such details as conductor size, color, and other data should be included on the drawing of the type shown in Fig. 12-23.

Cable or Highway Connection Diagrams

When it becomes impractical to draw individual connection lines because of diagram complexity, a number of them may be merged into single lines that run horizontally or vertically between the component symbols. These single lines are connected to the individual component terminals by short lines, 1 to $1\frac{1}{2}$ inches long, that are known as "feed lines." An example of feed-line data is shown in Fig. 12-24.

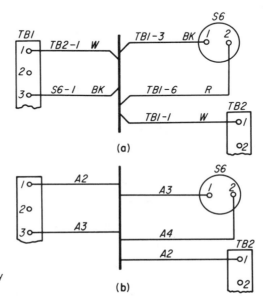

FIG. 12-24. Cable or Highway
Connection Diagram Data.

One method of feed-line information is represented in Fig. 12-24(a). The destination of the connection line or "feed line" from terminal #1 of *TB1* is shown as "*TB2-1*," or terminal #1 of terminal board *TB2*. Its destination is also indicated by the inclination of the feed line where it joins the heavier "cable" or "highway" line. Wiring color is indicated according to the color abbreviations or code number designations (see Table 12-3).

Another method of presenting information is shown in Fig. 12-24(b). The feed lines are identified at each point where they emerge from the cable line by a code letter, number, or a combination of both, and are drawn perpendicular to the cable line.

The wire data included on feed lines for Fig. 12-24(a) may include such information as wire color, type, or code; and the destination by component reference designation and its terminal identification.

The wire data for the method in Fig. 12-24(b) is included in a separate list on the diagram drawing or on a separate drawing made for this purpose.

Cable lines should be continuous and located in their relative positions to the component symbols. Wiring may be segregated for electrical reasons by drawing a separate cable line. Identification of cable lines that are part of a harness assembly may be made on the diagram by either a local or general note.

To simplify finding the feed-line terminations on connection diagrams that contain branch circuits, additional identification may be provided at branch points as in Fig. 12-25. Here, the direction of connection lines *A1* and *A2* may readily be followed through the branches.

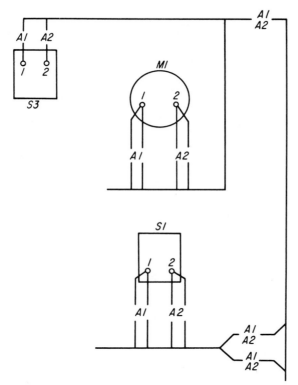

FIG. 12-25. Destination of Circuits at Branch Points.

Interrupted Line Connection Diagrams

There are two principal types of interrupted line connection diagrams—the base line, where the feed lines terminate in a short base line; and the feed line, where no base line is used and the feed lines from the component symbols terminate a short distance from the symbol. In both instances, the ends of feed lines have destination identification in the form of letters, numbers, or both, and in the base-line method, are identified by color abbreviations of the feed lines.

An example of the base-line connection diagram is shown in Fig. 12-26.

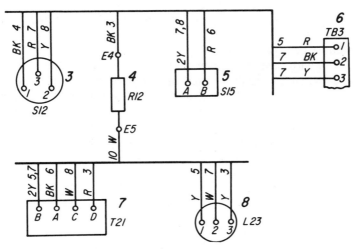

FIG. 12-26. Partial Base-line Connection Diagram.

Note that, in addition to its reference designation, each component symbol also has a location number assigned for wiring identification. These location numbers are assigned to component symbols, from left to right and from top to bottom, in the same manner as reference designations are assigned on many schematic diagrams. They are assigned to components that have feed lines extending to them (see *4* in Fig. 12-26) but not to components that are wired point-to-point.

Location numbers and letters should be at least twice as high as the other lettering used on the diagram and located next to the symbol or within it, if joined component symbols are shown.

Only a single feed line needs to be shown when more than one connection wire connects to a component terminal, as in the feed line connected to terminal "*B*" of *7*. Destination points are given as *5* and *7*, while *2 Y* represents two yellow wires. To avoid possible confusion, different colored wires connected to the same terminal should usually be indicated by separate feed lines. Although the base lines need not be continuous, they should extend far enough to connect feed lines of component symbols that are located on approximately the same reference level. The base lines are drawn as convenient rather than to indicate the actual cable routing.

The destination of each feed line in Fig. 12-26 can be readily traced. In location *3*, the black (*BK*) feed line connected to terminal #*1* extends to resistor *R12* at location *4*. Similarly, terminal #*3* at location 3 has a red (*R*) wire extending to terminal *D* at location *7* of transformer *T21*.

Feed lines should be drawn far enough apart to provide lettering space and be offset when drawn to the base line from opposite sides to avoid giving the appearance that they continue across the base line.

Although Fig. 12-26 is only a part of a base-line diagram, the same practices apply to the complete diagram. Conventional graphic symbols are used to indicate shielded wires, wires forming a common cable or harness, or splices, etc.

An example of a partial feed-line type of connection diagram is shown in Fig. 12-27. As with base-line connection diagrams, each component symbol is arbitrarily assigned a number, a letter, or a combination of the two in a logical sequence. The destination is indicated at the end of a short feed line from each component symbol terminal by the component number and the terminal to which it is connected (see *A*, terminal #*1*). Indication is also given if more than one connection extends from the component symbol (see *A*, terminal #*2*).

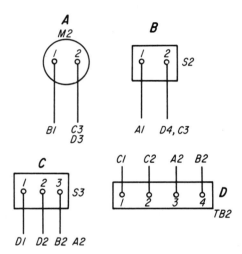

FIG. 12-27. Partial Feed-line Connection Diagram.

Multiwire Connection Diagrams

These are actually point-to-point interconnection diagrams developed for use on printed-circuit boards in place of the conventional plated conductors.

The documentation prepared for such a board establishes the interconnections between individual points and the paths they follow. The conductors are insulated and thus can cross each other. This eliminates the tedious development of plated interconnections that cannot cross each other at any point on the board. The conductors can form several layers of insulated wire placed in a definite pattern.

The connection diagram drawing should show the component side of the board. The identification of points on the diagram is done both alphabetically and numerically to simplify the procedure. Integrated circuits and other components should preferably be organized in columns and rows, Fig. 12-28, with .100-mil spacing and all connection points located on grid intersections.

Each point, identified as "Net No.," is listed on a "Multiwire Net List." Pins or points are identified by the numbers and areas of components. Such a

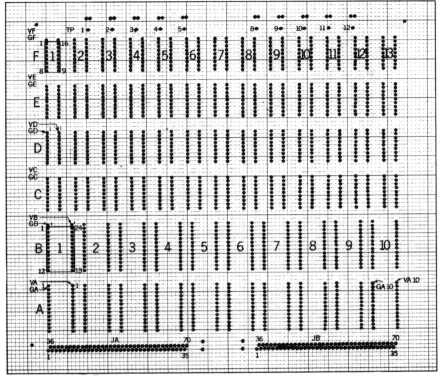

Typical component layout.

FIG. 12-28. Component Pin Identification.

list is to be supplied to the company that will do the actual multiwire wiring; see Fig. 12-29.

The list may also be prepared as a coordinate *X-Y* list. Thus there is no need for actual connection diagram. The schematic diagram can serve the purpose if all actual connection points are identified on it. They form the "Net List."

Tabular Connection Diagrams

The wiring information in this diagram type is given in a table or a tabulation that lists the connections between the component symbols but omits any connection lines between them. Such diagrams are generally used with industrial connection diagrams and are discussed in Chapter 14.

Route Diagrams

Occasionally it is desirable to present a circuit or a part of one by separating parts of the symbol and drawing the parts in the order they occur in the circuit. Thus, a complete section of the circuit can be seen at a glance, even though it

MULTIWIRE NET LIST   PART NO.   PAGE NO.

NET NO.	REF DESIG					
001	U005-07	C0003-1	R0003-1			
002	U005-05	R0002-1				
003	U005-40	R0002-2	VR001-2	C0004-1		
004	U005-25	U001-05	U002-04	U003-04		
005	U005-26	U001-06	U002-03	U003-03		
006	U005-27	U001-07	U002-02	U003-02		
007	U005-28	U001-08	U002-01	U003-01		
008	U005-29	U001-09	U002-21	U003-21		
009	U005-30	U001-10	U002-05	U003-05		
010	U005-31	U001-11	U002-06	U003-06		
011	U005-32	U001-13	U002-07	U003-07		
012	U005-33	U001-14				
013	U005-35	U001-02	U004-05	U004-04		
014	U005-36	U001-03	U002-17	U003-17	U004-09	
015	U005-19	U004-01	U004-02			
016	U005-18	J01-00H				
017	U005-01	R0001-1	U007-13	U002-20	U003-20	
018	U005-37	Y0001-1				
019	U005-38	Y0001-2				
020	U005-16	U006-02	U001-15	U002-10	U002-08	
021	U005-15	U006-04	U001-16	U002-12	U002-11	
022	U005-14	U006-06	U001-17	U002-14	U002-13	
023	U005-13	U006-08	U001-18	U002-16	U002-15	
024	U005-12	U006-12	U001-19	U003-09	U003-10	
025	U005-11	U006-14	U001-20	U003-11	U003-12	
026	U005-10	U006-16	U001-21	U003-13	U003-14	
027	U005-09	U006-18	U001-22	U003-15	U003-16	
028	U004-03	J01-00G				
029	U006-03	U007-03				
030	U006-05	U007-05				

MULTIWIRE
Kollmorgen Corporation

FIG. 12-29. Multiwire Net List.

may include switching, relay contacts, or widely separated components or their parts.

This procedure is useful in developing printed circuit and component board layouts, portions of connection diagrams, and control circuits in industrial control equipment. A typical route diagram is shown in Fig. 12-30.

In the circuit shown, terminal *E3* on one end of cable assembly *W5* connects through green wire (*G*) to receptable *D* of connector *J4*. Pin *D* of connector *P4* is mated with receptacle *J4* and in turn connects to fixed resistor *R22* and variable resistor *R32*. The movable contact #2 and end terminal #1 of the resistor connect to choke *L10*, while the circuit beyond is interrupted by switch *S2*.

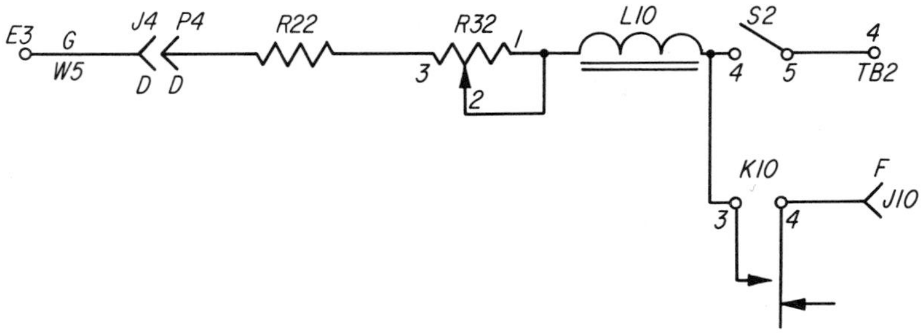

FIG. 12-30. Route Diagram.

When this swtich is closed, the circuit terminates at terminal #4 of terminal board *TB2*.

A side circuit connects from terminal #4 of switch *S2* to terminal #3 of relay *K10*. When the relay contacts are closed by relay coil (not shown), the circuit continues to contact *F* of connector *J10*.

If necessary, a route diagram may be split and continued on the line below or elsewhere on the diagram, or even on another drawing. However, the break-off point should be labelled with a letter or number to provide continuity of the circuit at another point.

Thus, a complex circuit may be illustrated by subdividing it into simpler route diagrams that show only the components entering into such subdivisions. Route diagrams are frequently used in instruction manuals to illustrate operational details.

Interconnection Diagrams

An interconnection diagram is used to show connections between multiple-unit assemblies that require several connection diagrams. Such a diagram depicts and suitably identifies only the cabling and the connector connections between the various units.

It also carries reference information to other drawings and diagrams and relevant mechanical assembly data. The cable and connector assemblies are identified by reference designations and referenced to the detail drawings of such assemblies. Units are represented by squares or rectangles and identified by a letter, a number, or a descriptive title. Connectors or terminal boards which use the same identification given on the connection diagram of each unit are the only terminations that should appear on the unit blocks. Terminals should be shown in the same sequence as on the connection diagram of the unit.

Examples of typical interconnection diagrams are shown in Fig. 12-31. Units *A* and *B* in Fig. 12-31(a) are interconnected by cable *W205*, which has connectors *J202* and *P207* at its respective ends, and by cable *W207*, which has connectors *P214* and *P203* at its ends. Details of these cables are referenced to their mechan-

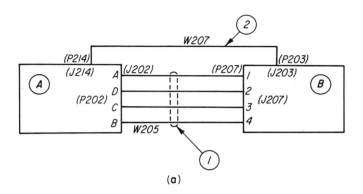

(a)

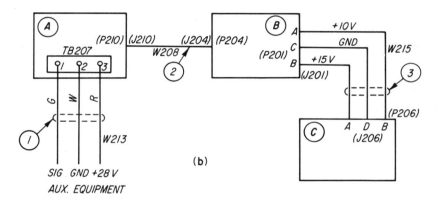

(b)

FIG. 12-31. Typical Interconnection Diagrams.

ical assembly drawings by "call-outs" *1* and *2* in the title block of the interconnection diagram drawing.

Cable *W213* in Fig. 12-31(b) connects between terminals *1*, *2*, and *3* of terminal board *TB207* to external auxiliary equipment. The colors and connections of this cable to the equipment are indicated. Cable *W215* between units *B* and *C* has the functions indicated on the cable.

Connector identification should be provided both at the unit and also at cable assembly ends.

SUMMARY

Various industry standards have set the practices involved in drawing connection or wiring diagrams. Such diagrams may be classified as point-to-point, highway, interrupted line, route, and interconnection type. The equipment components are interconnected by the various conductor types. The paths of these conductors are color coded to help trace them within the equipment. The components themselves may be represented in the form of circles or rectangles, depending upon the component's physical shape. Symbols are located as they appear in equipment assembly.

QUESTIONS

Sketches should be included with answers when necessary.

12.1 What kind of conductors are used for chassis wiring?

12.2 How are the wire sizes identified?

12.3 Give three methods used to list stranded wire.

12.4 What is bus wire? Where is it used?

12.5 Describe a coaxial cable.

12.6 What is the standard color code for chassis wiring? List three examples.

12.7 Abbreviate the following wire colors using the three prevalent codes: (a) white; (b) blue; (c) brown; (d) green.

12.8 List the various wiring techniques.

12.9 Name some of the methods used to mark interconnection wiring.

12.10 Describe the multiwire interconnection system.

12.11 Show three examples of individual conductor identification and describe the details.

12.12 Describe the following: (a) tracer; (b) braid; (c) wire marker; (d) jacket; (e) body color; (f) coaxial cable; (g) cable.

12.13 List the three basic components of a connection diagram.

12.14 Show six examples of semiconductor basing diagrams.

12.15 Illustrate several methods of showing symbol details on a connection diagram.

12.16 What is a pictorial connection diagram? Make a sketch of its basic features.

12.17 What are some of the points to be considered before making a layout of a connection diagram?

12.18 Describe some of the requirements and practices to be followed in laying out a connection diagram.

12.19 List the advantages of rearranging terminal board terminals on a connection diagram.

12.20 What steps are to be followed in preparing a multiwire connection diagram?

12.21 What are the various connection diagram types? What factors govern the type used?

12.22 Describe briefly: (a) a feed line; (b) a highway line; (c) a base line.

12.23 What is a tabular diagram?

12.24 Describe: (a) an interconnection diagram; (b) a route diagram. Sketch an example of each.

EXERCISES

Develop preliminary layouts for these exercises on cross-section paper with four divisions per inch. Refer to parts catalogs such as Electronic Engineers Master *and other catalogs for details of components specified on the drawings, such as sockets, connectors, switches, transistors, diodes, and so forth.*

12.1 Sketch a view of a chassis with components, and indicate how a connection diagram is developed from it.

12.2 Sketch the following symbols used on connection diagrams: (a) a three-conductor shielded cable; (b) a male and female coaxial connector; (c) a wiring-harness assembly; (d) a twisted pair; (e) a power transformer; (f) a transistor; (g) an integrated circuit; (h) a 16-lead ceramic (DIP); (i) a TO-72 transistor.

12.3 Make a sketch of a simple cable diagram terminating in male and female connectors.

12.4 Make a sketch of a simple point-to-point diagram.

12.5 Make a point-to-point connection diagram of the voltage regulator, described in Exercise 8-11, and connected as in Fig. 13-37. Draw the components in outline form double-size and use a connection line spacing of $\frac{1}{4}$ inch. Locate the terminals as shown in that exercise.

12.6 A five-position, six-pole rotary switch consists of three wafers (Fig. 12-32). Draw a base-line diagram with switch wafer *A* on the right and lines *A*, *D*, *E*, *F*, *J*, and *M* connecting to the upper base line. The remaining lines connect to

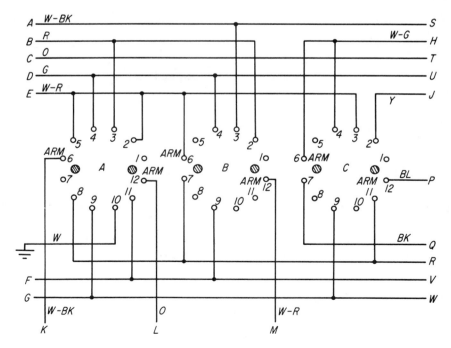

FIG. 12-32. Partial Switch Schematic.

the lower base line. Draw circular switch wafers with horizontal mounting holes. Draw five individual contacts on each half of the wafer extending toward the center rotor, which has two semicircular contact segments with a non-bridging contact extending at one end. Each switch arm contact extends to one of the segments and makes a continuous contact with the segment during switch rotation. Show the wire color and destination on feed lines, e.g., *W-R A2* on the feed line from terminal *5* on switch wafer *A* to terminal *2*. Assign consecutive numbers to line letter designations, e.g., *C = 3, F = 6*, and so on, so that the destination on the feed line from terminal 3 on wafer *A* to line *B* will be indicated as *R2*, from terminal 5 to line *E* as *W-R5*, and so forth.

12.7   Using data in Exercise 12.6, draw a point-to-point connection diagram showing the wafers separated and in line vertically with wafer *C* at the top. Wafer mounting holes are also vertical. Show wires *A, B, E, F*, and *K* vertically on the left and the remaining wires on the right-hand side. Show jumpers on switch wafers, for example, terminal *5* on *A* to terminal *2*, terminal *11* on *A* to terminal *9* on *B*, and so on. Identify jumpers and other connection wires by consecutive numbers starting with *1*, see Fig. 12-11(c), and include designating numbers, letters, and colors as given. Show terminals *A7, A9, A10, 10B,* and *10C* connected to a ground lug.

12.8   Revise the drawing in Exercise 12.7 as follows: remove the jumper from *A5* to *A2*, and a jumper from *A1* to *A5*; add a one-watt 1000-ohm composition resistor from *B1* to *B8*; and add a ground wire to *B11*. List these changes as revision *A* in the revision block of the drawing.

12.9   The connection diagram in Fig. 12-22(b) has the components arranged as in Fig. 12-33. Draw a highway type wiring diagram. Indicate the color and destina-

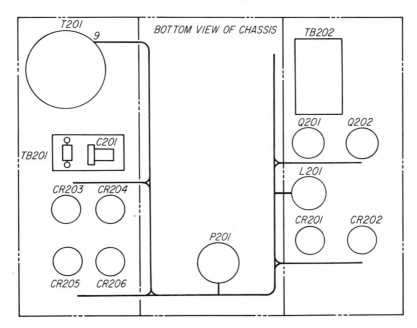

FIG. 12-33. Power-Supply Component Layout.

tion of each feed line, and indicate the turn-offs by diagonal lines. A suggested routing is shown by a heavy line in Fig. 12-33. Indicate wire sizes, excluding *T201* leads, on the feed lines by a code—type *A* for #22 *PVC* stranded on all connections except from terminals *E* and *B* on *P201*. These are type *B* or #18 *PVC* stranded. As part of the drawing notes, include color abbreviations and an example of coding.

12.10 Prepare connection diagrams in ink of the following items that may be assigned by the instructor: (a) electronic voltmeter; (b) signal generator; (c) phonograph preamplifier; (d) a regulated power supply unit; (e) transistor checker. Make these diagrams of the point-to-point and base-line types, and include reference designations, wire colors, and types, and destinations (on the base-line diagram type). Diagrams should be checked against schematics before inking.

12.11 A plug-in card is shown in Fig. 12-34, along with its schematic diagram. Make a point-to-point connection diagram, using the plug-in terminals for terminations, and locate components approximately as shown on the card. Show reference designations.

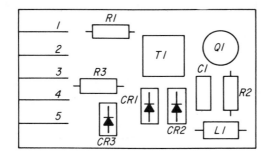

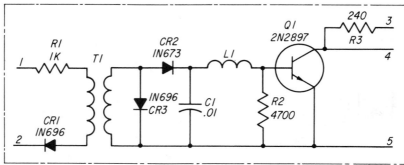

FIG. 12-34. Plug-in Card and its Schematic Diagram.

13

PRINTED CIRCUITRY

The use of miniaturization and subminiaturization in electronic equipment design has, to a large extent, been responsible for the introduction of a new technique in inter-component wiring and assembly that is popularly known as the "printed circuit." This process involves a conductive pattern, formed on one or both sides of one or more insulating laminates or substrates. The term "printed wiring" refers only to the conductive pattern that is formed on the laminate to provide point-to-point connections.

Printed circuitry has several advantages:

1. Circuit characteristics can be maintained without introducing variations in intercircuit capacitance.
2. Component wiring and assembly can be mechanized by wave soldering.
3. Quantity production can be achieved at lower unit cost.
4. The size of component assembly can be reduced with a corresponding decrease in weight.
5. Inspection time can be reduced because printed circuitry eliminates the probability of error.

The two general categories of printed circuitry are rigid printed-circuit board and flexible printed wiring. In addition, there is the multilayer type, which consists of a number of layers of electrical conductors separated from each other by insulating material and formed into a flexible or rigid body.

13.1 PRINTED CIRCUITRY PROCESSES

Several processes are in use to produce the desired conductive pattern on an insulating laminate or board. The required pattern may be obtained by spraying or stencilling deposits of metal powder or paint, by mechanical die stamping or embossing, or by photographic and chemical means, which is the most common process.

Etched Process

There are usually several steps involved in the so-called "etched" method for a laminate clad with copper foil or other conductive material. First, a drawing is prepared showing the pattern to be duplicated. This is done by drawing the desired pattern in ink or taping it on a stable drawing material two or four times the finished size, Fig. 13-1(a). The pattern is then reduced photographically to actual size, Fig. 13-1(b). After the clad laminate has been cleaned, the copper-foil surface of the laminate is coated with a light-sensitive emulsion. The pattern develops on this emulsion when it is exposed through the negative. An acid-resistive coating, Fig. 13-1(c), or "resist" is formed when

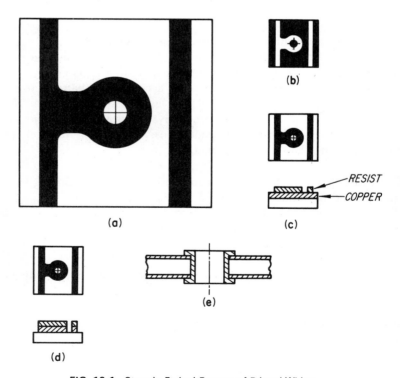

FIG. 13-1. Steps in Etched Process of Printed Wiring.

the emulsion hardens. The copper-foil coated laminate is then placed in a chemical solution, such as ferric chloride, that dissolves all of the copper foil, except those areas protected by the resist, Fig. 13-1(d), leaving only the copper pattern when the acid resist is removed. A water rinsing process is used after etching to remove all chemicals.

The connection between the conductive patterns on laminate material, with foil on both sides, is provided by means of plated-through holes. The bare holes in the laminate are coated with a conductive film for depositing the plating metal to a minimum thickness of .001 inch. A flange forms on each end of the plated-through hole to complete the connection between the two patterns, Fig. 13-1(e).

Additive Process

Very large scale integrated circuits (VLSI), such as small chip carriers of up to 80 terminations in a small square configuration, require a different approach in printed circuitry from the subtractive or etched method of board manufacture.

The "additive process," as its name implies, adds the copper conductors on the laminate or substrate material by depositing the desired conductor pattern. It allows plating of 5-mil lines and spaces in between, with a ± 1-mil tolerance, compared to a minimum limit of 10 mils for etched circuitry. This is the CC-5 process developed by the PCK Division of the Kollmorgen Corporation.

Figure 13-2 illustrates the various steps in this process. Holes are drilled or punched in the substrate (b), which is an unclad catalytic base material

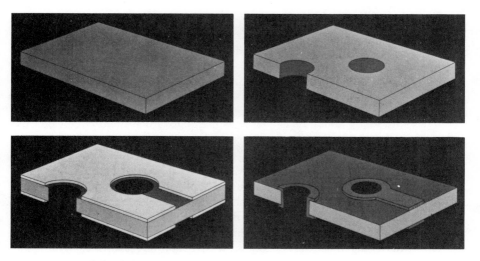

FIG. 13-2. Steps in Additive Process of Printed Circuit Boards.

FIG. 13-2. (Continued)

with adhesive added, Fig. 13-2(a). Then the desired copper conductor image is transferred to the substrate by silk screening or photoprinting (c), followed by depositing electroless copper, using a copper anode and a chemical solution. The plating resist is then stripped (d) and is followed by solder masking and protective coating (e).

Substrate materials commonly used in this process are FR-2, FR-3, and FR-4 (see page 401). There are several advantages to this process as compared to the etched method. One is that very fine conductor lines, .005 inch wide or even narrower, can be deposited because there is no undercutting of the conductor lines that occurs in the etched process. Secondly, thinner thickness of copper conductors can be deposited and also there is less waste of copper in this process and less undesirable residue.

The interconnections form between the two sides of the double-printed board during the plating process through holes previously drilled in the board.

13.2 MATERIALS

Printed-wiring materials consist primarily of the laminate base material and the various types of foils that are bonded to it or deposited on it by the additive process.

Base Materials

There are many materials available as laminates for printed-circuit boards. Thermosetting plastics, such as various grades of paper-base phenolics, melamine, silicone, glass-base epoxy, and Teflon, are used frequently. The reliability of the printed-circuit board depends upon the quality of the laminate used.

The less expensive boards are made of paper-base phenolic resin, type XXP or XXXP (NEMA), which is the equivalent of type PBE in Military Specification MIL-P-3115. The higher the number of X's the higher the dielectric strength. The "P" stands for punching grade.

Other materials in use are: glass base with melamine resin for high abrasion resistance (NEMA grade G-5); glass base with epoxy resin (NEMA grade

G-10) for high mechanical strength, low dimensional change, and fungus resistance; glass-base Teflon for microwave applications; glass-base silicone resin (NEMA grade G-7) for temperatures up to 177°C; and nylon with phenolic resin binder (NEMA grade N-1) for low moisture absorption and high insulation resistance.

Conductor Materials

The most widely used conductor material is high-purity electrolytic copper foil bonded to the base laminate. This foil comes in various thicknesses which are generally referred to on the basis of weight per square foot: .00067 inch thick or $\frac{1}{2}$ ounce; .00135 inch thick or 1 ounce; .0027 inch or 2 ounces; .0041 inch or 3 ounces.

Silver, brass, and aluminum are also used, while gold or solder plating is needed in some applications to cover the conductive pattern.

13.3 BOARD ASSEMBLIES

On single-sided boards, the components are assembled on the side opposite from the pattern. Their leads are cut to the correct length and shaped before they are inserted in the board. Then, they are either hand-soldered or wave-soldered to the conductor pattern. In wave-soldering, the pattern side of the boards is passed over a standing wave of molten solder which automatically solders all of the component leads to the patterns.

On two-sided boards, the two patterns are connected by either funnel eyelets or plated-through holes. After the components are mounted on one side, the board is submerged momentarily in the molten solder, which fills the holes with solder by capillary action and thus connects the component leads to both patterns.

Printed Board Assembly Protection

After the components have been assembled and soldered to the printed board, the finished assembly is sprayed with a clear resin or coated by dipping. This coating, known as "conformal coating," provides both mechanical protection and also protection from the effects of moisture and fungus. Various types of epoxies and polyurethane resins are used for this purpose. They may be of the air-dry or bake-dry varieties.

13.4 MILITARY SPECIFICATIONS FOR PRINTED CIRCUITRY

Since the introduction of printed circuitry into military electronic equipment, appropriate standards and specifications have been issued to achieve uniformity in definitions and applications.

Among these standards and specifications are: MIL-STD-275D, *Printed Wiring for Electronic Equipment*; MIL-STD-429C, *Printed Wiring and Printed Circuit Terms and Definitions*; MIL-P-13949E, *Plastic Sheet, Laminated, Copper Clad (for Printed Wiring)*; MIL-P-28809, *Printed Wiring Assemblies*; MIL-P-50884B, *Printed Wiring, Flexible, General Specification For*; and MIL-P-55110C, *Printed-Wiring Boards*.

Military Standard MIL-STD-275 specifies:

1. Parts shall be mounted on one side of the printed-wiring board.

2. All parts weighing over $\frac{1}{4}$ ounce per lead shall be mounted by clamps.

3. Two-part printed-circuit connectors, consisting of two plastic bodies, containing male and female quick-disconnect electrical contacts and integral aligning hardware, shall be used to integrate plug-in printed-wiring subassemblies.

4. All printed-wiring assemblies shall be supported within one inch of the board edge on at least two opposite sides.

Some of the definitions listed in MIL-STD-429 are:

Additive Process—a process for obtaining conductor patterns by the selective deposition of conductive material on unclad base material.

Annular Ring—a strip of conductive material completely surrounding a hole.

Interfacial Connection—a conductor that connects conductive patterns on opposite sides of the laminate.

Plated-through Hole—metal deposited on sides of the hole through the laminate for electrical connection purposes and extending to the conductive pattern on both sides of the laminate.

Printed-Circuit Assembly—a printed-circuit board on which separately manufactured components have been added.

Register Mark—a mark used to establish the registration, see page 405.

Resist—a material used to protect the desired portions of the printed-circuit pattern from the action of the etchant, solder, or plating.

Terminal Area—that part of a printed circuit that makes connection to the conductive pattern, such as the enlarged area of conductor material around a component mounted hole. This is preferred to such terms in use as "boss, land, pad, or terminal pad."

Military Specification MIL-P-55110 classifies printed-wiring boards according to the following types:

1. Single-sided board
2. Double-sided board
3. Multilayer board

Acceptable Base Materials

Various types of base materials listed in MIL-P-13949 are given here together with their NEMA grade equivalents:

Type	Material	NEMA Grade
	Paper base, phenolic resin	FR-2
PX	Paper base, epoxy resin, flame-retardant	FR-3
GF	Glass-fabric base, epoxy resin, flame-retardant	FR-4
GH	Glass, woven-fabric base, epoxy resin, temperature-resistant and fire-retardant	FR-5
GE	Glass-fabric base, epoxy resin	G-10
GB	Glass-fabric base, epoxy resin, temperature-resistant	G-11
GI	Glass (woven fabric base), polyimide resin	
GP	Glass, non-woven, fiber base, Teflon, flame retardant	
GR	Glass, non-woven, continuous fiber base, epoxy resin, flame retardant	
GT	Glass, woven fabric base, Teflon, flame retardant	

It should be noted that some of the base materials that are acceptable for commercial applications, such as XXXP, paper-base phenolic resin, and glass-fabric base melamine are not acceptable for military equipment.

Board Subdivision

MIL-STD-275D shows boards with interfacial connections, or connections that go from one board side to the other or boards with clinched jumper wire soldered to both sides, Fig. 13-3.

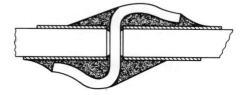

FIG. 13-3. An Interfacial Connection.

The same specification divides the boards according to the method of attaching components, Fig. 13-4(a) a plated-through hole; (b) standoff terminal.

Conductor Spacing

Larger spacings shall be used whenever possible and the minimum spacing between conductors, between conductor patterns, and between conductive materials, such as conductive markings or mounting hardware, shall be in accordance with Table 13-1.

For general use a minimum spacing of .031 inch should be used for a peak

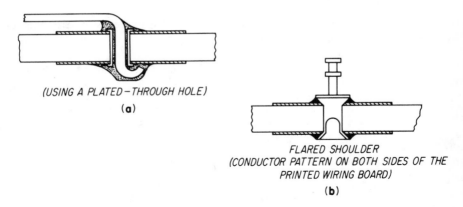

(USING A PLATED−THROUGH HOLE)

(a)

FLARED SHOULDER
(CONDUCTOR PATTERN ON BOTH SIDES OF THE
PRINTED WIRING BOARD)

(b)

FIG. 13-4. Component Lead Terminations.

value of 150 volts, and increased proportionately up to 500 volts. Above that, the spacing should increase .00012 inch per volt.

Table 13-1   Conductor Spacings (MIL-STD-275)

Voltage Between Conductors (dc or ac peak volts)	Minimum Spacing (in.)
0–15	.005
16–30	.010
31–50	.015
51–100	.020
101–300	.030
301–500	.060
Greater than 500	.00012 per volt

13.5 CONDUCTORS AND COMPONENTS

Certain conductor practices have been adopted on printed wiring boards for commercial and military use.

Conductor Size

The current-carrying capacity of the conductor depends upon its cross-sectional area, the allowable temperature rise, and the characteristics of the board laminate. Conductor widths of .010 inch or more are accepted for military requirements and for commercial use. Most conductors with a width ranging between .015 and .031 inch are acceptable for all connections, with the possible exception of power circuits, which may require wider or thicker conductors.

Military Specification MIL-STD-275 specifies conductor size based upon temperature rise above ambient (Fig. 13-5). The lower part of this figure converts the width of the conductor and its various foil thicknesses from inches into cross-sectional area in square mils. These are represented by the vertical subdivisions on the graph. By extending these subdivisions upward, it is possible to determine the temperature rise for a given current with a given cross section. For example, a conductor with a cross section of 50 square mils will have a rise of 45°C when a current of 2 amperes is passed through it. Such

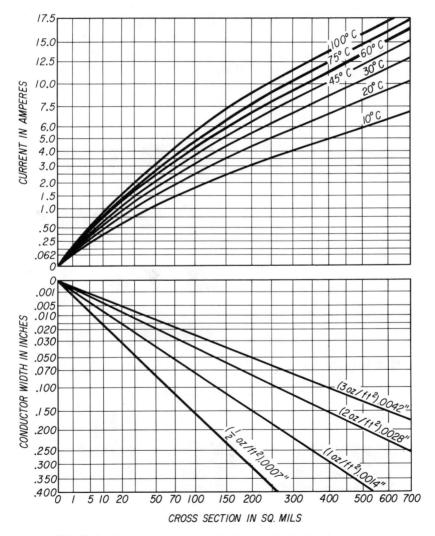

FIG. 13-5. Current-Carrying Capacity Compared with Conductor Size.

a conductor can be .020 inch wide if it is of 2-ounce copper or .038 wide if it is of 1-ounce copper, as determined from the lower part of this figure.

From this, it can be seen that a conductor, .062 inch wide and of 2-ounce copper, is more than ample for most applications.

When necessary, conductor widths of .005 inch can be specified if the board pattern is produced by the additive process.

Conductors and Termination Areas

Examples of conductor pattern shapes are shown in Fig. 13-6(a). The conductor termination areas, which are also known as "pads" or "lands," are given in Fig. 13-6(b) and (c). Note that the conductor width, Fig. 13-6(d), extends smoothly into the pad area by radial fillet on the left or by straight side fillet teardrop on the right. Fillets should be included at all intersections to avoid sharp corners. Some additional termination area shapes are shown in Fig. 13-6(e).

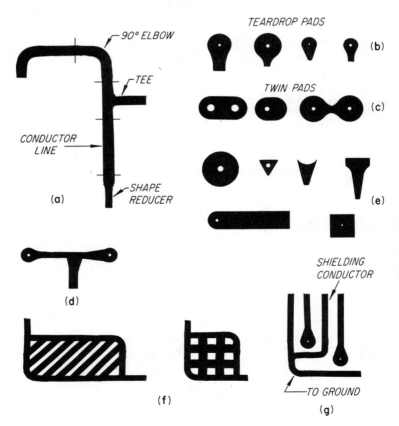

FIG. 13-6. Basic Conductor and Termination Shapes.

If a large area is required for shielding or some other reason, the area should be broken up into a striped or checker pattern, Fig. 13-6(f). This prevents overheating of the laminate when the board assembly is wave-soldered.

The shielding required between adjacent conductors on the laminate may be produced by inserting a shielding conductor, Fig. 13-6(g), between the conductors and connecting it to a grounded conductor path on the board.

Specialized Conductors in Termination Tapes

The increasing complexity of printed wiring and semiconductor components has resulted in the development of specialized tapes and preassembled multiple terminations in opaque black and transparent blue and red. They simplify the draftsman's task in laying out single, double-sided, and multilayer circuitry.

The precision-cut tapes are available in such widths as .015, .020, 026, .031, .040, .046, .050, .062, .090 and up to 2 inches. Thus, if a 4:1 pattern is made the actual board line width would be .004, .005, etc.

Some of the connector artwork patterns that are available are shown in Fig. 13-7(a). They form the multiple connector pads on the end of the printed-circuit board for insertion in connectors with .050, .100, .125, .150, .156, and .200 inch center-to-center spacing.

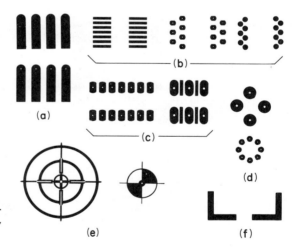

FIG. 13-7. Artwork Patterns. (Courtesy Bishop Graphics, Inc.)

In Fig. 13-7(b) are shown the multiple artwork patterns for flat packs, in (c) the patterns for dual-in-line packages (DIPs), and in (d) the patterns for the TO-type transistors.

There are several types of targets (fiducials) or registration marks used in artwork preparation. Two of them are shown in Fig. 13-7(e). The larger target is located over the intersecting grid lines on the artwork drafting film. They appear through the peripheral spokes, vertical and horizontal, when properly aligned.

Board corner marks, Fig. 13-7(f), are required to establish the board's outline in relation to the registration marks in the artwork pattern. The inner edges of these marks establish the board's outline.

Drafting Films and Grids

Drafting films used in the preparation of printed-circuit artwork may be polyester or clear Mylar®* .003, .005, or .007 inch thick. Mylar®* .0075 inch thick is also available with printed grid patterns, with the grid lines spaced .050, .100, or .125 inch apart. Grid patterns are also available in metric, with lines spaced 1, 2, 4, and 5 mm. The lines may be opaque black, blue, or brown. The last two colors are "dropout" colors, i.e., they will not reproduce photographically.

The artwork may be laid out directly on the grid patterns or a sheet of this grid Mylar®* may be slipped below a sheet of transparent Mylar®* on which the artwork is laid out. Whenever possible, holes for components should be laid out at the intersections of the grid lines for easier board drilling layout.

Tape Application

Once the pads have been attached in the artwork area, tape is attached to the area with an overlap, Fig. 13-8(a). The cut end of the tape should be held with the forefinger, and the tape unrolled. The tape should be laid down, and pressed in place *but not stretched*. This is important. For 90-degree turns, corners and nested elbows are available, Fig. 13-8(b), or they may be formed from tape by crimping it on the inside radius.

Dual-in-line packages (DIP) and flat pack terminal configurations, Figs. 13-7(c) and (b), can be secured in tape packages with scale sizes of 1:1, 2:1, and 4:1. Use a film with a suitable grid pattern and lightly draw the outlines

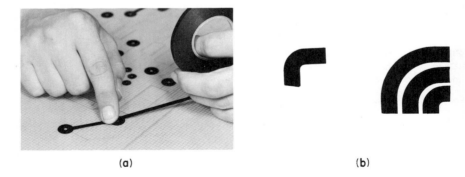

(a)   (b)

FIG. 13-8. Tape Application. (Courtesy Bishop Graphics, Inc.)

*® du Pont registered trademark.

of each DIP or flat pack in accordance with the scale selected and the dimensions given in the manufacturer's catalog. Two nonadjacent pins are located
for reference. After the protective transparency over the conductor tapes is
removed, the pattern is laid down using the component outline and the two
marked pins as reference points, Fig. 13-9(a). This method permits precise
conductor tape placement so 10, 12, 14 or more pads can be laid down simultaneously for DIPs or flat packs. Pads can be trimmed to desired length to
fit the case and lead length of the component.

A somewhat similar procedure is followed in transferring TO-type configurations. Determine from the manufacturer's catalog the diameter of the
transistor pin circle, multiply this by the scale of the artwork (4:1, 2:1, etc.),
and draw this diameter on the artwork layout in the space selected for the

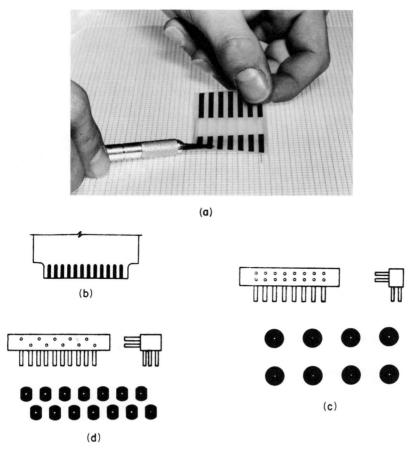

FIG. 13-9. Laying out Multiple Patterns. (Courtesy Bishop Graphics,
Inc.)

component location. Locate one pin on this circle diameter and then transfer the pads from the "TO" multiple pad all at once to the artwork.

Multiple connector artwork patterns are also applied in a similar manner. The connector to be used determines the application. There are three connector types in use: the insertion type, which utilizes the multiple pads shown in Fig. 13-7(a) that form terminations of the printed wiring pattern, Fig. 13-9(b); multiple pads for a separate right-angle connector plug with contacts in line, Fig. 13-9(c); and multiple pads for a separate right-angle connector with the contacts staggered, Fig. 13-9(d).

Two-Sided Printed-Circuit Boards

One of the chief problems encountered in preparing artwork for the two-sided printed-circuit boards has been in securing accurate registration between the two sides.

Two methods are in use. One, where the artwork patterns are drawn or taped separately on the same drawing, with "Front" and "Back" notations and at least three registration marks or "bull's eyes," such as shown in Fig. 13-7(e); and second, where blue and red transparent tapes are used on one film sheet to produce a blue pattern on one side and a red pattern on the reverse side.

Correct registration is of utmost importance in multilayer boards where shorting between layers may occur because of improper alignments when the board assembly is drilled for plated-through holes.

In the blue-red artwork layout, Fig. 13-10(a), separation of the two colors is accomplished photographically by using suitable filtering of blue and red colors as required, Fig. 13-10(b).

Tape and Terminal Pad Selection

Several factors control the width of tape used in the artwork pattern: the current to be carried by the printed wiring, the space available for the conductors, and the tolerances specified for conductor width, which in turn depend upon conductor thickness.

A tape width of .062 inch, reduced by a 4:1 ratio, will produce a .015-inch-wide conductor on a small printed-circuit board. This has proven to be a practical minimum for a majority of applications. For military applications, MIL-STD-275D governs the spacing in relation to the voltages applied (see Table 13-1).

The size of the termination pad depends upon the application. For single- and double-sided boards, the pads or annular rings should be as large as possible if they are for military applications. Their size is governed by the diameters of the component leads attached through plated-through holes, voltage difference between adjacent pads, and expected tolerances on double-sided boards if the patterns are formed separately.

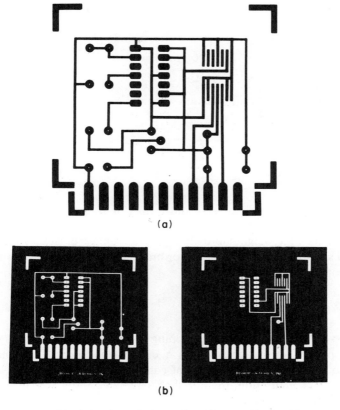

FIG. 13-10. Color Artwork Separation Photographically. (Courtesy Bishop Graphics, Inc.)

In multilayer boards, with possible misalignment between layers, the width of the annular ring (a pad with a hole in it) ideally should not exceed .010 inch in width. For example, with a component lead diameter of .030 inch and lead clearance allowance of .010 inch, the required hole diameter would be .050 inch. If an annular ring .010 inch wide is required, its overall diameter for .050 inch diameter hole would be .050 plus $\pm$.005 tolerance, plus a factor of .020. This equals a .080-inch-diameter minimum pad.

All of the above dimension references are for actual board dimensions. The artwork pattern must be multiplied four times if the scale used is 4:1. Greater accuracy is obtained with the 4:1 scale because artwork tolerances are reduced by a factor of four.

Printed-Wiring Components

Two types of components are used in printed-wiring assemblies—conventional components and specially designed components for inserting or mounting on such assemblies. The latter class includes RF, IF, and audio transformers;

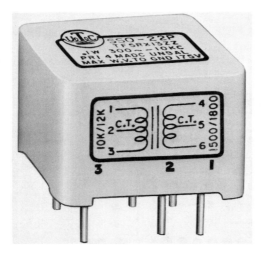

FIG. 13-11. Printed Circuit Board Plug-in Inductor.

connectors; relays; potentiometers; and transistor sockets. A plug-in component is shown in Fig. 13-11.

Components that weigh more than half an ounce must have additional mechanical support provided by epoxy or brackets, rather than being suspended only by their leads or terminals.

13.6 PRINTED BOARD DESIGN

There is no simple formula for developing a printed board assembly and its associated pattern or patterns if it is a double-sided board. A compromise will have to be reached in the choice of component sizes and layout, the conductor pattern configuration, the board shape and size, the hole locations and sizes, the clearances between components, the conductor widths and separation, the one- or two-sided pattern, the shielding and grounding requirements, production methods, and other considerations.

Some of the guidelines followed in actual practice are:

1. Single-sided rectangular boards, $\frac{1}{32}$ to $\frac{1}{16}$ inch thick, should be used whenever possible. Boards for miniaturized equipment, which commonly use thicknesses of $\frac{1}{32}$ inch or less, are an exception.

2. Double-sided boards should have plated-through holes, funnel eyelets, or jumper wires.

3. The board area should be restricted to less than 50 square inches and a maximum length of 10 inches in any direction in most applications.

4. All holes should be located at intersections with a grid of .025 or .050 inch.

5. A minimum spacing of one-and-a-half times the board thickness should be maintained between hole edges for paper-based laminates and one board thickness for other materials.

6. A minimum distance of one-and-a-half to two times the board thickness should be allowed between the edge of the holes and the outer edge of the board.

7. There should be termination areas on both sides of the board for plated-through holes. The hole diameter should be at least $\frac{1}{64}$ inch larger than the inserted lead size.

8. Termination areas of eyeleted holes should be from $\frac{5}{32}$ to $\frac{3}{16}$ inch in diameter.

9. Termination areas of plain holes should be about $\frac{5}{32}$ inch in diameter, and the holes in the board between .010 and .015 inch larger than the component lead size.

10. The edge of the conductor should be at least $\frac{3}{32}$ inch in from the edge of the board.

11. The distance between component mounting holes (Fig. 13-12) should be based upon the following minimum, with the X dimension chosen to the nearest grid intersection. The X dimension consists of two dimensions, each $\frac{1}{16}$ inch, for the straight portions of leads extending from the body of the component or the end leads, plus two radii dimensions of lead diameter doubled. For example, a component with a body length of $\frac{1}{4}$ inch and lead diameters of $\frac{1}{32}$ inch should be mounted in holes spaced $\frac{1}{2}$ inch apart.

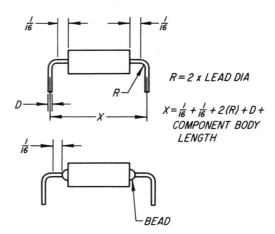

$$R = 2 \times LEAD\ DIA$$

$$X = \tfrac{1}{16} + \tfrac{1}{16} + 2(R) + D +$$
COMPONENT BODY LENGTH

FIG. 13-12.   Component Mounting Distance.

12. Components should be mounted flush with the surface of the board and ceramic capacitors or other similar components perpendicular to the board.

13. Only one lead wire should be inserted in a hole.

14. Heavy components should be located close to the supports on the board, and all components should be mounted parallel to the edges of the board, not angularly unless it is required by circuitry.

Several sketches or layouts will probably be necessary to reach a workable solution. The draftsman is expected to develop these sketches as well as the necessary drawings and artwork.

The required drawings include conductor pattern, master artwork, hole layout, board trim details, marking details, and assemblies.

Printed Board Layout

EIA has adopted a grid line system to simplify printed board layout. It is based upon grid lines spaced .025 inch apart for the subminiature system and .100 inch for the miniature system. This modular dimensioning system applies to hole spacing in printed boards, test point locations, connector terminal spacing, overall board dimensions, and so forth. It simplifies locating the individual board holes because a tabulated coordinate system can be used for all hole positions and referenced to a common point at the intersection of the vertical and horizontal axes.

Printed-Circuit-Board Design

There are a number of steps involved in designing a printed-circuit artwork layout.

The first is to obtain the schematic diagram of the circuitry that is to be assembled on the printed-circuit board or card. Next, a complete list should be made of all components that are to be mounted on the board. Information for each component should include: manufacturer; part number; voltage rating, if applicable; tolerance, mechanical and electrical; and temperature limitations.

A decision must be made as to the board terminations—the number and type, whether integral with the board or included by a separate connector plug. The schematic diagram will determine the number of external connections and the voltage and current values likely to be encountered. The voltage and current will influence the spacing between the board conductors and the selection of the foil thickness on the laminate. If the terminations are integral with the board, a connector receptacle should be selected at this time so the edge of the board at the termination end is suitably notched to match the receptacle.

The next problem is to determine the approximate board size. One of the chief factors is the number of external connections to be made from the board. Connectors are available with center-to-center contact spacing of .050 to .200 inch, so connector length can be reduced somewhat by using connectors with the closer contact spacing. On double-sided boards, termination pads can be plated on both sides of the board, thus reducing the number of terminations per side and the length in half.

Once the approximate length of the board along the connector edge has been established, or board length has been predetermined on the basis of available space, the next step is to make a rough layout of the components. This will establish the approximate width of the board.

Commercially made templates, without adhesive backing and two or four times size, are available with outlines of such components as transistors, flat packs, DIPs, and so forth. They can be used to make an initial component layout on a clear, nonmatte finish drafting film as they readily adhere to its surface, Fig. 13-13(a), and yet can be readily moved.

An estimate of the board's overall size can be made by laying out all of

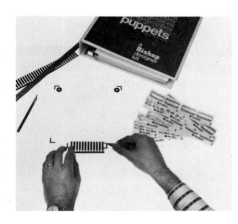

FIG. 13-13. Steps in Using Commercially Made Templates. (Courtesy Bishop Graphics, Inc.)

the component templates required for the printed-circuit board, as determined from the schematic diagram. Then, starting with the board's outline and delineation marks, proceed by laying down the component pattern, Fig. 13-13(b), and when the component layout is finished, add the circuit interconnections, Fig. 13-13(c), by laying a matte surface overlay on the component layout. The next step is to make solid connection lines for the component side of the board and dash lines for the circuit side, if the board is of the double-sided type.

Once the component layout and wiring appear to meet the circuit requirements, Fig. 13-13(d), the final artwork is ready for layout. Using a print made from the interconnection overlay and component layout in place, position over it the film with grid lines on which the final artwork is to be laid out. Component terminations can be shifted on grid line intersections if necessary to facilitate locating the interconnection holes on the drilling drawing layout. Add the interconnection tapes to the layout, blue on the component side and red on circuit side.

In laying out the conductor pattern, conductor crossover must be avoided if a one-sided board is desired. Several trial layouts of component arrangement may be necessary to achieve this, and some of the conductors may have to take a circuitous path to eliminate a crossover. Such a trial-and-error method becomes especially necessary with complex wiring patterns.

If it is impossible to avoid one or two crossovers without creating a complex circuit pattern, these crossovers should be treated as another component on top of the board and two termination areas provided for each crossover on the artwork pattern. A piece of insulated wire is soldered between these areas on the actual board.

If it becomes evident that a number of crossovers cannot be prevented, a double-sided board is required, and a rough diagram for each board side has to be drawn, with a final component layout made to locate the components accurately.

Single-sided boards also require a final board layout, with connection or mounting points located at the intersections of the grid lines whenever possible. Component layout has to be made concurrently with conductor pattern layout to allow sufficient space for conductors.

Once the board configuration has been established, the artwork pattern layout can be started. If the component termination and mounting points have been located at grid intersections, their locations can be readily determined on the artwork layout. The latter is generally made 4:1 size to reduce tolerance buildup.

Start the artwork pattern by taping the board outline, locate the registration marks in relation to it, and establish one scale dimension between the registration marks or other points as a guide for the photographer in securing exact reduction. The dimension should carry a tolerance as a further stress on accuracy. The various single and multiple termination pads should be located as the next step.

Double-sided boards can be laid out with either separate artwork patterns with registration marks accurately located to align the two patterns, or single pattern made on both sides of a polyester or Mylar®* film with special transparent blue and red tape used on both sides as explained previously.

After all of the termination pads have been located on the artwork, the next step is to apply the tape interconnections in the paths selected on the board wiring diagram. Tape .062 inch wide is suitable for conductors not carrying any power, or wider tape may be used if space permits.

Whenever voltages of any magnitude exist, spacing between the edges of pads or taped circuits should be checked as the artwork pattern layout proceeds. There is generally sufficient space between the two rows of DIP terminals for two or three conductor tapes.

The wiring pattern should be carefully checked against the schematic diagram once it has been established, following the procedure outlined in Chapter 16 for checking of wiring diagrams.

Component Mounting

Transistors, flat packs, and DIPs may be mounted by various methods, as shown in Fig. 13-14.

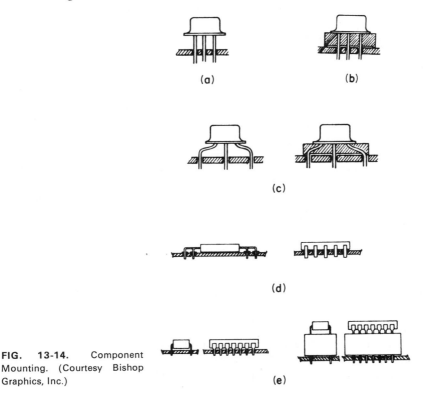

(a)

(b)

(c)

(d)

FIG. 13-14.   Component Mounting. (Courtesy Bishop Graphics, Inc.)

(e)

*® du Pont registered trademark.

Transistors may be mounted with the leads straight through the board as in Fig. 13-14(a) or with a spacer between the board and the transistor to eliminate possible shorting, Fig. 13-14(b). In (c), the transistor leads are offset, which is desirable when transistors with 6, 8, 10 or more leads are used. Some of these transistors come with preformed offset leads or the leads may be formed at assembly. Again, a spacer may be used to provide mechanical support for the transistor.

Flat packs may be surface mounted by forming and trimming the leads to size or by forming alternate leads in offset fashion and inserting them through the board, Fig. 13-14(d).

Dual-in-line packages (DIPs) may be mounted with the terminals going through the board, Fig. 13-14(e). A minimum hole diameter of .066 inch is required to clear the flat section of each terminal. They may also be plugged into DIP sockets mounted on the board, and their terminals soldered to the wiring pattern.

Component Lead Attachment

Component leads pass through holes drilled in the printed-circuit board and are attached by soldering as shown in Figs. 13-3 and 13-4.

Heat Sinks

Consideration should be given to the inclusion of heat sinks during the mechanical layout of the printed-circuit board. See Chapter 4 for additional cooling means of such semiconductor components as power transistors, diodes, and rectifiers. Additional board space will be required by heat sinks, depending upon their configuration, and thus disposition of components will be directly affected by the heat sinks. This makes it necessary to include the heat sinks in the initial design.

Some components may require forced cooling in addition, to maintain their safe operating temperature. This is especially true when many printed-circuit boards or cards are assembled together in a card cage with ventilation restricted by the assembly configuration.

Printed-Wiring Layout

Although layouts for printed-wiring boards vary considerably because of the different circumstances, certain steps are common in most layouts.

The schematic diagram of a transistorized preamplifier circuit for a photomultiplier tube shown in Fig. 13-15 is to be laid out as a printed-wiring-board assembly. For mechanical and electrical reasons the board has to be circular in shape, be of a definite maximum diameter, and have the photomultiplier tube socket located at the center of the board. The power input and amplifier output connections are to be RF connectors of the BNC connector type.

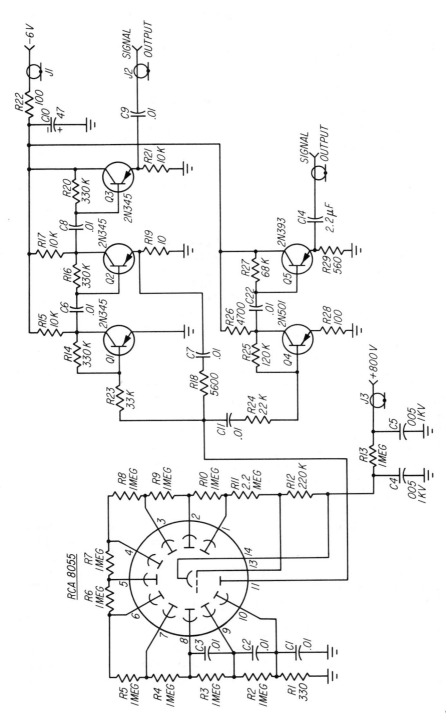

FIG. 13-15. Schematic Diagram of a Photomultiplier Amplifier.

417

The specified board diameter is $3\frac{1}{2}$ inches, single-sided type, $\frac{1}{8}$ inch thick, with 2-ounce copper bonded to glass-fabric base, epoxy resin laminate.

An examination of the schematic diagram shows that the following components have common connections:

R1-R2-C1-C2-V1/10	R16-C6-Q2/B
R2-R3-C2-C3-V1/9	R16-R17-C8-Q2/C
R3-R4-C3-V1/8	R20-C8-Q3/B
R4-R5-V1/7	R22-J1
R5-R6-V1/6	C9-J2
R6-R7-V1/5	R19-C7-Q2/E
R7-R9-V1/4	R21-C9-Q3/E
R8-R9-V1/3	R18-C7
R9-R10-V1/2	C11-R24
R10-R11-V1/1	R24-R25-Q4/B
R11-R12-V1/13	R25-R26-C12-Q4/C
R12-R13-C4-V1/14	R27-C12-Q5/B
R13-C5-J3	R29-C14-Q5/E
R23-R18-C11-V1/11	C14-J4
R23-R14-Q1/B	R1-C1-C4-C5-J3-Q1/E-R19-R21-J2-
R14-R15-C6-Q1/C	R28-R29-J4-J1-C10
R15-R17-R30-R22-C10-R26-R27-	(all common ground returns)
Q3/C-Q5/C	

A list such as this is useful for reference during the layout process. It may further be expanded by identifying such common connection points as high voltage, circuit paths to be shielded, and other specialized items by means of various colors on the schematic diagram.

The physical requirements mentioned earlier establish the placement of some of the components on the board. For example, resistors *R1* through *R12* and capacitors *C1* through *C4* must be located near the center of the board to have short connections to tube *V1* elements. A rough trial wiring diagram (Fig. 13-16) should be drawn freehand with the components in their approximate physical locations.

Because the majority of the components are located near the periphery of the board, the ground return also is located there so that it will provide grounding of the coaxial connectors without interfering with the wiring pattern. Because the component leads are to be hand-soldered, the wide grounding strip at the connector area does not present a possible blistering problem.

Such a preliminary diagram, drawn several times as large as its physical counterpart, may need to be altered several times to find a workable circuit pattern that does not have crossovers. The conductor paths run across the components, between their termination areas or lands. Several path layouts may have to be tried, and considerable judgment used, to achieve this goal.

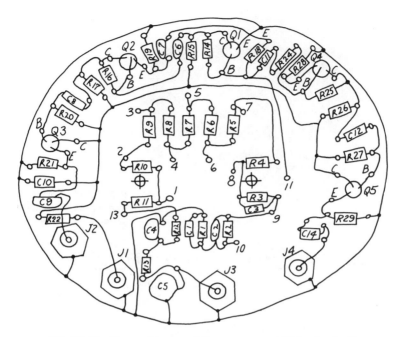

FIG. 13-16. Proposed Layout of Components and Wiring for the Circuit in Fig. 13-15.

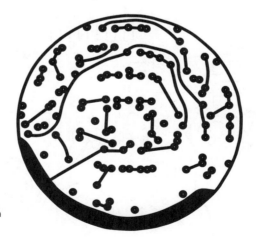

FIG. 13-17. Conductor Pattern Artwork.

The final arrangement of the wiring pattern is shown in Fig. 13-17, while Fig. 13-18 shows the opposite side with all of the components wired in.

The assembly drawing of the photomultiplier printed-wiring board is given in Fig. 13-19. It is similar to the conventional assembly drawing and includes all of the mechanical and electronic components in the material list.

FIG. 13-18. Components Assembled on the Printed Circuit Board.

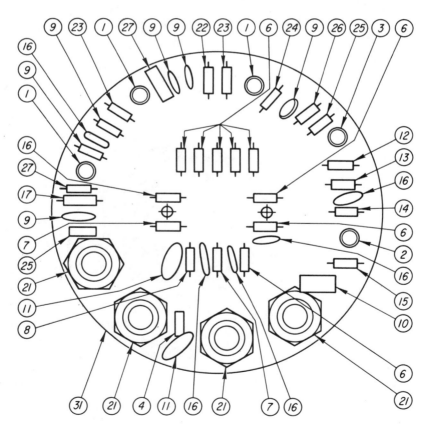

FIG. 13-19. Assembly Drawing of the Component Side of the Photo-multiplier Board.

13.7 PRINTED-WIRING-BOARD DRAWINGS

Drilling Drawing

This drawing type is made from a transparency of the artwork, a procedure that is both economical and accurate. The drawing must include the following information:

1. A reference to the artwork
2. Specifications of the material from which the printed board is fabricated
3. Directions for finishing the pattern, such as solder plating or baking
4. The size and shape of the finished board
5. The diameters of all holes with tolerances.

The grid intersection in the center of the termination areas or lands will generally locate hole centers with a tolerance of $\pm.005$ inch. The holes are identified on the drawing by letter designations at the pattern and a tabular listing of hole sizes.

Marking Drawing

This drawing lists the reference designations for each component, the external connection terminal designations, etc. It is made on plastic drawing film using "stik-on" or pressure-sensitive lettering similar to the tape and pads on the artwork pattern. It represents the view as seen from the component side of the board. Markings should be placed where they will not be covered by components. Leader lines may also be added to indicate the designated component (Fig. 13-20). The drawing should not show any holes, location

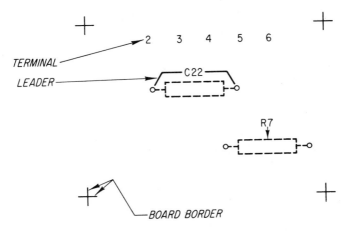

FIG. 13-20. Marking Drawing Details.

dimension lines, or other dimensioning because it is the master drawing for the screen process of marking identification.

All lettering, condensed or extra-condensed, should be in capitals that will be at least $\frac{3}{32}$ inch high when reduced. Components with definite polarity terminations that are mounted on the board, such as diodes, should have their polarities indicated on the marking drawing. The conventional negative, positive, or schematic symbols, such as for a diode, are used for this purpose.

Printed-Wiring-Board Assembly Drawing

This drawing is made on a standard drawing sheet and includes the following:

1. Hardware components assembled on the board, such as brackets, clamps, standoffs, etc.
2. The parts list including such items as printed-board artwork, drilled board drawing, and specifications for the marking process
3. A few typical markings to indicate orientation
4. Electronic components, such as transistors, capacitors, dual-in-line, etc., available as "stik-on" symbols, that save considerable drafting time, and are made in 1, 2, and 4 times component size.

13.8 MULTILAYER PRINTED-WIRING BOARDS

The problem of increasing density of intercircuit wiring, compounded by the decrease in the size of the circuit components down to the integrated circuit size, has necessitated a new approach to wiring. This development is the multilayer printed-wiring board, which interconnects the many components or printed-circuit cards. It may also serve as an enlarged printed-circuit card on which the integrated circuits (either TO-type packages or flat packs) are interconnected permanently. The interconnections between such assemblies are completed within the board, with only the desired circuit points for external connections to other parts of equipment brought out.

Construction

Multilayer boards actually consist of from two to as many as fourteen printed-circuit boards with a thin layer of what is known as "prepreg" material placed between each layer, thus making a "sandwich" assembly (Fig. 13-21). The printed circuit on the top board is similar to a conventional printed-circuit board assembly except that the components are placed much closer together or they may have many terminals, which necessitates the use of the additional board layers for the required interconnections.

This technique presents problems in interconnections between layers. Some

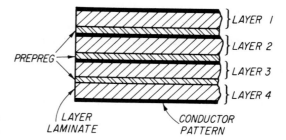

FIG. 13-21. Cross Section of a
Multilayer Printed Circuit Board.

methods used to avoid these problems are: plated-through holes, as done with conventional double-sided boards; metallic pins fused in place for interconnections between layers; and eyelets fused in place for the same purpose.

One of the internal board layers can be what is known as a ground plane, which is drilled out before assembly to clear the interconnections between the other layers. The ground plane provides capacitance to ground from the various board conductors and also acts as a ground shield.

Such a layer may also be a voltage plane or bus to provide required potential to the semiconductors or integrated circuits mounted on the top layer board.

After the individual board patterns have been fabricated from the artwork, the boards are assembled, with great accuracy taken in aligning the layers and layers of prepreg material inserted between the layers. The assembly is then placed in a heated press with a 50-ton capacity or higher. The heat and the pressure fill the voids around the conductor pattern on each layer and bond the layers into one solid assembly. The accuracy in locating each layer in relation to the other determines the width of the annular rings (conductor pads drilled out after forming the multilayer board) that are used to interconnect the various points in each layer. Layer misalignment, on the other hand, could result in short circuits within the multilayer board.

A number of multilayer boards may be laminated as one large assembly, and thus, the individual boards must be cut to their finished size. Small board quantities are routed or machined to size while larger board quantities are blanked to size in a die.

Board Design

When a complex schematic diagram indicates the need for a multilayer board, one of the first problems to solve is the wiring distribution between the layers and the number of layers that will be required. Since there are many possible wiring distribution patterns, it becomes a trial-and-error method of design requiring considerable time on the part of the draftsman to achieve the design with a minimum number of layers. To determine the simplest multilayer pattern design approach rapidly, computers have been utilized to save time.

A preliminary layout of the components is made according to the practices previously described for the single- and double-sided boards. A rough connection diagram is also made from a print made from the layout. This connection pattern should show all of the termination points numbered, which can then be transposed into a set of wiring tables listing the interconnections between the various points.

The connection pattern is then subdivided into the expected number of layers, and the pattern for each layer is developed in rough form according to these tables.

Some connection points on the individual sections of the multilayer board layout may have to be rerouted as the layout work nears completion. The pencilled wiring layout of each section is then transposed into the conventional taped artwork pattern for each layer. The important point is to achieve an almost perfect layer-to-layer registration so that the interconnecting holes for plated-through holes or pin connections will coincide in the finished board assembly. Hence, the importance of accurately locating each board layer relative to its registration marks on the artwork pattern cannot be overstressed. Also the registration marks on one layer must coincide with the marks on every layer. These marks must be positioned on each layer so they will match only when the various patterns are placed in their correct relationship.

Examples of multilayer boards are shown in Fig. 13-22.

Artwork

The artwork required for a multilayer board consists of an extremely accurate pattern layout for each layer. To obtain such accuracy, the artwork is made four times the finished conductor pattern size, with the registration or tool marks and board outlines located with extreme accuracy for each board layer, and the dimension to which the board artwork is to be reduced specified on the artwork for each layer.

One of the prime requisites in multilayer board design is the orientation and location of each board layer in respect to the other layers. To achieve this, tool holes, preferably three in number, should be located relative to the outer edges of the finished board (Fig. 13-23).

These holes should be located close to the edge of each layer outline on the artwork. The circuit pattern is located in relation to the three holes. In the above figure, note that these holes are dimensioned relative to each other rather than from the board edges. This eliminates the tolerance buildup that would otherwise contribute to inaccuracy in the relative placement of board layers.

In addition to the three tool holes, the artwork should also include two reduction targets with the exact reduced toleranced dimension between them serving as a photographer's guide in reducing the artwork to correct size (Fig. 13-24).

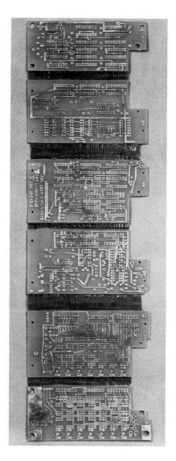

FIG. 13-22. Multilayer Circuit Boards

Figure 13-25 shows a completed multilayer board, while Fig. 13-26 illus-trates the artwork required for each layer. Note the three tool or registration marks on each artwork pattern. It is of utmost importance that these marks be located exactly the same distance apart on each layer of artwork and that the board outlines are located in an identical position in relation to these marks on each layer. Each layer of artwork should clearly show the part outline and the corner markers.

Conductor terminations (also known as pads or lands) are generally round in shape, and they should be at least .050 to .060 inch larger in diameter than the diameter of the hole drilled through them. Upon drilling, they become annular rings .025 to .030 inch wide with exact layer alignment, or narrower with misalignment.

The interconnections between the various layers of a multilayer board are

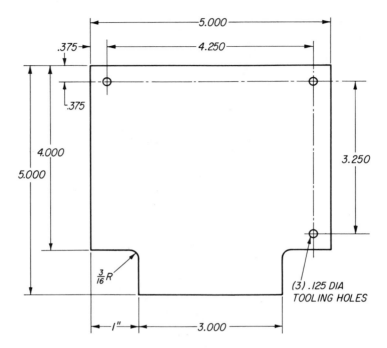

FIG. 13-23. Typical Tooling Hole Dimensioning.

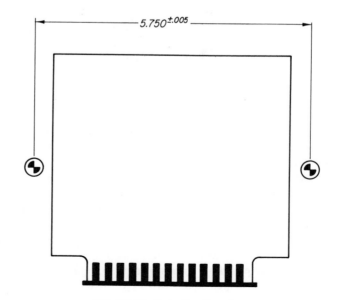

FIG. 13-24. Reduction Targets.

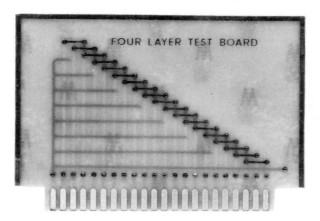

FIG. 13-25. Completed Multilayer Board.

made by plated-through holes as done with the double-sided boards (Fig. 13-27). It is necessary to have an accurately dimensioned drilling drawing to drill the various holes in the multilayer board.

Artwork Layout

In laying out multilayer artwork, several steps are necessary:

1. Determine the schematic diagram for the complete multilayer board assembly.

2. Make a rough layout on grid paper of the components that are to be mounted on the top layer of the board assembly and determine the approximate board size and shape. If internal corners are involved, they should be rounded rather than square to avoid the development of stress. Components can be located close together because the additional board layers provide space for the connections between the components. Place all the components at 90 degrees to each other rather than at odd angles so the component intersections occur at grid intersections.

3. Make a rough connection diagram (see Chapter 12) for the complete assembly.

4. Subdivide the connection diagram into smaller connection diagrams for each board layer.

5. Lay out the artwork for the first or top layer with its own dimensional pattern, tool marks, and reduction dimension.

6. Lay out the conductor patterns for the remaining conductor layers and any voltage and ground plane layers. Number the individual layers starting with the top layer.

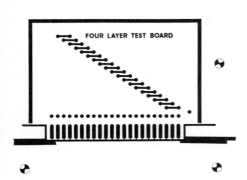

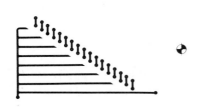

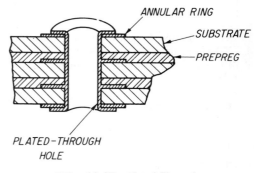

FIG. 13-27. Plated-Through
Hole in a Multilayer Board.

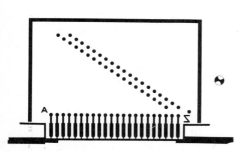

FIG. 13-26. Individual Layer
Artwork for Multilayer Board.

Multilayer Board Detail Drawings

A set of such drawings consists of the following:

1. An outline drawing of the finished multilayer board that shows all the necessary dimensions together with their tolerances; a cross-sectional view of the board should be included, with each layer numbered for identification purposes.

2. Conductor pattern artwork for each layer with conductor and laminate thicknesses specified; each conductor layer should be identified numerically.

3. Drilling detail drawing showing the location of all drilled holes, their sizes, and whether clearance or for plated-through holes; their location can be by the standard grid pattern, or by pad centers or dimensioning. If they are located by the grid pattern, the holes are dimensioned by the X-Y coordinates from the lower left corner of the board outline.

4. A marking drawing showing such markings as reference designations, nomenclature, part number of the board, part number of the board assembly, and so forth; these markings can be printed on the top board with special ink, or they can be etched simultaneously with the conductor pattern. In the latter case, they should be included on the artwork for the outside layer conductor pattern.

5. A ground plane drawing showing the drilling details for clearance holes, since they have to be drilled before the multilayer board assembly is laminated.

6. A voltage plane drawing showing drilling details because it also must be drilled prior to laminating.

Materials for Multilayer Boards

Epoxy glass, type GE or GF, is generally used for multilayer boards as are polyester film, polyimide film, and Teflon. The material and its type should be specified on the drawing.

The prepreg or "B" stage material is epoxy cloth, .0025 to .006 inch thick, commonly used between the board layers. The thickness used depends upon the thickness of copper on each board laminate and should be left to the discretion of the multilayer board manufacturer. Thus, only the completed board thickness should be specified.

Multilayer Board Tolerances

Due to the complexity of multilayer boards, their tolerances have to be greater than for the single- or double-sided printed circuit boards. Such tolerances must be indicated on the board detail drawings and closer tolerances may be obtained at an increased or premium cost.

For example, hole locations can be held to a standard tolerance of $\pm.002$ or to a premium tolerance of $\pm.001$ inch. Holes of less than $\frac{1}{4}$ inch in diameter can be drilled to a tolerance of $\pm.002$ inch. Plated-through hole tolerance depends upon the laminate thickness and the hole diameter. The standard tolerance for holes .037 to .060 inch in diameter in $\frac{1}{16}$ inch laminate is $\pm.003$ inch, or $\pm.002$ inch for premium tolerance. For thicker laminates, the tolerance is $\pm.005$ inch or $\pm.002$ inch for premium boards.

Conductor width tolerance varies with the conductor thickness. It is $\pm.001$ inch for $\frac{1}{2}$-ounce copper conductor, $\pm.003$ inch for 1-ounce copper, and $\pm.005$ inch for 2-ounce copper.

Pattern location to outside board edge tolerance is standard at $\pm.015$ inch or $\pm.010$ inch for premium boards.

Standard tolerance for layer to layer registration can be $\pm.010$ inch or $\pm.005$ inch for premium boards.

The thickness of multilayer boards with connector tabs integral with the board pattern must be specified to a nominal thickness with a $\pm.007$ inch tolerance to ensure adequate external connector contact.

Conductor Pattern Plating

In many instances, the conductor pattern is plated with an additional metal or alloy, such as tin-lead for easier assembly soldering, with gold on conductor terminations on the board, or with tin-nickel.

The minimum recommended plating thickness is from 80 millionths of an inch for gold to 300 millionths of an inch for tin-solder and tin-nickel.

Multilayer Board Substitute

In many instances the Multiwire®* wiring method (page 386) can substitute for multilayer boards. It provides a means to interwire many connection points by wiring that is composed of several insulated conductor layers as compared to the use of some boards in the multilayer assembly whose sole purpose is to provide the interconnections. The ease of changing the connections involved in circuit modifications is another factor in its favor.

13.9 FLEXIBLE PRINTED CABLES

The production of flexible printed cables is an adaptation of the printed circuit process. These cables are flat and consist of a number of conductors of a desired pattern that are enclosed between layers of insulating material with only the terminations at each end exposed.

The conductors can take any shape within close limits to fit the configuration of the connector terminals that connect to each end of the cable. The cable can be single layer or multilayer. It can be twisted, bent, or accordion-pleated if the cable has to contract or expand with the motion of the equip-

*® Registered trademark of Kollmorgen Corporation.

ment to which it is attached. Because the individual terminations of the cable are specifically positioned, they do not need to be color-coded for circuit tracing, which simplifies assembly and prevents possible connection errors.

Military Specification MIL-P-50884B, *Printed Wiring, Flexible, General Specification For*, covers the requirements of these cables for military use.

Base Materials

FEP Teflon, glass cloth, polyimide (Kapton H®*), Nomex®*, and Mylar®* are commonly used as insulating base materials or laminates. Cables range in thickness from .007 inch and up when the conductor is .001 inch. Additional special stiffening materials such as epoxy may be specified to provide rigidity to the laminate at each cable end or over the entire cable area.

Conductors

Rolled copper foil that weighs 1, 2, 3, 4, or more ounces per foot is used for the conductor. Two-ounce copper is the preferred thickness.

The required conductor thickness and width are determined by the current-carrying capacity and/or resistance requirements and the allowable temperature rise. A minimum conductor width of .020 inch or a conductor width ratio to thickness of ten or more is preferred. A minimum spacing of .015 inch should be allowed between adjacent conductors, with greater spacing allowed at exposed areas such as connector pins. This recommendation does not apply, however, to high-voltage applications.

Because the printed flexible cable has a large flat radiation area, the conductors are able to carry heavier current with the same cross-sectional area conductors than the conventional round wire cable. The other limiting factors are the resistance per foot or the voltage drop in each cable conductor with a given current, and allowable temperature rise.

Table 13-2 lists the various conductor thicknesses and widths, along with recommended maximum current for each conductor and the resultant voltage drop per foot at a temperature rise of ten degrees C. Current carrying capacity of flexible cable at higher rise temperatures is shown in Fig. 13-28. The indicated conductor current capacity should be reduced 10 to 25 percent if there is little opportunity for the cable to radiate its heat.

Connectors

These can be of the printed circuit, removable crimp, or solder cup types that are available commercially! Exposed drilled termination areas or pads are required at each end of the conductor to assure a dependable soldered connection at each termination point. Pad diameters should be at least three times the diameter of the hole that they surround.

*® du Pont registered trademarks.

Table 13-2   Teflon FEP Laminate-10°C Temperature Rise
(Courtesy Teledyne Electro-Mechanisms)

Conductor Thickness Inches	Conductor Width	Current Amps.	Voltage Drop Per Foot at Recommended Current
.00135	.020	1.2	.36
(1 oz.)	.025	1.4	.35
	.050	2.3	.29
	.075	2.9	.25
	.100	3.6	.22
.0027	.020	1.5	.23
(2 oz.)	.025	1.8	.23
	.050	2.8	.17
	.075	3.7	.15
	.100	4.6	.14
.004	.020	1.8	.18
(3 oz.)	.025	2.1	.17
	.050	3.3	.13
	.075	4.3	.12
	.100	5.4	.11
.0054	.030	2.6	.13
(4 oz.)	.050	3.7	.11
	.075	4.9	.10
	.100	6.2	.09
.0067	.035	3.2	.11
(5 oz.)	.050	4.0	.10
	.075	5.6	.10
	.100	7.0	.08

Cable Design

The first step in designing a flexible printed cable is a rough sketch of the conductor layout. Then, a model of the cable assembly is developed in paper, cardboard, or transparent plastic. This can be shaped to fit the components, and it finally evolves into a finished product as shown in Fig. 13-29.

Master Artwork

The procedures outlined for the artwork for printed-circuit boards and multilayer circuitry also apply to the artwork for printed cables. When selecting conductor tape width, allowance should be made for the undercutting that occurs during the etching process and amounts to at least .005 inch reduction in conductor width. Thus, a conductor tape width on a 4:1 artwork layout should be .020 inch wider than the required conductor width. A typical cable artwork is shown in Fig. 13-30.

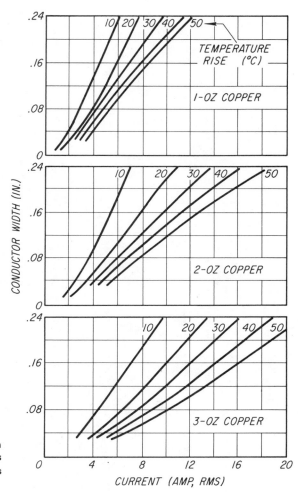

FIG. 13-28. Current-Carrying Capacity of Printed Conductors with Kel-F Board Insulation. (Sanders Associates, Inc., Flexprint Products Div.)

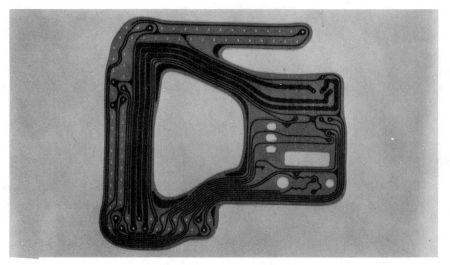

FIG. 13-29. (a) A Flexible Printed Cable.

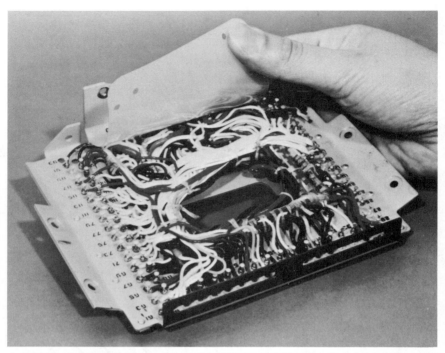

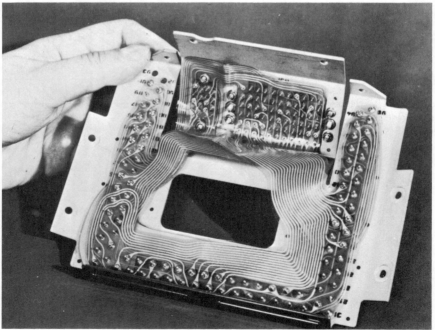

FIG. 13-29. (b) "Before" and "After" Conversion to Flexible Printed Cable.

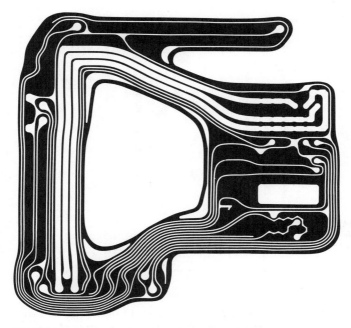

FIG. 13-30. Artwork for Flexible Printed Cable Shown in Fig. 13-29(a).

Termination Pads

These should be of the teardrop type with a 90-degree elbow or ear, see Fig. 13-31(a), and should be placed on top of a conventional teardrop pad as shown in (b). This provides better pad anchoring when the pads are bared for joining to connector terminals. Preferably, the pads should be bared on one side only. One end of the cable can be folded over if necessary to avoid reverse side baring. Pads with a center hole should be used to act as a guide for drilling.

Conductor Placement

The conductors and pads should be located so their edges are at least .060 inch away from the edge of the cable and .015 inch from the nearest clearance hole. On the 4:1 artwork layout these distances would be multiplied by a factor of four, Fig. 13-31(c).

Electrostatic Shielding

Several approaches are possible when such shielding is required. One is insertion of a grounded counductor, to separate conductors connected to a disturbing circuit from the conductors of a circuit to be protected. A further

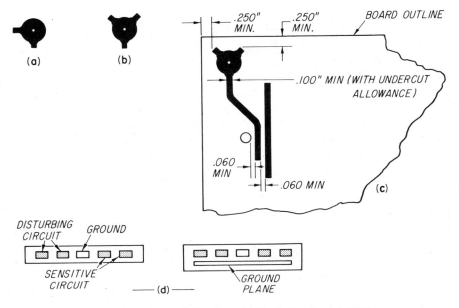

FIG. 13-31. Flexible Cable Artwork (4:1) Details. (a) 90-Degree Elbow. (b) Termination Pad with Anchors. (c) Conductor Details. (d) Conductor Shielding Methods.

improvement is obtained by including a solid ground plane and extending it across all the conductors, in addition to the grounded conductor, see Fig. 13-31(d).

Combining Cables

It is often better to substitute several smaller or narrower cables for one large or wide cable. The cables can be nested together, which results in a small increase in the overall cable thickness.

Cable Connections

Tubelets or eyelets are used to strengthen cable ends when the cable termination points have to be removed a number of times from the connector or other termination points. These eyelets have inner diameters ranging from .023 to .110 inch and corresponding external diameters of .046 to .218 inch.

Other cable terminations take the form of horizontal pins, silver-brazed in place, and projected from the end of the cable. These pins are encapsulated to form a connector that is assembled into a mating printed-circuit board.

Another form of cable termination are the vertical pins, .021 inch in diameter, silver-soldered in place at the end of the cable.

Both these termination types conserve space and reduce weight by eliminating external connectors.

Cable Tolerances

These should be as liberal as possible because, in normal applications, the flexibility and "give" in the cable are sufficient to overcome tolerance buildup. For example, a normal tolerance, hole to hole center, is $\pm.005$ inch.

Cable Forming

With cable thicknesses as low as .008 to .020 inch, it is possible to form cables into various required shapes or fold over their ends.

Other Details

A dimension should be given along the longest distance on the artwork to indicate the required reduction. The artwork is generally enlarged 4:1, as with other printed-circuit products.

If the cable is quite long and repetitive except at terminations, the artwork can be photographed in sections and the negatives spliced together. Lettering, numbers, or other markings for circuit and conductor identification can be added to the artwork.

Cables with a "Memory"

By utilizing cable insulation with a "memory," it is possible to make a multiconductor flexible cable that will expand to three or four times its compressed length (Fig. 13-32). As one end of the cable is moved, the accordion-type folds expand or contract automatically.

A "window-shade" or roll-up flexible cable (Fig. 13-33) has been developed

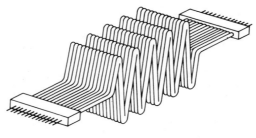

FIG. 13-32. Flexible, Accordion Fold-Type Printed Cable.

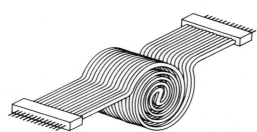

FIG. 13-33. Roll-up Power Cable.

with the same type of insulation. This cable type provides a compact cable package in applications where space is at a premium and automatic cable take-up is required.

SUMMARY

With the development of transistors and integrated circuits came a corresponding decrease in the interconnection wiring between such components.

Printed-circuit wiring has been developed using the etched and additive processes which make it possible to mass produce printed-wiring boards at a reasonable cost.

Because of the close proximity of the conductors and semiconductor components assembled on the boards, extreme accuracy and close tolerances are a requisite for manufacturing drawings. This is accomplished by making the drawings many times larger than the actual product.

Printed-circuit boards can be of the rigid or flexible types depending upon their use, and multilayer boards have become commonplace.

QUESTIONS

Sketches should be included with answers when necessary.

13.1   Describe a printed circuit, giving some of the major advantages and features.

13.2   What are some of the processes used to produce a printed circuit? Which are most commonly used?

13.3   Describe the additive process of printed-circuit-board manufacture.

13.4   What is copper-clad laminate? Describe the designation method for the conductor material.

13.5   Describe the following: (a) a plated-through hole; (b) dip soldering; (c) dip coating; (d) land; (e) an interface connection; (f) resist; (g) a radial fillet.

13.6   Sketch several of the more commonly used termination shapes on printed-circuit boards and identify each by name.

13.7   List some of the materials used for printed-circuit boards.

13.8   Describe some of the factors governing the conductor widths and spacings on a printed-circuit board.

13.9   List some of the important electrical and mechanical factors that should be considered in a printed-circuit-board layout.

13.10 What determines the distance between component mounting holes? Illustrate by sketching a typical pigtail lead-mounted component.

13.11 What is the grid-line system?

13.12 Describe: (a) single pattern; (b) double pattern; (c) edge distance; (d) artwork pattern.

13.13 Explain: (a) adhesive aids; (b) trim outline; (c) register marks; (d) board outline.

13.14 Outline the various steps to be followed in typical printed-circuit layout.

13.15 What identification information appears on the artwork for a printed circuit?

13.16 What is a drilling drawing? List the information that should appear on it.

13.17 Describe a typical marking drawing, and itemize the information that shouʌɑ be included.

13.18 Give a brief description of the construction of a multilayer board.

13.19 What is a voltage plane?

13.20 How is the accuracy of a multilayer board artwork achieved?

13.21 What is a tool hole?

13.22 What method is employed to obtain exact artwork reduction?

13.23 Describe briefly the steps in artwork layout of multilayer boards.

13.24 List some of the drawings required for multilayer board details.

13.25 What is a prepreg?

13.26 List three multilayer board tolerances.

13.27 What materials are used in plating multilayer board patterns?

13.28 What is the dual purpose of a ground plane in multilayer printed-circuit boards?

13.29 What is a flexible printed cable? Discuss its construction and advantages.

13.30 On the basis of the current capacity of a flexible printed cable as shown in Fig. 13-28, what minimum conductor width would be required for the following conditions: (a) 2 amperes, 2-ounce copper, 20°C rise; (b) 8 amperes, 2-ounce copper, 40°C rise; (c) 6 amperes, 1-ounce copper, 40°C rise?

13.31 How is a flexible cable layout developed?

13.32 How is the artwork developed for long, flexible printed cables?

13.33 Make a list of the drawings that may be required for a typical circuit-board assembly.

13.34 What would be the minimum spacing between two printed-circuit conductors connected to the following: (a) 50 volts dc; (b) 150 volts ac (RMS); (c) 250 volts dc; (d) a peak of 500 volts ac.

EXERCISES

Develop freehand layouts for the following exercises with exception of Exercises 13.4 and 13.5, on cross-section paper with four divisions per inch. Refer to parts catalogs when necessary for details of sockets, connectors, switches, and other components.

A fine-pointed 4H pencil or the special types of drawing pencils used with plastic drawing film should be used for outlines of circuit shapes. The conductor pattern should be uniform in width and drawn with smooth curves. If India ink is used for the pattern, the pattern should be filled smooth without pin holes or gray areas. Different drawing media may be used—vellum, bristol board, or polyester film.

13.1 A plug-in card, shown in Fig. 13-34, is used for mounting and to bring out all connections of four flat packs to external circuitry. The plug connections on the

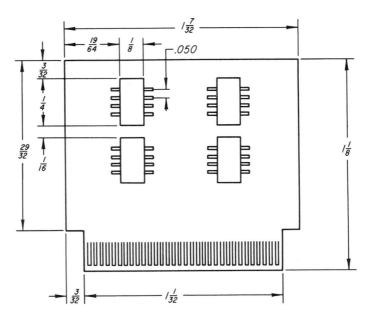

FIG. 13-34. Plug-in Card for Flat Packs.

component side of the card extend $\frac{1}{8}$ inch and terminate $\frac{1}{32}$ inch from the bottom edge. They are spaced on .030-inch centers. The conductor width is $\frac{1}{64}$ inch throughout. Locate the flat packs in the positions shown and develop the conductor layout to within $\frac{1}{16}$ inch of the flat pack outline. Use a scale of 10: 1 and make a master layout in ink on vellum or polyester film, accurate to within $\frac{1}{32}$ inch. Include the card outline, the $1\frac{1}{8}$-inch dimension as reference, and the material. Show the board outline and outline the corners with wide lines or tape. Add three registration marks and one overall reference dimension.

13.2   Repeat the layout in Exercise 13.1 using adhesive drafting aids shown in Figs. 13-6 and 13-7 for the conductor pattern.

13.3   Make an assembly drawing of the card in Exercise 13.1. List the four flat packs to be welded to the circuit pattern. Show overall dimensions as in Fig. 13-34. Card material is $\frac{3}{64}$-inch glass-epoxy laminate, with 1-ounce copper foil.

13.4   Make a full-size drawing of the board shown in Fig. 13-23. Convert all dimensions to metric, and use a metric-size drawing sheet. Specify the metric size drill nearest to that shown in this figure. Consult the tables in the Appendix to make the conversion and add suitable tolerances.

13.5   Make a double-size assembly drawing of the board shown in Fig. 13-34. Convert all dimensions to metric and use a metric-size drawing sheet. Consult the conversion tables in the Appendix.

13.6   Complete the conductor pattern in Fig. 13-35 by drawing it on $\frac{1}{8}$-inch grid paper and by scaling the figure. The radii are represented by crosses, the centers of lands by circles, and the strips by parallel lines. Pad diameters are as follows:

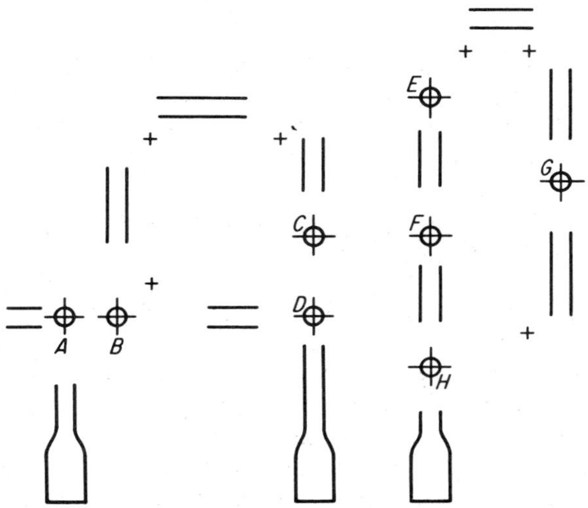

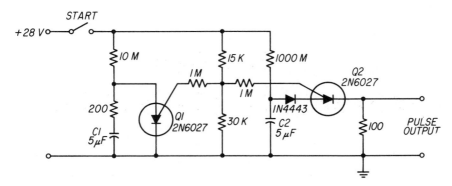

FIG. 13-35. Conductor Shapes to be Filled In.

A, $\frac{3}{8}$ inch; B, $\frac{5}{16}$ inch; C, $\frac{1}{2}$ inch; D, $\frac{3}{8}$ inch; E, $\frac{5}{16}$ inch; F, $\frac{3}{8}$ inch; G, $\frac{1}{2}$ inch; H, $\frac{3}{8}$ inch. Use $\frac{1}{16}$- or $\frac{1}{8}$-inch fillets as appropriate and radii of $\frac{1}{4}$ inch.

13.7 The long delay timer, shown in Fig. 13-36, is to be laid out as a printed-circuit plug-in card. Its size should be kept to a minimum, and its shape should be rectangular. External connections are to be located at one end, with a notch added to prevent incorrect card insertion. The resistors are to be $\frac{1}{2}$-watt composition type, and the capacitors are to be solid tantalum, rated at 50 volts WVDC. Conductor width is to be $\frac{1}{32}$ inch, with a minimum separation of $\frac{1}{32}$ inch between conductors and the same distance from the edge of the card. Terminal pads are to be $\frac{1}{8}$ inch in diameter, located on .100-inch grid, and have center holes .040 inch in diameter. Use foil dimensions of .100 by $\frac{1}{4}$ inch for terminal pads. Draw the pattern in pencil with smooth curves and with fillets

FIG. 13-36. Transistor Circuit. (Courtesy Semiconductor Department, General Electric Company.)

at the terminal pads. Develop the circuit pattern layout 4:1 to correspond with the schematic diagram and include a reference dimension. Use pencil on vellum for the original layout and transfer it to drafting film on which the pattern should be done in India ink. As an alternate method, the layout can be done with adhesive aids in the form of precut shapes.

13.8   Make an assembly drawing of the plug-in card in Exercise 13.7 on a *B*-size sheet and list all of the components in the parts list.

13.9   Make a 4:1 layout of a connection end of a flexible cable assembly. The connector is MS3100A-20-9S with 8 contacts. The end of the cable is to fit over the contact ends and is to be soldered. The conductor widths are to be $\frac{1}{16}$ inch, with $\frac{1}{32}$-inch separation. Land diameters at the connector end are to be $\frac{3}{16}$ inch. Make a pencil layout on vellum and show the cable extending 4 inches from the center of the connector. Prepare another layout on plastic film over the pencil layout, and lay out the circuit pattern with self-adhesive tape and shapes. Provide a reference dimension for photographic reproduction.

13.10   Figure 13-37 is a schematic diagram of a voltage regulator circuit. Its component board details have been given in Exercise 8.11. Make a 4:1 layout for a printed-circuit board using the same dimensions and locate the components for simplest intercomponent wiring without crossovers. Make a finished artwork drawing on plastic film, using India ink or adhesive aids for circuit wiring paths and lands. Holes for all component leads, with the exception of the variable resistor, are to be .031 inch in diameter. Holes for the variable resistor are .055 inch in diameter, for seven terminal stud terminals .062 inch in diameter, and for four mounting

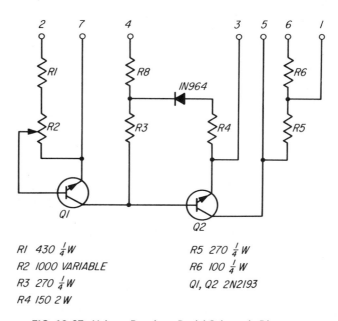

RI 430 $\frac{1}{4}$ W   R5 270 $\frac{1}{4}$ W

R2 1000 VARIABLE   R6 100 $\frac{1}{4}$ W

R3 270 $\frac{1}{4}$ W   QI, Q2 2N2193

R4 150 2 W

FIG. 13-37. Voltage Regulator Partial Schematic Diagram.

screws $\frac{5}{32}$ inch in diameter. Use suitable notes as required. Provide one reference dimension and specify 1-ounce copper foil.

13.11 Another double-sided board example is shown in Fig. 13-38. This is a decoder board on which three integrated circuits are mounted. The board measures $4\frac{1}{2}$ inches long by $2\frac{3}{16}$ inches maximum width. Lay out both sides of this board, scaling the figure to obtain other dimensions. Use conductor tapes, teardrop shapes, and 14-terminal ICs for terminal positions. Select the corresponding connector for this board and list it on the drawing. Identify the connector and IC terminals. Dimension the board outline and add three registration marks on the outside of the board outline.

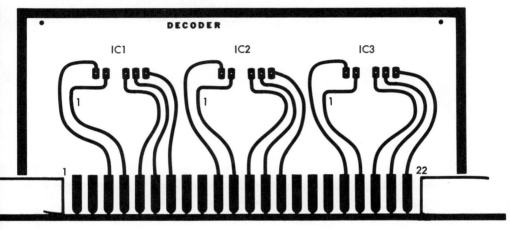

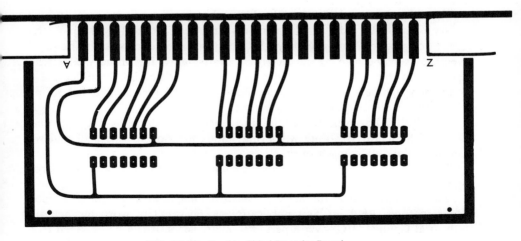

FIG. 13-38. Double-Sided Decoder Board.

14

INDUSTRIAL ELECTRONIC
DIAGRAMS

Industrial electronic diagrams are considerably different in many respects from the diagrams for electronic communication equipment.

One of the major differences is that the principal component types are in the control category such as contactors, solenoids, actuators, push buttons, timers, and similar items. In addition, the component parts are generally heavier and bulkier and may feature mechanical linkages, cams, and time controls, and, to a lesser extent, such purely electronic components as transistors. Finally, the circuits are mainly electrical rather than electronic.

As a result, the older graphic symbols have been retained by the industrial control companies. Unfortunately, this situation is sometimes confusing to the eventual user and also to the electronic draftsman if his primary experience has been in the field of communication electronics. Besides the difference in component symbols, there also are differences in the general diagram format and in the reference designation practices.

Eventually, however, the continuing work being done by the American National Standards Institute will bring standardization to this field and will lead to a more uniform set of practices.

Some of the current standards in the industrial control field and sponsoring organizations are:

Electronic Standards for Mass Production Equipment and General Purpose Machine Tools, EL-1-71. This publication is the result of work by the Joint Industrial Council (JIC) over a period of many years. It can be obtained from the National Machine Tool Builders' Association (NMTBA).

444

General Standards for Industrial Control and Systems, 1CS-1-1978. This is a publication of National Electrical Manufacturers Association (NEMA).

Electrical and Construction Standards for Numerical Machine Control, RS-281B. This is a publication of Electronic Industries Association (EIA), and has been adopted by the Department of Defense.

14.1 DIAGRAM DEFINITIONS

There are two basic diagram types used in industrial control circuitry—the elementary, which is the equivalent of the schematic diagram in communication electronic equipment, and the connection, which is the equivalent of the wiring diagram in the communication electronic equipment. These diagram names are used interchangeably throughout this chapter.

The elementary diagram has three basic parts: the graphic symbols of the components, which are arranged to suit the circuit; device designations, which distinguish similar components from each other; and connection lines, which interconnect the graphic symbols and are usually drawn in a progressive or "ladder-like" manner.

The connection (or wiring) diagram also consists of these same basic parts. The components, which are represented by geometrical figures, graphic symbols, or basic physical outlines, are shown in their general relationship within the equipment. The device designations used to identify the component symbols are the same as on the elementary or schematic diagram. The connection lines represent every individual connection wire in the equipment.

The so-called "lineless" diagram is a somewhat simplified type of a connection diagram. The lines representing the connection wires are replaced by a tabulation that indicates the individual connections between the components.

Another similar diagram type employs the "destination wiring" technique instead of a tabulation. In this diagram type, which was described on page 385, the destination point for each terminal connection is indicated at the terminal.

The interconnection diagram discussed in Chapter 12 may also be considered as a special type of a connection diagram. It shows external connections between the equipment and such associated components as controls, power input, load connections, and so on.

The abbreviations used in industrial control diagrams are the same as in communication electronic equipment diagrams and were listed in Chapter 3.

14.2 ELEMENTARY DIAGRAM PRACTICES

Many industrial applications involve timing processes for automatic sequencing operations. These processes are used for controlling motor speed, electrical welding, positioning of machine tools, as well as controlling temperature, flow, pressure, and other functions.

Since time, sequence, and motion may all be involved, the elementary diagram must be laid out according to these requirements.

Most of the circuits in industrial electronics are electrical, and consequently many of the component graphic symbols and reference designations appearing in them will not be found in electronic communication diagrams.

Graphic Symbols

The graphic symbols for industrial control circuits that appear in the NMTBA standard, Appendix A, are shown in Fig. 14-1, together with basic device designations above each symbol. This standard specifies that other symbols should be selected from ANSI Standard Y32.2-1975, *Graphic Symbols for Electrical and Electronics Diagrams*. The variation in symbols can be seen by comparing Fig. 14-1 and the corresponding symbols in Chapter 9.

Some of these differences are: the use of the alternate rectangle symbol for the resistor symbol instead of the zigzag; the use of two parallel lines for the contact symbol, with a diagonal line added for closed contacts; and the use of a circle for the alternate relay coil symbol instead of the rectangle.

Device Designations

The device designations used in industrial diagrams, comparable to reference designations in electronic diagrams, are a combination of consecutive numbers and assigned letters. The numbers, which differentiate between functions or components of the same class, *precede* the letter designations, as in 3CR, 2FU, etc.

As many as four capital letters may be used for the letter portion of the device designation. They are assigned on the basis of function or component type (Table 14-1). They directly follow the numerical designation without a hyphen or space and are placed on the same level.

A suffix letter is used to identify component subdivisions. For example, 3CAP-A, 3CAP-B, and 3CAP-C identify the three sections of capacitor 3CAP.

Device designations and component values are placed next to their graphic symbols as is done in electronics communication diagrams.

DC Motor Control Drive

The diagrams in Figs. 14-2 and 14-3 show a circuit of a fractional-horse-power dc motor drive. A potentiometer *1POT* (*R5*) provides continuous speed adjustment.

Figure 14-2 is drawn according to the schematic practices for communication electronic equipment while Fig. 14-3 illustrates the same circuit drawn according to industrial diagram practices. Note the differences.

During half of the voltage cycle, the motor shunt field is supplied from the ac line through part of rectifier *1REC*. During the other half of the cycle,

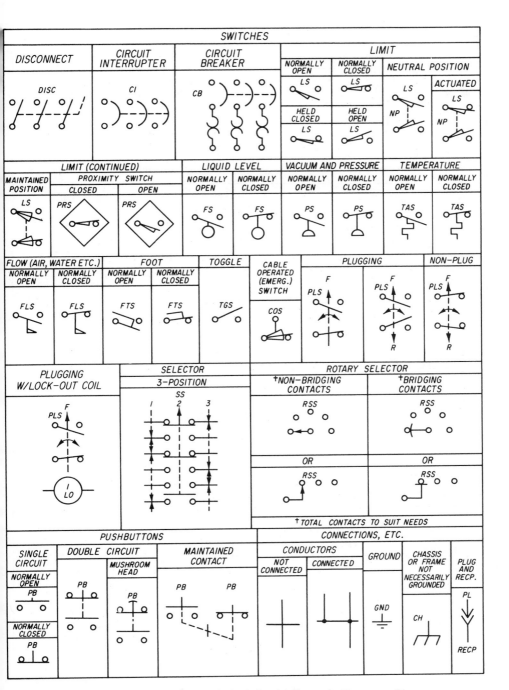

FIG. 14-1. Graphic Symbols for Industrial Electronic Diagrams with Basic Device Designations Above Each Symbol.

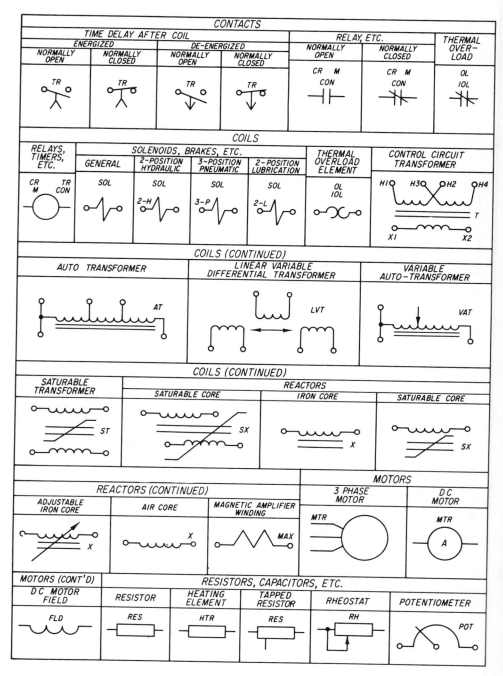

FIG. 14-1. (Continued)

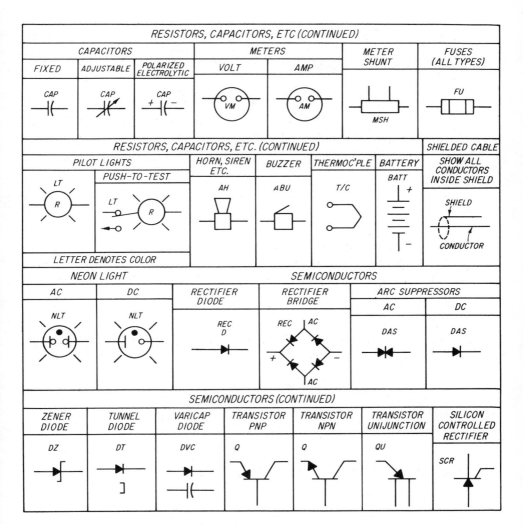

FIG. 14-1. (Continued)

the other part of rectifier *1REC* provides a path for the current that results from the collapse of flux lines in the motor field.

Resistor *4RES*, capacitor *2CAP*, resistor *5RES*, and capacitors *3CAP* and *4CAP* provide a 90-degree phase-shifted voltage between the grid and the center of filament of thyratron tube *1ET*. A dc reference voltage is obtained from potentiometer *1POT* which is connected in series with resistor *2RES* and rectifier *2REC*. The negative bias voltage below bus *1* obtained by resistor *1RES* makes it possible to adjust the motor speed down to zero. A combination of three factors controls the firing of the thyratron: (1) The voltage lags 90 degrees behind the voltage of the thyratron plate; (2) voltage lags behind the

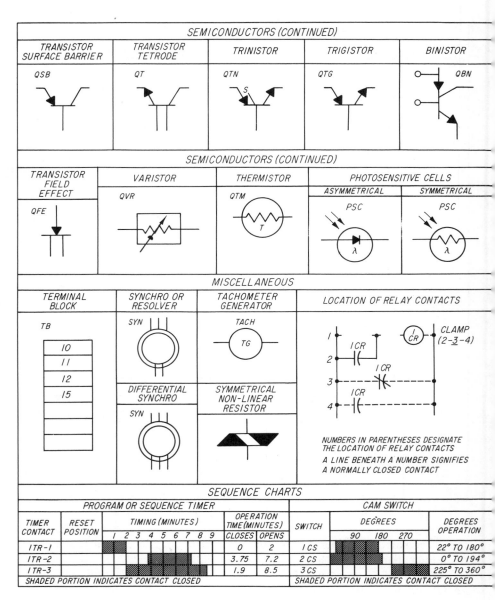

SEMICONDUCTORS (CONTINUED)				
TRANSISTOR SURFACE BARRIER	TRANSISTOR TETRODE	TRINISTOR	TRIGISTOR	BINISTOR
QSB	QT	QTN	QTG	QBN

SEMICONDUCTORS (CONTINUED)				
TRANSISTOR FIELD EFFECT	VARISTOR	THERMISTOR	PHOTOSENSITIVE CELLS	
			ASYMMETRICAL	SYMMETRICAL
QFE	QVR	QTM	PSC	PSC

MISCELLANEOUS			
TERMINAL BLOCK	SYNCHRO OR RESOLVER	TACHOMETER GENERATOR	LOCATION OF RELAY CONTACTS
TB 10 11 12 15	SYN	TACH / TG	CLAMP (2-3-4)
	DIFFERENTIAL SYNCHRO SYN	SYMMETRICAL NON-LINEAR RESISTOR	NUMBERS IN PARENTHESES DESIGNATE THE LOCATION OF RELAY CONTACTS / A LINE BENEATH A NUMBER SIGNIFIES A NORMALLY CLOSED CONTACT

SEQUENCE CHARTS

PROGRAM OR SEQUENCE TIMER						CAM SWITCH				
TIMER CONTACT	RESET POSITION	TIMING (MINUTES) 1 2 3 4 5 6 7 8 9		OPERATION TIME (MINUTES) CLOSES	OPENS	SWITCH	DEGREES 90 180 270			DEGREES OPERATION
1TR-1				0	2	1 CS				22° TO 180°
1TR-2				3.75	7.2	2 CS				0° TO 194°
1TR-3				1.9	8.5	3 CS				225° TO 360°
SHADED PORTION INDICATES CONTACT CLOSED						SHADED PORTION INDICATES CONTACT CLOSED				

FIG. 14-1. (Continued)

adjustable dc reference voltage on the grid of the thyratron; and (3) the armature of the motor develops back emf voltage.

Motor control is initiated by pressing the "*start*" button, which closes the main contactor coil (circled *1M*) circuit. To stop the motor, the "*stop*" button

Table 14-1 Device Designation Letters

AM	Ammeter	PC	Printed circuit
AT	Autotransformer	PL	Plug
CAP	Capacitor	POT	Potentiometer
CB	Circuit breaker	PS	Pressure switch
CH	Chassis or frame (not necessarily grounded)	Q	Transistor
		QBN	Binistor
CON	Contactor	QFE	Transistor, field-effect
CR	Control relay	QSB	Transistor, surface-barrier
CRE	Control relay, electronically energized	QT	Transistor, tetrode
		QTG	Trigistor
CRM	Control relay, master	QTM	Thermistor
CT	Current transformer	QTN	Trinistor
D	Diode	QU	Transistor, unijunction
DISC	Disconnect switch	QVR	Varistor
DT	Tunnel diode	REC	Rectifier
DVC	Varicap diode	RES	Resistor
DZ	Zener diode	RH	Rheostat
FB	Fuse block	S	Switch
FTS	Foot switch	SCR	Silicon-controlled rectifier
FU	Fuse	SOC	Socket
GND	Ground	SOL	Solenoid
HTR	Heating element	SYN	Synchro or resolver
IL	Pilot	T	Transformer
LS	Limit switch	TB	Terminal block
M	Motor starter	T/C	Thermocouple
MB	Magnetic brake	VM	Voltmeter
MTR	Motor	WM	Wattmeter
PB	Pushbutton	X	Reactor

is pressed which opens the main contactor coil circuit. This in turn opens the "*1M*" contacts in series with the motor armature and, thus, removes the power from the armature circuit. Contacts "*1M*" in series with resistor *6RES* then close, connecting the resistor across the motor armature. The coasting armature, acting as a generator, brakes itself to a stop.

Control Diagram Practices

A comparison of Figs. 14-2 and 14-3 points up several basic differences:

1. In the electronic communication schematic diagram, the circuit starts at the left or the input end, and extends across the diagram to the right, or the output end. In the elementary diagram, input is at the top, and the circuit progresses downward between two vertical feed lines.

2. In the electronic communication diagram, the relay or contactor symbol *K1* has all of its component parts together, while in the elementary diagram, the symbol for the main contactor "*1M*" is subdivided throughout the diagram. A circled "*1M*" represents the coil. The normally open and closed contacts are indicated by the same capital letter and located for circuit convenience.

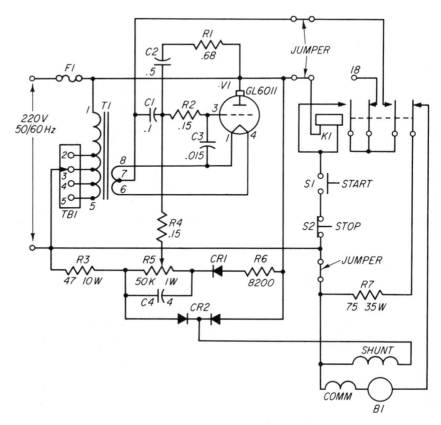

FIG. 14-2. Schematic Diagram of a DC Motor Drive.

3. The elementary diagram is subdivided into several ladder-like circuit levels between the vertical feed lines, following a definite sequence. The control circuits, such as the *START* and *STOP* push buttons, and the primary of the input transformer (*1T*), if one is used, are placed at the top of the ladder. The other transformer windings are located for circuit convenience (see *SIT*).

The circuits are arranged at each successive level in the order in which they are energized. These circuit levels are comparable to the route diagrams described in Chapter 12, with each route diagram starting at one vertical feed line and extending across to the other.

4. The left-hand vertical feed or supply line is often considered to be the "high, live, or positive" bus. Coils of contactors, relays, solenoids, and other control mechanisms are usually connected to the right-hand or "negative, neutral, or low" potential bus. The contacts of relays, contactors, and switches, or other circuit opening or closing devices are connected to the left-hand or positive bus as shown in Fig. 14-4.

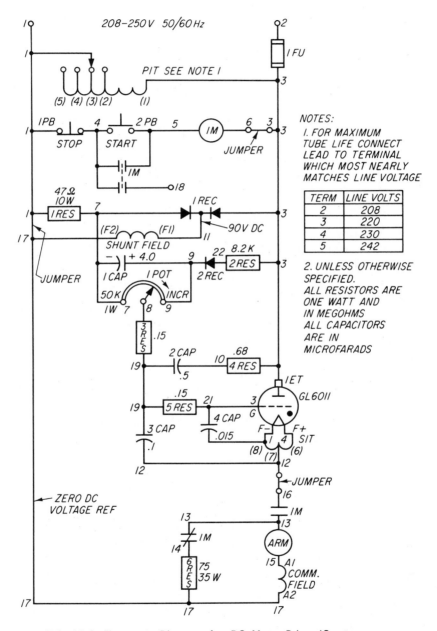

FIG. 14-3. Elementary Diagram of a DC Motor Drive. (Courtesy General Electric Company.)

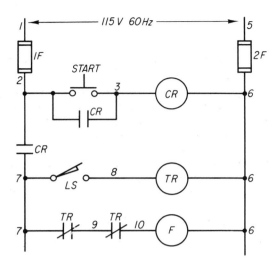

FIG. 14-4. Typical Elementary Diagram Practices.

5. It is common practice to assign numbers to the connection lines until they are interrupted by a symbol as illustrated in Figs. 14-3 and 14-4. The same number carries throughout the length of each vertical feeder (see *1, 3, 12, 16* in Fig. 14-3) until the feeder line is terminated in a transistor or some other symbol or interrupted.

6. It is also common practice to include reference designations within the symbols for resistors, relay and contactor coils, and any others that have sufficient area. See *1RES, 1M*, etc. Vertical symbol designations read from the right of the drawing.

7. Another practice is to put parentheses around the transformer termination numbers and the other component terminals to distinguish them from the numbers used to identify the wires between components.

8. General drawing notes may include such information as:

RESISTORS ARE 1 WATT UNLESS OTHERWISE SPECIFIED.
NUMBERS IN PARENTHESES ARE TERMINAL OR LEAD NUMBERS ON COMPONENT.

9. The diagram itself may include such local notes or information as the voltages at various points, references to general notes, the ratings of fuses and contactors, the identification of component parts, and the functions of various components.

A reference parts list listing the details of the various components and their identification by symbol and drawing number may also be included.

Diagram Layout

The layout of an industrial elementary diagram is much simpler than an electronic communication schematic diagram, which is often quite complex, and in most cases, the diagram area will be quite small by comparison.

However, the following points should be established before making the actual drawing:

1. The sequence of operations and thus the number of steps in the circuit ladder must be established.

2. A rough layout should be prepared on grid paper, allowing sufficient space for symbols, device designations, lettering, local notes, and component identification.

3. The horizontal level line with the most symbols will determine the spacing between the vertical feed lines.

4. The length of vertical feed lines will be governed by the number of horizontal levels or circuits, symbol sizes, and lettering. In long, complex elementary diagrams, it may be necessary to split the circuit at the bottom of the drawing sheet and to continue it on the right.

14.3 CONNECTION DIAGRAM PRACTICES

The details concerning views, terminations, drawing notes, and wiring presented in Chapter 12 also apply to connection diagrams of industrial electronic equipment.

Component Representation

In most cases, components are represented by their physical outlines and drawn proportionately to size, or they may be shown by graphic symbols. Another method is the geometric shape presentation, which uses either a circle or a rectangle depending on the physical appearance of the component on its wiring side.

Device Designations

These designations correspond to the elementary diagram designations and are stamped or silk-screened on the equipment next to the component.

The practices outlined in Chapter 12 regarding terminal identification, pigtail mounted components, component ratings, wire identification, and other data also apply to industrial electronic diagrams.

Connection Lines

In the industrial control circuits, the connection lines on the elementary diagram (Fig. 14-3) are numbered for both identification purposes and also for connection diagram use. This numbering generally starts at the top of the diagram and proceeds across sequentially, left to right and down.

When the circuit path passes through a component symbol, the number changes (see *4, 5, 6* in Fig. 14-3).

Connection lines are drawn straight, either horizontally or vertically, and along the shortest path on point-to-point diagrams.

Conductors

Usually, the conductors must be identified by the same numbers used in the elementary diagram and should be color-coded for wiring identification.

As a rule, this color identification follows the JIC *Electronic Standards for General Purpose Machine Tools*, EL-1-71, for circuit identification given in Table 14-2.

Table 14-2   Circuit Identification Color Code

Circuit	*Color*	*Abbreviation*
Line, load, and control circuit at line voltage, AC or DC	Black	BK
AC control circuit	Red	R
DC control circuit	Blue	BL
Interlock control circuit on panel energized wired from external source	Yellow	Y
Equipment grounding conductor	Green	G
Grounded circuit conductor	White	W

In most cases, complex equipment or panel assemblies are wired with cable or harness assemblies that have their conductors suitably color-coded, and separated from the remainder of the wiring on the drawing.

Diagram Notes on Sequencing

As mentioned previously, industrial electronic control equipment may involve the operation of relays, contactors, limit switches, push buttons, solenoids, time relays, and other controls. The sequential operation of these control elements may be listed by either a drawing note or a tabulation on the diagram, beginning with the first operational step and progressing through the operation of each control component.

Point-to-Point Connection Diagrams

This diagram type is used for simpler industrial control equipments. The conductors, represented by the connection lines, are all shown, although not necessarily in their exact location in the equipment. The details of this diagram type, given in Chapter 12, also apply to industrial connection diagrams.

Cable or Highway Connection Diagrams

Complex equipment with numerous connections is best represented by a cable or highway connection diagram. This diagram is similar to the point-to-point diagram except that the connection lines are merged into cables or highways (see Fig. 14-5). The direction of each conductor run is indicated by slanting the end of the connection line near its junction with the cable line.

Each connection line in Fig. 14-5 has a coded designation, either a letter, a number, or both, to facilitate locating its termination points. The coded

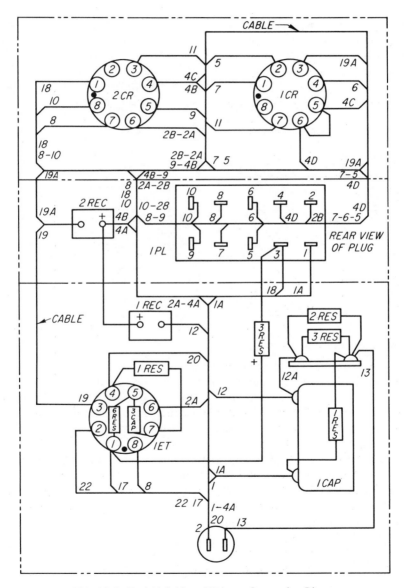

FIG. 14-5. Typical Cable or Highway Connection Diagram.

designations are repeated at branch cable junctions to indicate the direction in which the conductors extend along the cable (see *2B-2A*).

Considerable diagram space is saved by merging the connection lines into a single-line cable with no noticeable decrease in diagram readability.

Lineless or Tabular Connection Diagrams

In this diagram type the component symbols are identified by device designations and connection line termination points by coded designations, which consist of numbers, letters, or both. Although the connections are omitted on the diagram, they are listed individually in tabular form in the "*FROM*" and "*TO*" columns (Fig. 14-6). To provide full wiring information, such additional information as wire size, color, etc. is included in the tabulation.

TRANSFORMER CONNECTIONS

LEAD NO.	CKT NO.	CONNECT FROM	TO	VIA CUTOUT
1	3	(1)	1FU	A
2	1	(2)	1TB-2	B
3	1	(3)	1TB-3	A B
4	1	(4)	1TB-4	B
5	1	(5)	1TB-5	A B
6	F+	(6)	1ET	A
7	12	(7)	PCB	A
8	F-	(8)	1ET	A

WIRE TABLE

CKT NO.	WIRE SIZE	CONNECTIONS FROM	TO
2	14	2TB-12	2TB-16
1-2	14	PCB *#1*	PCB *#17*
1-4	18	PCB *#1*	1TB (CUT. "A")
6-3	18	PCB *#3*	PCB *#6*
3-3	14	1FU	ANODE 1ET
3-4	14	1FU	PCB *#3*
2-2	14	1FU	PCB *#2*
11-2	18	1ET	PCB *#11*
14-2	14	6RES	PCB *#14*
17	14	6RES	PCB *#17*

FIG. 14-6. Tabulation for an Industrial Tabular Diagram. (Courtesy General Electric Company.)

Interconnection Diagrams

This diagram type, illustrated in Fig. 14-7, shows the connections from the equipment in Fig. 14-6 to external equipment. The coded designations *8, 13, A1, F1*, etc. denote each connection wire that forms part of the cable harness, which extends from the panel assembly to the external controls and the dc motor drive. The external controls include push buttons and the speed adjustment potentiometer.

To supplement the diagram information, the general drawing notes include operating instructions and installation notes, information that will be useful later for instruction manual purposes.

Programmable Controller Diagrams

Stepping switches, cam programmers, and relay logic have been the common industrial control methods. Their drawback has been mechanical breakdowns and wear, as well as contact problems, enhanced by difficult industrial environment.

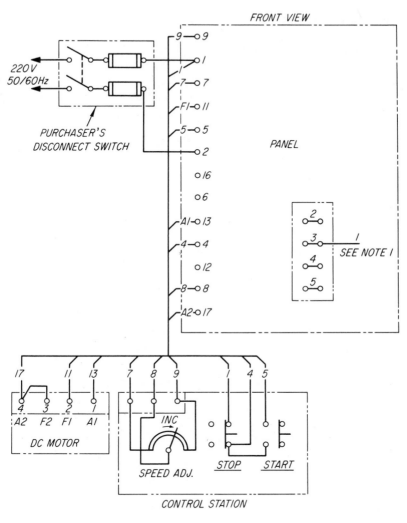

FRONT VIEW

220V
50/60Hz

PURCHASER'S
DISCONNECT SWITCH

PANEL

SEE NOTE I

9—○9
○1
7—○7
F1—○11
5—○5
○2
○16
○6
A1—○13
4—○4
○12
8—○8
A2—○17

2
3
4
5

17 11 13 7 8 9 1 4 5

4 3 2 1
A2 F2 F1 A1
DC MOTOR

INC

SPEED ADJ.

STOP   START

CONTROL STATION

NOTES:
1. CONNECT LEAD TO TERMINAL NEAREST TO LINE VOLTAGE

TERMINAL	LINE VOLTAGE	TERMINAL	LINE VOLTAGE
2	208 V	4	230V
3	220 V	5	242V

2. THE DISCONNECT SWITCH MUST BE CLOSED FOR AT LEAST 30 SECONDS
BEFORE STARTING THE MOTOR.

FIG. 14-7. Typical Industrial Interconnection Diagram. (Courtesy General Electric Company.)

A new control system has been developed, a direct replacement for the devices previously used, which has several advantages:

1. Immunity from EMI, RFI interference
2. Visual monitoring of output
3. Instantaneous reset.

It employs a control module which includes two memory programmers for a stored program set up at the factory and which can also be replaced at the user's site. The system has a number of inputs, outputs, and independent time delays which can override the master clock.

To develop a program for the control module, the task to be performed is divided into a series of time intervals or "steps," which can differ in length. As the controller progresses from one step to the next, the outputs will change

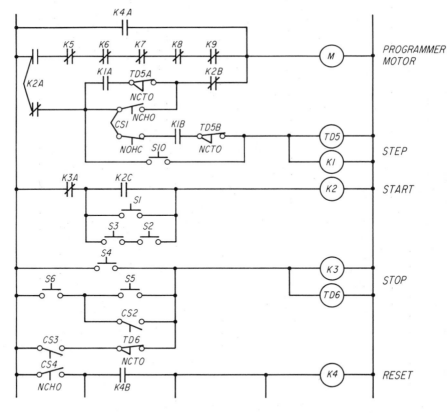

FIG. 14-8. Typical Programmable Control Elementary Diagram Practices.

in conformance with the information programmed at that step. Special instructions such as Reset, Stop, and others will be executed by the controller at the same time. Figure 14-8 is a partial ladder diagram, similar to the conventional relay diagram, Fig. 14-4.

SUMMARY

Industrial electronic diagrams require a somewhat different approach, and consequently the electronic draftsman must acquire additional knowledge to draw these diagrams in accordance with industry standards. Graphic symbols, format, and reference designations differ from their electronic communication counterparts, and the circuits are primarily electrical rather than electronic.

QUESTIONS

14.1 What is an elementary diagram?

14.2 What are some of the industrial applications of electronics?

14.3 Describe destination wiring.

14.4 What standards should be referred to when preparing industrial circuit diagrams?

14.5 How do the device designations on industrial diagrams differ from reference designations used on electronics communication diagrams? Give several examples, including multiple unit components.

14.6 What devices are represented by the following designations: (a) RES; (b) CT; (c) LS; (d) TB; (e) PS; (f) CAP; (g) FU; (h) FTS; (j) X; (k) PB; (l) REC. Show a typical symbol for each.

14.7 Discuss the general arrangement of an industrial schematic diagram and describe the circuit level system.

14.8 What is one outstanding connection-line practice followed in both elementary and connection diagrams?

14.9 What special drawing notes may be required on a connection diagram of industrial electronic equipment?

14.10 List some of the circuits and the identification colors to be found on a connection diagram.

14.11 Describe a tabular connection diagram.

14.12 What methods are used to indicate the progressive operation of controls on industrial circuit diagrams?

EXERCISES

Develop freehand layouts on opaque cross-section paper before making finished drawings on plain vellum or vellum with disappearing grid lines. Include junction dots, polarity signs, shielding, abbreviations, lettering, and device designations. Use symbol templates whenever possible. Refer to transistor manuals for correct designations and element connections.

14.1 Sketch some of the industrial graphic symbols that differ from electronics communication symbols.

14.2 Sketch a typical industrial control circuit and indicate: (a) the live side of a power supply; (b) levels; (c) wire numbers; (d) jumper; (e) potentiometer; (f) closed contacts; (g) the parts of a divided transformer symbol; (h) a typical device designation.

14.3 Sketch the methods of showing device designations on an industrial schematic diagram.

14.4 A schematic circuit of a motor speed control is shown in Fig. 14-9. Draw an elementary diagram of this circuit arranged in accordance with industrial diagram practices. Include device designations and component values on the diagram. Develop it in the ladder-type circuit and assign wire numbers and level designations. Make a list of all components by name, values, and designations.

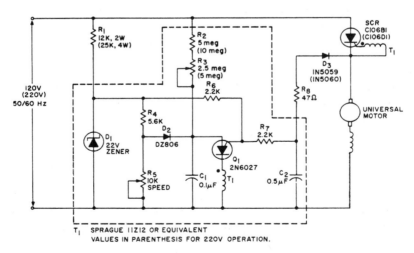

FIG. 14-9. Motor Speed Control. (Courtesy General Electric Company.)

14.5 Make drawings of a chassis 2 inches high and a component board on which all the components in Exercise 14.4 will be mounted, except the motor. Provide a small, 4-terminal block on the chassis for connections to the ac supply voltage and to the motor. The board is to be $\frac{1}{16}$-inch thick phenolic with components laid out for maximum space utilization. Use 150WVDC capacitors and 2-watt resistors except for R5, which is rated at 5 watts. Use the chassis to house the component board.

14.6 Make a lineless or tabular connection diagram of the circuit in Exercise 14.4. List all conductors, their sizes, colors, and destinations. Include any necessary notes.

14.7 An outline drawing of an electronic resistance-sensitive relay assembly is shown in Fig. 14.10. The chassis has: (1) an electron tube and socket; (2) magnetic

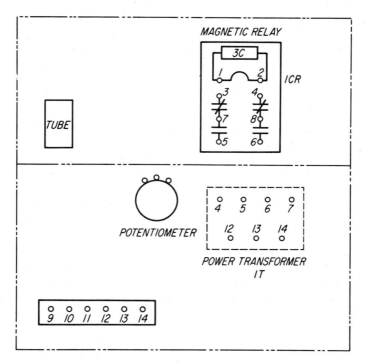

FIG. 14-10. Electronic Resistance-Sensitive Relay Assembly.

relay 1 *CR* with screw terminals *1-8*; (3) potentiometer 1*P* to regulate sensitivity; (4) power transformer 1*T* mounted inside the chassis; (5) a 150,000-ohm resistor mounted on tube socket; (6) two 18,000-ohm resistors connected from tube socket to terminal strip; (7) .22-μF capacitor from tube socket to terminal strip; (8) 1.0-mF capacitor across terminals 1 and 2 of relay *1CR*. The circuit is shown in Fig. 14-11. Relay terminals *3-8* are used for external control only. Draw a ladder-type elementary diagram and identify the levels and individual conductors by numbers. Identify components by industrial device designations.

14.8 Draw a point-to-point connection diagram of the relay assembly in Exercise 14.7 following industrial diagram practices. Assign wire colors as listed in the color code table given in this chapter.

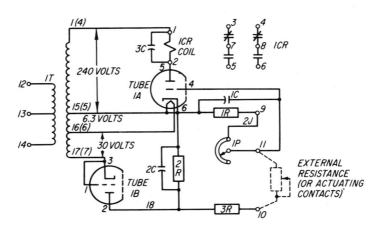

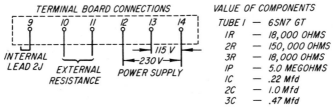

FIG. 14-11. Elementary Diagram of the Assembly in Fig. 14-10. (Courtesy General Electric Company.)

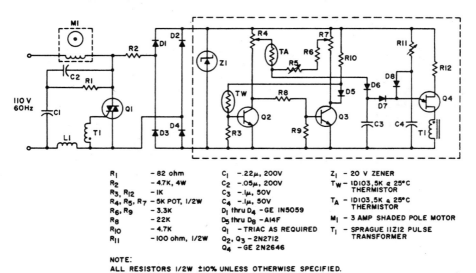

R₁	– 82 ohm
R₂	– 4.7K, 4W
R₃, R₁₂	– IK
R₄, R₅, R₇	– 5K POT, 1/2W
R₆, R₉	– 3.3K
R₈	– 22K
R₁₀	– 4.7K
R₁₁	– 100 ohm, 1/2W

C₁	–.22μ, 200V
C₂	–.05μ, 200V
C₃	–.1μ, 50V
C₄	–.1μ, 50V
D₁ thru D₄	– GE IN5059
D₅ thru D₈	– AI4F
Q₁	– TRIAC AS REQUIRED
Q₂, Q₃	– 2N2712
Q₄	– GE 2N2646

Z₁	– 20 V ZENER
T_W	– ID103,5K a 25°C THERMISTOR
T_A	– ID103,5K a 25°C THERMISTOR
M₁	– 3 AMP SHADED POLE MOTOR
T₁	– SPRAGUE 11Z12 PULSE TRANSFORMER

NOTE:
ALL RESISTORS 1/2W ±10% UNLESS OTHERWISE SPECIFIED.

FIG. 14-12. Fan and Coil Blower Motor Speed Control. (Courtesy General Electric Company.)

14.9 Prepare a lineless wiring diagram for the conductors required in Exercise 14.8. List wire numbers, wire colors, and end connections to and from the respective components. Include a suitable table and any notes that may be required.

14.10 An elementary diagram for a fan and coil blower speed control is shown in Fig. 14-12. Draw this as a ladder-type elementary diagram and identify all wires by numbers. Include a terminal board for 110-V terminals. Make a list of all components by name, value, and designations.

15

WIRING HARNESSES

To satisfy both military specifications and many commercial applications, components must be rigidly mounted and identified, equipment must be subdivided into subassemblies, and the connection wiring must be in an orderly form, such as cabled assemblies.

This last requirement is met by forming the connection wires between components, terminal boards, connectors, and component subassemblies into groups of wires that are bound mechanically by a lacing cord or special clamping devices. Such a wiring assembly is popularly known as a harness, a harness assembly, or a cable assembly. The last term generally applies to the interconnection cables between equipment units.

A harness extends throughout the equipment assembly, with branches or arms that reach the smaller subassemblies or individual components. Care must be taken, however, to route the wiring away from such items as fasteners, hinges, terminals, and sharp corners and edges. Eyelets, grommets, or rounded holes are used to protect the harness assembly when it passes through partitions. Although harness assemblies are secured to the equipment with metallic or plastic clamps, they have slack or are "looped" at such points as hinges and shock-mounted assemblies to eliminate strain on the wires. To facilitate soldering and to reduce the strain at connector pin joints, the individual wire connections are looped at connectors.

In some industrial equipment, the conductors on control panels and between the equipment sections are routed in troughs or channels of wire mesh or sheet metal. These troughs have side openings for the branch connections. The conductors are laid in the troughs loose rather than laced or tied together.

All of the wires for the wiring harness may be pre-cut, with terminals attached to the conductor ends and the terminations coded by color, number, or other means; this both reduces the possibility of wiring errors, and cuts costs by permitting the use of unskilled personnel to assemble harness wiring.

Because the wiring of complex electronic equipment represents a considerable part of the total cost of the equipment, the savings that result from the use of harness assemblies more than compensate for the cost of preparing the harness assembly drawings.

15.1 CONDUCTORS

Many of the conductor details are common to all harness assemblies.

Size

The size of the conductor depends upon the current and temperature requirements of the harness assembly. If it is appreciable, the conductor length determines the voltage drop. To avoid such a drop, long conductors should use the next larger conductor size. Some of the conductor details are listed in Table 15-1.

Table 15-1   Conductor Details

Wire Size (AWG)	Recommended Max Current (amp)	Ohms per 10 ft	Voltage Drop per 10 ft at Max. Current
22	1.5	.160	.24
20	3	.100	.30
18	4.5	.064	.28
16	7	.040	.28
14	12	.025	.30
12	18	.016	.28
10	30	.010	.30

Insulation

The conductor insulation should be suitable for the voltages and temperatures encountered in the harness installation. Among the various insulation types in use are polyvinyl chloride (PVC) and Teflon.

The use of polyvinyl chloride (PVC) as insulation on harness conductors or other applications is prohibited in military airborne equipment.

Color Coding

Several color-coding methods are used to identify each conductor in a harness assembly. These methods were described in Section 12.2.

Conductor Details

Any conductors that require individual or group shielding should be indicated on the connection diagram, the harness assembly drawing, and the wire list, which is generally a part of the harness drawing.

Conductors for ac line wiring should be twisted in pairs if they are included in the harness. This minimizes the chance of a pick-up by other conductors inside or outside of the assembly. Again, such conductors should be indicated on the connection diagram, the harness drawing, and the wire list.

High-voltage conductors, or conductors that carry RF or pulse circuits should not be included in the harness assembly.

Conductor Lacing

The conductors that form a harness assembly have to be held together mechanically. Ordinarily, this is accomplished by lacing the assembled conductors with a flat lacing tape of nylon, Dacron, and for higher temperatures, Teflon-finished fiberglass. The lacing tape can be applied manually or with a special cable-lacer tool. As a rule, the harness is laced so the loops are on top and the loop ties on the bottom, nearest to the chassis surface.

Plastic ties, zipper tubing, and spiral plastic binding are also used to hold conductors together, but no matter what method is used, the material should be specified on the material list and identified on the harness outline.

Conductor Terminations

Conductor ends may be stripped to a length of $\frac{1}{4}$ to $\frac{1}{2}$ inch; they may have various types of crimped or soldered terminal lugs; or they may have other termination means. These mechanical details are generally listed on the harness assembly drawing and may also be indicated symbolically on conductor ends

15.2 HARNESS DRAWING DEVELOPMENT

Although the process of drawing a harness assembly may appear to be complicated, it actually can be subdivided into several steps.

Harness Outline

The mechanical assembly drawing of the electronic equipment should be studied carefully before the harness assembly drawing is started. The assembly drawing that shows all of the mechanical, electronic, and other components and subassemblies should be examined to determine the path or paths that the harness assembly will have to follow to avoid such obstructions as partitions. An actual mechanical assembly of the equipment is even better for such an examination.

A print of this mechanical assembly should be covered with a blank drawing sheet that is transparent enough to show the details of the assembly. This sheet is then used for the harness assembly drawing.

The connection diagram of the equipment, which shows the connections within the equipment in relation to the various components, acts as the guide in wiring harness layout. The number of conductors in various "runs" between components can be determined from it, and this, in turn, determines the harness width at various points and the location of the harness branches.

The student should draw the basic outline of the harness first. The number of conductors and their approximate diameters determine its width at various points. For example, if there are 15 conductors at a given point and the cross section of the harness is to be square, the outline will have to be wide enough to accommodate four conductors. Then, assuming an average conductor has a diameter of $\frac{3}{32}$ inch, the harness width at that point will be four times $\frac{3}{32}$, or $\frac{3}{8}$ inch. This method of approximation provides a quick approach to developing harness width. There will, of course, be variations in the conductor diameter or the number throughout the harness or its branches, so that the width at various points will have to be redetermined. If twisted pairs or shielded conductors are included in the harness, an allowance will have to be made for them in estimating harness width.

In certain areas, component disposition may limit the harness width, and require building the harness up to a greater height.

To assist in harness layout, each conductor on the connection diagram print should be marked in advance with such information as approximate diameter, wire size, insulation, color code, and many other details that will eventually appear on the harness assembly drawing. The beginning of a harness layout is shown in Fig. 15-1. It is not necessary to indicate lacing because this can be included in a drawing note along with other special details, nor it is necessary to show individual conductors, except at the termination points.

Harness Drawing Layout

The foregoing description of harness layout practices applies to the layout of the whole harness outline, including the branches. Once the main harness path has been selected and the harness widths established, only the various branches and individual connections need to be laid out to complete that part of the harness assembly drawing. Although some of the wires might "run" perpendicular to the main body of the harness when they are connected to the components, they are shown in the plane of the drawing. The harness drawing thus represents a plan view of the chassis with its sides folded out, the same as the practice followed in connection diagrams (see Fig. 12-12).

An example of a typical harness assembly is given in Fig. 15-2. To locate both ends of each conductor, numbered balloons or call-outs are assigned to each point where individual conductors terminate. The balloons are numbered

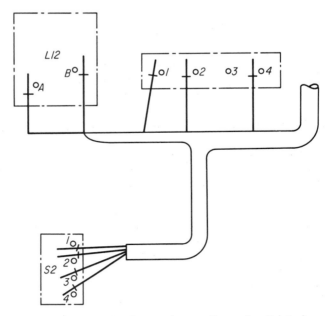

FIG. 15-1. Beginning of a Harness Layout. Connection Points Are Shown for Reference.

from left to right and from top to bottom. Each conductor termination has the color abbreviation of the conductor and the number of the balloon indicating the conductor termination at the opposite end.

Therefore, the run of each conductor can be found from either end, and the entire harness should have an even number of terminations, thus providing a simple check on conductors in the harness layout. The color coding or numbering of conductors was given in Section 12.2.

The extent of the harness lacing is indicated pictorially (Fig. 15-1) up to the point where the individual conductors emerge from the body or branches of the cable. Rounded corners, with as large a radius as possible, help to shape the larger conductors.

The appropriate symbols are used to identify harness terminations and other harness details. These symbols, shown in Fig. 15-3, should be included in the drawing notes.

Estimating Termination Lengths

It is necessary to determine the length of the individual conductors from the point where they break out of the laced harness to the point where they connect to the component.

The approximate position of the harness break-out point must be determined on the chassis layout and the length of each conductor measured to its connec-

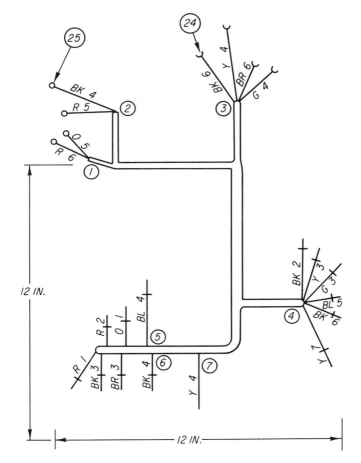

FIG. 15-2. Typical Harness Assembly Layout.

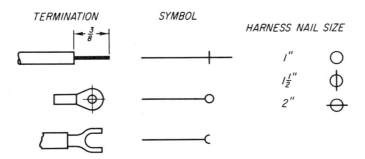

FIG. 15-3. Termination and Harness Nail Symbols.

tion point, with an allowance for the differences in height. This can be done by using a piece of wire to measure the distance and allowing an extra $\frac{3}{4}$ inch for stripping and tinning the conductor end and to permit some slack at the connection point. An even greater allowance should be made for looping the individual connections at multiple-pin connectors.

A complete mechanical assembly of the equipment is very helpful in estimating the connection lengths if one is available. The conductor terminations should be shown in their exact locations on the harness assembly drawing and with their lengths corresponding to the measurements obtained as described above.

15.3 HARNESS PROBLEMS

Harness assemblies present some layout problems, and thus, certain precautions must be taken to prevent shop difficulties.

Harness Obstructions

Once a harness has been laced it cannot be reshaped to avoid unforeseen obstructions or for installation on the chassis.

The closed loops in the harness are another source of difficulty. If possible, the conductors should be rerouted to keep loops out of the harness layout. They certainly cannot be included in the layout if they have to pass through an obstruction. When it is impossible to avoid such loops, they should be large enough to slip over any obstruction in their path when the harness is being placed in position. If there is any doubt about the harness installation, a sample harness, made from the harness layout drawing, should be used to check the harness positioning on an actual chassis assembly. This will also help to check conductor lengths.

Additional openings may be required in the chassis, rack, cabinet, or their partitions to permit passage of the harness branches to connection points. Grommets, eyelets, or rounded hole edges must be added to the mechanical chassis detail drawings for harness protection.

Complex Harnesses

So far, this description has been confined to harness layouts in one plane, that of the drawing paper. However, many harnesses are more complex, with the body or branches of the harness extending perpendicular to the rest of the harness or in more than one plane.

Harnesses that require a right-angle bend along the body of the harness may be formed without constructing a special form for this purpose. This is done by a notation on the harness drawing giving instructions to fasten a rod, $\frac{3}{16}$ inch in diameter, at the bend point to provide the desired bending relief (see

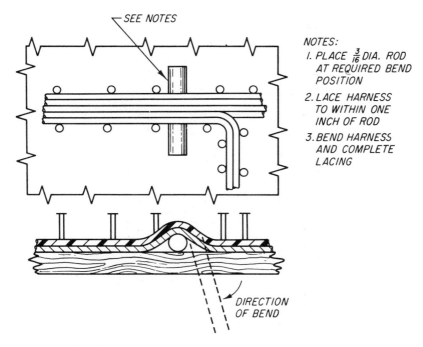

NOTES:
1. PLACE $\frac{3}{16}$ DIA. ROD AT REQUIRED BEND POSITION

2. LACE HARNESS TO WITHIN ONE INCH OF ROD

3. BEND HARNESS AND COMPLETE LACING

FIG. 15-4. Method Used to Provide Bending of Harness Assembly.

Fig. 15-4). As each conductor is placed in the harness jig, it follows the rise created by the rod. Thus, the hump formed at that point increases with each additional conductor. Such a harness assembly is laced to within an inch or so of the hump. When the harness is removed from the harness jig, it is bent at that point and the lacing is then completed. In making the harness layout, care must be used to select the correct bend direction. If the bend has to be made in the opposite direction of that shown in Fig. 15-4, a reverse print of the harness should be made and used for the harness layout.

The draftsman should examine the wiring diagram carefully to decide whether two or more harness assemblies might be preferable to one harness, which could become quite complex. Such a procedure will simplify manufacturing the harness and wiring it on the production line.

The wiring diagram should indicate the individual wiring harnesses and be developed along with the harness design even though it cannot be completed until all of the harness designs have been finished.

When a multiplane harness assembly is required, the various bends are indicated on the harness drawing and, if necessary, dimensions included. A suitable form is built to manufacture the harness, using the drawing as a guide. One or more views may be included on the drawing to show the relationship of the various harness sections to the plane of the assembly (Fig. 15-5).

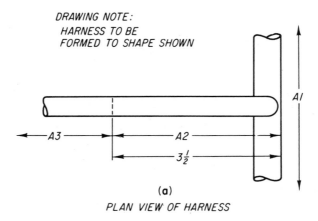

(a)

PLAN VIEW OF HARNESS

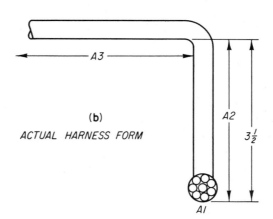

(b)

ACTUAL HARNESS FORM

FIG. 15-5. Drawing Details for Developing Complex Harnesses.

15.4 HARNESS MANUFACTURING DETAILS

These manufacturing details may help the student in designing harness assemblies.

Harness Drawing Application

For manufacturing purposes, a print of the harness drawing is cemented to a plywood base and protected with a coat of clear varnish or lacquer. Nails, or pins made especially for this purpose, are driven along the outline of the harness, up to the termination points where various individual conductors emerge. The nails are spaced an inch or two apart to keep the individual con-

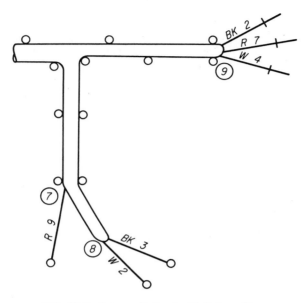

FIG. 15-6. Harness Outlined with Nails or Pins.

ductors in place. They are also located at all branch points (Fig. 15-6) and should project to at least twice the harness height.

Another harness cable technique utilizes a jig board made of layers of metallic screen separated by layers of honeycomb material. The harness drawing print is taped to the board (Fig. 15-7). The board construction allows ready attachment of wiring accessories at various harness points in accordance with the layout.

Special elastic retainers hold the cable wire and automatically form round

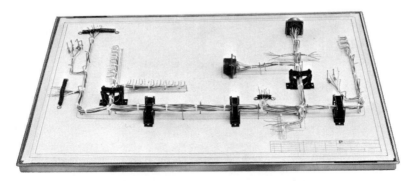

FIG. 15-7. Wiring Harness Jig Board.

Table 15-2 Typical Wiring List

	Conductor					From				To			
Wire no.	Size AWG	Item no.	Code	Color	Length (in.)	Circuit	Strip Wire (in.)	Tinned	Terminal Lug, Item #	Circuit	Strip Wire (in.)	Tinned	Terminal Lug, Item #
1-4	20	4	42-020-2	W	14	K2-3	3/8	Yes	17	T3-1	3/8	Yes	22
1-7	20	4	42-020-2	W	18	K3-5	1/2	Yes	14	T3-3	3/8	Yes	22
1-9	18	6	42-018-1	R	20	P2-3	3/8	Yes	—	TB2-1	3/8	Yes	21
2-5	22	3	42-022-3	O	9	P2-6	3/8	Yes	—	TB2-2	3/8	Yes	19
2-7	18	6	42-018-1	R	11	S1-2	1/2	Yes	—	S2-2	3/8	Yes	—
2-10	20	7	42-020-6	Y	14	S2-2	1/2	Yes	—	S2-3	3/8	Yes	—
3-6	22	2	42-022-4	BK	8	XQ1-1	1/2	No	—	XQ2-4	1/2	No	—
3-7	18	5	42-018-5	BR	9	XQ1-2	1/2	No	—	XQ2-5	1/2	No	—

Note: 1. Lengths specified are cutting lengths.
2. Coat all soldered joints with Item No. 27.
3. Item Nos. refer to Item Nos. in Parts List.

wire bundles as the wires are inserted. Other cable accessories include: harness springs for locating conductor ends at specified places, and straight and bent harness pins to locate and hold down the harness.

Harness Wiring Lists

In harness manufacture, the individual conductors are precut to the required length by hand or by automatic machinery, depending upon the quantity involved. Each conductor of a given size and color combination is stripped for a definite distance and tinned, if required. Then the various types of terminal lugs are attached as specified by crimping or soldering. To complete the individual conductor preparation, coded sleeves or insulation markings are applied as required, and such items as conductor shielding and pigtails are added.

Such extensive details require a systematic tabulation of all conductor requirements, connection points, and so forth. This may be given as a wiring list included on the harness assembly drawing (Table 15-2) or as separate sheets prepared by the manufacturing department.

The partial wiring list in Table 15-2 is cross-referenced to the parts list included with the wiring harness assembly drawing. Individual harness conductors are identified by the dual numbers they have on the harness drawing, as for example the *2-4* black conductor in Fig. 15-2. This conductor should be identified in a similar manner on the connection diagram print to facilitate checking later.

Details for each conductor are listed as shown. For example, conductor *1-4* is *#20 AWG*, white, 14 inches long, and is item 4 in the parts list. The insulation and other details follow code 42-020-2. Both ends are stripped $\frac{3}{8}$ inch and tinned. One end is soldered to terminal *#3* of relay *K2* in the electronic assembly, while the other has a terminal lug, item 22 in the parts list, and connects to terminal *#1* of transformer *T3*.

Similar lists are prepared for jumper wires and other individual conductors that are used for miscellaneous inter-component wiring but are not part of a harness assembly.

SUMMARY

Wiring harnesses or cabling assemblies require special drawings, with these assemblies laid out to scale and the various conductors identified in accordance with established practices. Harness conductors must be carefully selected, keeping current and voltage requirements in mind, and wiring lists prepared for shop use.

QUESTIONS

15.1 Describe a harness assembly. How does it differ from a cable assembly?

15.2 List some of the factors that govern the size of conductors in a wiring harness.

15.3 How are harness wires held together?

15.4 Where is the use of polyvinyl-chloride insulated conductors prohibited?

15.5 Outline the process involved in starting a harness layout.

15.6 How are the lengths of the individual conductors that extend outside of the harness drawing determined? What precautions must be taken?

15.7 How are conductors indicated on a typical harness drawing? Make a sketch of a simple 12-wire harness assembly with three branches. Color code the individual conductors.

15.8 What are some of the precautions to be taken in laying out a complex harness?

15.9 How is the approximate width of a harness run determined? Determine the approximate width of a harness with the following conductors: six wires, $\frac{1}{8}$ inch in diameter; ten wires, $\frac{3}{32}$ inch in diameter; and four wires, $\frac{5}{32}$ inch in diameter.

15.10 What is a harness wiring list?

15.11 What precautions must be taken in routing a harness assembly? Describe harness protection methods.

15.12 What is a cable tie?

EXERCISES

Make preliminary layouts freehand on grid paper, 4 by 4 divisions per inch. Refer to products catalogs for details of various components indicated.

15.1 Sketch some of the special symbols used on harness drawings.

15.2 A wiring harness is needed for the chassis shown in Fig. 15-8. The harness is to connect between terminal block (A), component boards (B) and (C), and terminal boards (D), (E), and (F). The chassis shown is 12 by 17 by 3 inches deep. The conductor numbers and their termination points are: (1) $A1$-$B2$; (2) $B2$-$D4$; (3) $A6$-$F1$; (4) $A4$-$D1$; (5) $A4$-$F3$; (6) $B5$-$F3$; (7) $F3$-$A3$; (8) $B1$-$C7$; (9) $B4$-$E3$; (10) $E2$-$A2$; (11) $E2$-$D3$; (12) $C2$-$B9$; (13) $B3$-$C3$; (14) $F2$-$C4$; (15) $D3$-$E4$; (16) $A5$-$F4$; (17) $F5$-$C5$; (18) $C6$-$B8$; (19) $C1$-$B7$; (20) $F6$-$B10$; (21) $B6$-$C8$. Conductors 1, 9, 10, 15, and 18 are #18; 5, 6, and 8 are #20; the rest are #22. List their equivalents in stranded wire on the harness layout. Allow an extra $\frac{1}{2}$ inch at all terminal points. Conductor terminations at the terminal block consist of #6 crimped terminal lugs; the remainder are stripped $\frac{3}{8}$ inch. The conductors are laced with nylon lacing cord, with ties spaced $\frac{1}{2}$ inch apart, and tied at each break-out point. Make a full-size drawing of the harness assembly according to the above data and to the scale shown. Include a reference dimension and such notes as may be necessary.

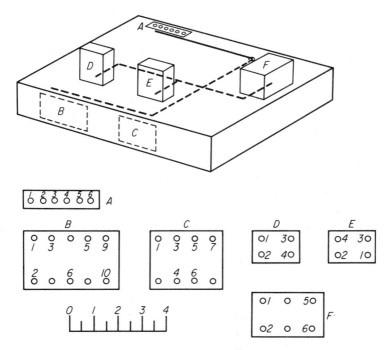

FIG. 15-8. Chassis and Wiring Harness Layout.

15.3 Make a wiring list of the individual conductors in Exercise 15.2. Individual conductor colors, with tracers if necessary, should be selected so that they do not repeat at the various boards or the terminal block, except where the circuit continues from the same terminal to another board or block. Include such notes as may be necessary and a reference dimension.

15.4 A junction box, 7 by 9 by 3 inches deep (Fig. 12-20) has the various terminal boards located one inch from the bottom of its inside surfaces. The terminal boards have #8 terminals, spaced $\frac{3}{8}$ inch apart. Make a wiring harness drawing, positioning the harness on the sides of the box below the terminal boards and keeping the bottom of the box clear of all wiring. Conductors connected to *TB1* are #16; all remaining wires are #20. All conductor terminations have #8 crimped lugs with insulating sleeves. The conductors are tied with nylon lacing cord, with ties $\frac{1}{2}$ inch apart. They are also tied at each break-out point. The harness should be laid out full-size and to scale. Include a reference dimension and any notes that may be required.

15.5 Make up a wiring list of the various conductors in Exercise 15.4. Select white conductors with a single tracer so that the tracer colors do not repeat at the various board terminals. Include such notes as may be necessary.

15.6 A wiring harness is needed for the chassis shown in Fig. 15-8, with the terminal board *F* omitted. The chassis shown is 10 by 17 by 3 inches deep. The conductor numbers and their termination points are: (1) *A1-D3*; (2) *A2-C1*; (3) *A3-E4*;

(4) *A4-B1*; (5) *B5-C8*; (6) *C3-E3*; (7) *B8-D3*; (8) *B3-C5*; (9) *B2-C2*; (10) *C4-D1*; (11) *B7-D2*; (12) *A5-B4*; (13) *B6-C7*; (14) *A6-B9*; (15) *B10-E1*; (16) *C6-E2*. Conductors 2, 6, 7, and 10 are #*16*; 1 and 5 are #*18*; the rest are #*22*. The remaining information is given in Exercise 15.2.

15.7 Make a wiring list of the individual conductors in Exercise 15.6. Individual conductor colors, with tracers if necessary, should be selected so that they do not repeat at the various boards or the terminal block, except where the circuit continues from the same terminal to another board or block. Include a reference designation, such notes as necessary, and assign reference designations to components.

15.8 A wiring harness is needed for the chassis shown in Fig. 15-8, with the terminal boards *D* and *E* omitted. The chassis shown is 12 by 15 by 2 inches. The conductor numbers and their destination points are: (1) *A1-C2*; (2) *A2-B2*; (3) *A3-F2*; (4) *A4-C2*; (5) *B3-C6*; (6) *C1-F4*; (7) *A6-C7*; (8) *B10-C4*; (9) *B1-F5*; (10) *A5-C5*; (11) *B5-C3*; (12) *B6-F1*; (13) *B9-F6*. Conductors 4, 8, and 9 are #18; 1 and 5 are #20; the rest are #22. The remaining information is given in Exercise 15.2.

15.9 Make a typical wiring list of the individual conductors in Exercise 15.8. Select individual conductor colors, with tracers if necessary, so that they do not repeat at the terminal block or the various boards. Include reference designations to these components, and such notes as necessary.

16

CHECKING ELECTRONIC DRAWINGS

The completed drawing, no matter whether it is a detail, assembly, or circuit drawing, cannot be considered finished and ready for release until it has been completely checked by an experienced checker. The drawing must not only be accurate and complete, it must also agree with the latest company and government standards, if military equipment is involved. Production requirements and shop capabilities must also be considered during the checking process. Errors must be found at the drafting room stage, not in the shop where they are costly and time-consuming. One small dimensional mistake, overlooked by the checker, causes extensive and costly rework in the shop that produces the item in large quantities.

The amount of checking personnel varies with the size of the drafting room, the equipment or parts being produced, and the level of experience of the drafting personnel. Large companies have a group of checkers in the checking department to scrutinize every drawing completed in the drafting room. In smaller firms, this job may be done by the group leader, the head of the drafting section, the project engineer, or even another draftsman—not a desirable practice, however, because of the possibility of mistakes not being discovered. A careful selection of checking personnel is a requirement generally met by transferring highly experienced draftsmen to the checking department. These trained men are thus capable of noting questionable fabrication designs or practices and of making supplemental layouts to check dimensional calculations.

A checker in a company producing electronic equipment is faced with the additional responsibility of checking circuitry drawings as well as mechanical drawings. If he fails to notice that two wires are joined in a wrong spot on the circuit diagram, he may contribute to extra expense and delays in the manufacture of the electronic equipment.

The draftsman may, of course, spot-check his own work when he completes the drawing and discover some of the obvious errors before he turns it over to the checking section, but the responsibility is still with the checker. Once the drawing has been checked and any corrections made, it is signed by the checker who then assumes full responsibility for its accuracy. The signatures of other personnel involved, such as the head of the drafting room and the project engineer, may also be added before the drawing is officially released.

16.1 CHECKING DRAWINGS: GENERAL INSTRUCTIONS

Only three items are required to check drawings: a copy of the original drawing, in the form of a black-on-white print, and a red, and a yellow pencil. Pencils of other colors may be found desirable for making special notations on the check prints. A black-on-white print allows checking calculations to be made of mechanical details, and connection lines on circuit drawings are easy to trace.

Such items as dimensions, notes, reference designations, and component values are checked off with a check mark ($\checkmark$) as the work progresses. The yellow pencil is used to note the correctness. Errors are circled with the red pencil to "spotlight" them against the drawing background. If necessary, the draftsman should be consulted to resolve any questionable dimensions, notes, or other data.

If inked drawings are to be checked, the pencilled version should be checked first and rechecked after inking. Costly corrections can be avoided by eliminating them in the pencil stage.

Once checked, the checking prints should be signed and dated by the checker and turned over to his supervisor or to the head of the drafting room.

The drawing types encountered in checking are: mechanical, either component, chassis, or a complete equipment; circuitry, either a schematic (elementary) or a wiring (connection) diagram; harness or cable assembly; or printed circuit master drawing. The checking of each drawing type follows a different pattern and, therefore, is discussed separately.

16.2 CHECKING MECHANICAL DRAWINGS

Since mechanical drawings are more complex by nature, the procedure for checking them is presented first.

As with all drawings to be checked, a set of prints is obtained for use exclusively for checking. These prints are then kept for future reference.

In addition, all material that pertains to the drawings being checked should be obtained. These include catalogs, calculations, prints or reference drawings, and the drafting standard book, etc.

When all of this material has been gathered, the checking should be done as follows:

1. If more than one drawing is to be checked, the simpler drawings, such as those of minor parts, are checked first for correct dimensioning, tolerances, references to other drawings, and other information. The red pencil is used to note errors and the yellow pencil to check off correct items. Dimension calculations can be made quickly on the margin and thus become part of the record on the checking print.

The less important dimensions are checked first, followed by dimensions referenced to other drawings of parts and assemblies, and finally hole dimensions and locations, notes, title block information, etc.

2. The ability to recognize dimensions that appear to be out of scale is very helpful in spotting the obvious errors on sight. Such errors could easily be overlooked if the given dimensions are accepted at face value. Drawings that are drawn out of scale should have the dimensions underlined by a straight or a wavy line.

3. Prints of details that were drawn accurately to scale are likely to shrink or stretch during printing, and thus they should not be scaled to check dimensions. Instead, all important dimensions should be calculated, with tolerances taken into consideration.

4. Shop personnel are not expected to calculate any required dimensions, and thus all drawings and views should be checked for complete dimensioning.

5. Drawings should show all necessary surface finish information and material specifications in detail and give references to material standards when required.

6. When required by shop use, drawings should include all necessary views, including sectional views, to clarify the detail or assembly.

7. The clearances between adjacent parts should be carefully checked for interference, and supplementary layouts made, if required.

8. If necessary, layouts should be made of the moving components to check for clearance between these parts throughout the complete motion. Such layouts should be enlarged to determine whether the movements operate correctly.

9. Since close tolerances increase the cost of a product, particular care should be taken to see whether they are necessary.

10. After all detail parts have been carefully checked, the subassembly and assembly drawings should undergo the same critical examination. The electrical and electronic components and such mechanical fasteners as screws, bolts, nuts, and washers should be carefully checked for reference to standard drawings, for catalog information, for ready availability, and for stock size. Additional expense is likely to result if attention is not given to such items on the assembly drawings.

11. The parts list should be checked to make sure it complies with the

detail and assembly drawings, the material specifications, the sizes of parts, and so on.

12. Any calculations furnished with the drawings should be checked.

13. Standard parts, materials, and finishes should be specified on the drawings whenever possible.

14. Any questionable fabricating practices should be checked out with the shop personnel. The limitations of the shop facilities should also be considered. Dimensions should be placed for ease of manufacturing.

15. If it is possible to simplify construction and consequently to effect a savings, a note should be made and brought to the attention of the proper personnel.

16. Designs involving equipment supported on shock mounts should be checked for adequate clearances during the maximum excursion of such equiment.

17. Subassemblies involving soldering or welding of such heat-treatable parts as springs should be checked for the effect of such joining methods on these parts.

18. Equipments designed for military applications should be checked to make sure they are within the maximum limits specified for entrance doors and hatches, as on vessels and submarines. Such equipments should be checked for compliance with applicable military specifications and standards.

16.3 CHECKING METRIC DRAWINGS

There are two basic types of metric mechanical drawings to be checked: a converted inch drawing and a metric drawing that was originally drawn in metric dimensions.

The first type involves checking every dimension to its exact metric equivalent, using the conversion tables in this book or some other source.

Check the following:

1. Tolerances translated into metric dimensions.

2. Surface finish information.

3. Mechanical fasteners, such as screws, bolts, nuts, washers, etc., should be converted to stock-size metric fasteners and checked for ready availability and stock size, if required by the specification.

4. Calculations furnished with the drawing should be checked for possible change to metric-size fasteners.

5. Equipment designed for military applications should be checked to make sure the drawings are within the metric requirements of specifications involved.

6. The notation METRIC should be added above the title block.

7. The tolerance block should be crossed out and a tolerance block in millimeters added.

8. The drawing should comply with the details outlined in Chapter 1, page 36.

9. A drawing note stating that the drawing is drawn in third-angle projection, a U.S. standard.

If metric drawings are made originally, check the following:

1. The drawing size should conform to the standard metric sheet sizes.

2. The drawing scale should be full, $\frac{1}{10}$, $\frac{1}{20}$, or $\frac{1}{50}$ size. The common drawing scales of $\frac{1}{2}$, $\frac{1}{4}$, $\frac{1}{8}$, etc., used for inch drawings are not used for metric drawings.

3. The drawings should be in the third-angle projection, the same as for inch drawing in the United States and Canada, and identified by the standard ISO symbol, shown in Fig. 1-23 (b).

4. Other details should follow the practices outlined in Chapter 1, page 36, and ASTM Standard E380-76, *Standard for Metric Practice.*

16.4 CHECKING CIRCUIT DRAWINGS

One problem likely to be encountered in a checking group is a lack of personnel capable of checking circuit diagrams. In some instances, it is necessary for the original draftsman to do his own checking of schematic, wiring, and other diagrams. If several draftsmen are employed to draw diagrams, they can cross-check each other's work as a substitute for a trained checker.

The need for accuracy cannot be over-stressed in checking schematic diagrams because any errors overlooked in checking will be magnified in the wiring diagram and harness drawings, which are based on the schematic diagram. An extra dot on two lines crossing on a schematic diagram could result in an interconnection on the wiring diagram that, in turn, might be difficult to locate on the completely wired equipment.

As with mechanical drawing, a black-on-white print is required to check the circuit diagram and harness. The following order must be adhered to in checking: (1) schematic diagram; (2) wiring diagram; (3) harness drawing.

Schematic Diagrams

Schematic diagrams must be checked against the original circuit or the sketch made by the electronic engineer or scientist. To prevent possible errors any questionable detail should be resolved before the actual checking is started, or notes should be kept of questionable items as the checking progresses.

The line-by-line, point-by-point checking method eliminates possible errors and provides a rapid self-checking method for the original draftsman.

Any desired connection line between two points on the original circuit sketch can be the starting point. It is drawn over with a yellow pencil or some other contrasting color that is readily visible on the original drawing. Exactly the same portion of the circuit should be drawn over on the schematic diagram with a colored pencil. It should extend only as far as the points on the original circuit sketch and, if necessary to prevent error, looped around points on the print that do not connect on the original circuit sketch.

Taking only one connection line on the sketch at a time eliminates the chances for error. The checking process of tracing each connection line on the original sketch to the checking print is repeated until every connection line and point are eliminated on both. The checker should not attempt to remember any circuit connection to be added later. It should be included at once on both the sketch and the print. Again, a red pencil is generally used to indicate errors.

Other Check Points

The following items also are checked during the connection line checking process:

1. Correct symbols for components
2. Circuit symbols for values and tolerances, reference designations, polarity markings, arrowheads when required, and other data
3. Terminal numbering and transistor and integrated circuit identification
4. Identification of terminal blocks, units, etc.
5. Possible ambiguity of reference designations or component values because circuit symbols are crowded
6. Drawing title, punctuation, and spelling
7. Notes relating to capacitor, resistor, and other values

All junction points should be rechecked against the original because they are the most common source of error.

Wiring Diagrams

The wiring or connection diagram is checked by following the corrected schematic diagram as the master and using a black-on-white print of each.

The checking procedure for wiring diagrams is similar to checking schematic diagrams. Again, a yellow or other color pencil is used to check off correct information and a red pencil to note errors.

The same line-by-line, point-by-point method is used. A connection line on the wiring diagram is selected between two points and drawn over with the yellow pencil. The same connection, from identical point-to-point, is located on the schematic diagram print and also drawn over with a yellow pencil. Points connected on the wiring diagram are included on the schematic diagram print. Any others are eliminated by looping around them. This process is repeated until all connection lines and all points on both the wiring and schematic diagram prints have been checked.

A similar procedure is followed in wiring diagrams of the cable type or of the base-line type by checking the connection points in these diagrams.

Tabular type wiring diagrams can be readily checked by noting and "to" destinations for the individual conductors. See also Fig. 12-29

Connecting leads have to be carefully checked in all wiring diagra color designations and continuity.

Any expected harness wiring should be shown drawn together on the w diagram to assist in making the harness or cable drawings later.

The other check points listed for schematic diagrams apply equally w in checking wiring diagrams.

Industrial Circuit Diagrams

Both the elementary and connection types of industrial circuit diagrams are similar in most respects to the conventional electronic schematic or wiring diagrams and the checking procedures are also similar. However, such items as device designations and graphic symbols must be checked carefully because they vary from conventional electronic circuit diagrams.

16.5 CHECKING HARNESS DRAWINGS

These drawings are checked against the wiring or connection diagrams on which they appear and, therefore, a black-on-white print of each is required. The harness drawing is checked using the wiring diagram as the master.

The same point-to-point procedure process should be used in checking the harness drawing. A lead on the harness assembly print should be selected, its ends checked off with a colored pencil, and the same lead traced with the colored pencil on the wiring diagram. The process is repeated until all of the harness wires have been located and marked over on the wiring diagram print. Care should be taken to locate the exact termination points on both the diagram and harness prints.

Each harness lead should be checked for continuity of color as well as for correctness of connection points, which should correspond with the wiring diagram.

If a wire list is a part of the harness drawing, it should be checked to make sure the wire type, insulation, shielding, and other data conform to the drawing.

Unless a sample wiring harness has been made, the length of each connection lead extending from the harness must be checked.

Harness conductors should be checked for such additional information as shielding if it is indicated on the wiring diagram.

The harness drawing also serves as a cross-check on the wiring diagram, with harness conductors shown tied together for each harness and each individual harness identified.

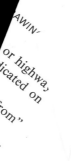

actually a wiring or connection diagram of a part
t card, or other electronic unit.

isolate that portion of the circuit on the schematic
rinted-circuit wiring. The procedures already outlined
g diagrams are followed, using this part of the dia-

d boards, care should be taken to include the interwiring
es in the checking process.

should be checked for the following:

awing number should be included within the board outline.

scale to which it is drawn should be checked.

ll holes outside of the wiring pattern should be indicated by round
spots.

4. Trim marks, dimensions between register marks, and FACE and BACK designations should be checked.

5. Reference designations, polarity markings, diode orientation, and terminal markings should be given when required.

Multilayer Boards

On the multilayer boards each double-sided board should be checked first, then the interwiring between the layers should be added on the schematic diagram as the checking proceeds.

16.7 FINAL CHECK

The check print and the original drawing is returned to the drafting room for correction.

As the corrections are being made the draftsman will mark out the red errors with a yellow pencil or a felt pen to indicate that the required corrections have been done. Upon completion of this work a new checking print is made and is sent along with original checking print to the checker for his final check.

SUMMARY

Electronic drawings must be carefully checked to achieve error-free drawings and eliminate the possibility of expensive shop rework to correct the overlooked deficiencies. Checking personnel must be capable of noting not only dimensional errors but also conformance with applicable standards, shop capabilities, etc. In small firms, the draftsman may be called upon to do his own checking, although such a practice is not desirable.

Checking metric drawings requires exceptional care when converting inch-pound dimensions and quantities on mechanical drawings to metric.

QUESTIONS

16.1 Why is checking so important?

16.2 What materials are required for checking purposes?

16.3 Describe the general factors to be considered in checking, and outline the procedures.

16.4 What are some of the important factors that should not be overlooked in checking mechanical drawings?

16.5 List some of the important items to be checked when a mechanical inch drawing has been converted to metric dimensioning.

16.6 What are some of the practices that should be checked on metric drawings?

16.7 What traits should a checker develop to assist him in his work?

16.8 Why is accuracy so important in checking schematic diagrams?

16.9 Name the method that should be followed in checking a schematic diagram.

16.10 List some of the important points to be checked on a schematic diagram.

16.11 What is one of the most common sources of error on a circuit diagram?

16.12 Discuss the method of checking a connection diagram.

16.13 How do industrial circuit diagrams differ from the communication electronics type?

16.14 Describe some of the important features in checking drawings.

16.15 Name some of the items to be observed when checking drawings of printed-circuit artwork.

17

GRAPHIC DRAWINGS

Technical data may be presented by using charts, graphs, or diagrams, which make a far more impressive presentation than a written description or a set of figures. These drawings are used for a variety of purposes: to compare values; to give design calculations; to present test results or data to management; to determine the values of variable quantities in relation to fixed quantities; and to present statistical data.

17.1 CHARTS

Several types of charts are in use. These include bar, circular, and organization charts. The type that is selected depends upon the application.

Bar or Display Charts

In this type of chart, quantities, relationships, and so forth are represented by vertical or horizontal bars that represent quantities drawn to a specific scale. Usually, these bars are also drawn against a time factor—the month or the year, for example.

Charts of this nature are frequently prepared with the same type of adhesive tape that is used in printed circuitry work and is available in various widths and colors.

Bar-type charts are also used effectively to illustrate progress reports, man-hours, cost, and similar items.

Circular Charts

These are also known as area charts and may be rectangular or of some other geometrical shape as well as circular. These charts are just as readable as bar charts and are used to indicate such quantities as percentage breakdowns. The individual segments may be shaded or colored. This technique is almost always used in advertising presentations.

Organization Charts

These charts show the duties, responsibilities, and authority of company management or specific departments, such as engineering, arranged in a "chain-of-command" order.

In drawing an organization chart, the exact responsibilities of each position and the name of the person holding it are given in a rectangular block. These blocks are arranged in descending order and have connecting lines to indicate the successive steps in responsibility.

17.2 GRAPH PAPERS

Graphs are often found in catalogs, technical magazines, equipment manuals, progress reports, and other technical publications. Special graph papers are available, printed on heavy drawing paper, tracing paper, or reproducible paper. The graphs themselves may be drawn in pencil or India ink depending on the final use.

Grid Patterns

These patterns are printed on the graph paper in green, orange, blue, or black ink. Orange and black grid patterns will appear on photographic or black-line reproduction prints. To prevent this, some grids are printed in special inks that will not reproduce.

The grid pattern is positioned on the graph sheet with a wider margin at the top or on the binding or punched edge, and the graph itself should be drawn with this wider margin at the top or on the left-hand side.

The two most common sizes of printed graph sheets, $8\frac{1}{2}$ by 11 inches (letter size) and 11 by $16\frac{1}{2}$ inches (double size), have corresponding grid pattern sizes of 7 by 10 and 10 by 15 inches, respectively.

The heavy drawing grid papers are generally used for graphs that are to be reproduced photographically or for rough graph layouts.

Square Grid-Type Paper

The square or cross-section grid (Fig. 17-1) is the most familiar type of graph paper. Because the vertical and horizontal divisions are of equal size, this type of grid is used to show quantity changes.

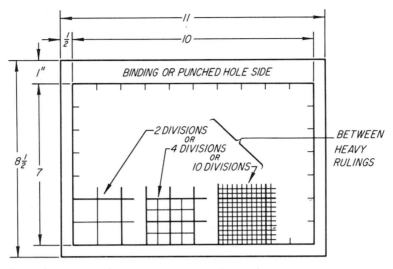

FIG. 17-1. Typical Square Grid Pattern Paper.

Square-section grids generally have a notation above the grid pattern to designate the number of divisions per inch. The most common divisions are: 4 × 4, 5 × 5, 6 × 6, 8 × 8, 10 × 10, 12 × 12, 16 × 16, and 20 × 20 per inch. A 10 × 10 sheet may have 10 divisions per inch in each direction, with every fifth line accented and every tenth line of a heavier weight. Other patterns are available with all the lines of equal weight or with only every tenth line accented.

Square grid paper is also used for engineering sketches and design. Grids with 16 × 16 divisions are good for such purposes because the division lines are $\frac{1}{16}$ inch apart.

Semilogarithmic-Type Paper

This grid pattern is a combination of equally spaced divisions in one direction and logarithmic divisions in the other, generally vertical.

If a logarithmic scale is drawn in comparison to the uniform or equal division scale (Fig. 17-2), it must be subdivided in accordance with the common logarithm (base 10) of the number. For example, log 2 equals .301, log 3 equals .477, and so on. The log 1 division is at the zero point on the conventional or arithmetic scale. [One log cycle covers a power of ten, 10 to 100 (10^1 to 10^2), 100 to 1000 (10^2 to 10^3), and so on.]

Semilogarithmic grid sheets (Fig. 17-3) usually have such designations at the top as "3 cycles × 70 divisions." This means there are 3 log cycles of 10 or a ratio of 1:1000 in one direction and 70 equal divisions in the other direction with heavy lines at each tenth subdivision. Every fifth line may or may not be accented.

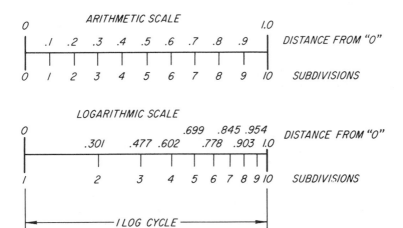

FIG. 17-2. Comparison of Arithmetic and Logarithmic Scales.

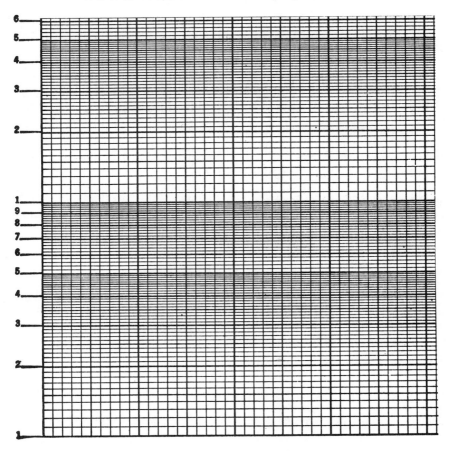

FIG. 17-3. Semilogarithmic Graph Paper.

Logarithmic-Type Paper

This paper type has logarithmic subdivisions both horizontally and vertically (Fig. 17-4) and may be designated "3 × 3 cycles," which indicates 3 log cycles in each direction or a ratio of 1:1000.

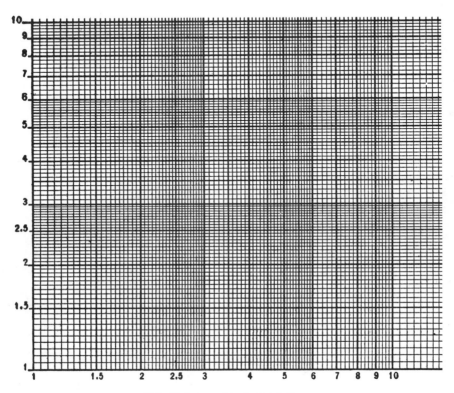

FIG. 17-4. Logarithmic Graph Paper.

Polar Graph Paper

This grid pattern (Fig. 17-5) is available in $8\frac{1}{2}$- by 11-inch sheets, with the pattern grid size of 7 by 10 inches. The angular scale is divided into single degrees and numbered both clockwise and counterclockwise every ten degrees. The ordinate scale, which consists of concentric circles equally spaced ten to the inch, is used to plot linear magnitude.

17.3 SCALE LAYOUT AND LETTERING

Selecting the correct layout and lettering scale is of the utmost importance because they determine the appearance and legibility of the graph. Figure 17-6 illustrates the effect of scale proportions. For example, the curve in Fig. 17-6(a)

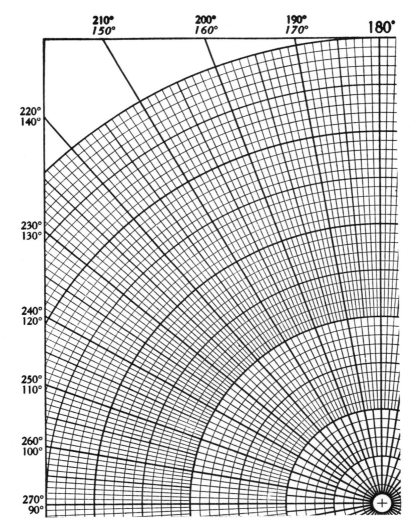

FIG. 17-5. Polar Graph Paper.

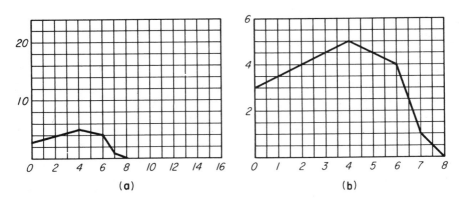

(a)   (b)

FIG. 17-6. Selection of Correct Scale Proportions.

covers only part of the available graph space, making it difficult to read the values at intermediate points. In Fig. 17-6(b), the scales were selected so that the curve extends over almost the entire graph area and results in a better balanced graph.

Graphic Scales

These scales have a straight or a curved line, known as the *stem*, with calibration marks over its entire length, Fig. 17-7(a). The scale length may be in inches or centimeters, and the zero end or origin may be on the extreme left as in Fig. 17-7(a), in the center as in Fig. 17-7(b), completely off the stem as in Fig. 17-7(c), or on a logarithmic scale.

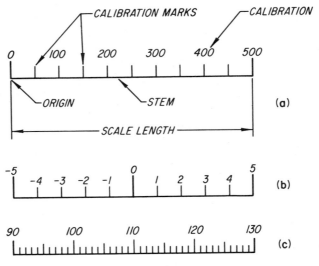

FIG. 17-7. Graphic Scales.

The scales shown in Fig. 17-7 are known as uniform or arithmetic scales and have equal spaces between the calibration marks. Other scales may be nonuniform, such as the logarithmic scale in Fig. 17-2, where the calibration marks follow a mathematical equation.

Lettering

Scale divisions on graphs should be well-separated, preferably on the major division lines which are normally one inch apart.

The division designations should be about $\frac{1}{16}$ inch away from the vertical and horizontal axes and centered on the major division lines, Fig. 17-8(a). They should be drawn about $\frac{1}{8}$ inch high. If both scales start at zero, only one cipher is used.

Although both scale division designations should read from the bottom

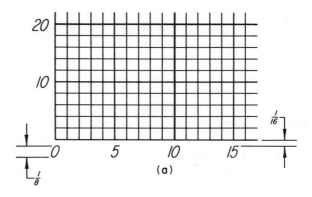

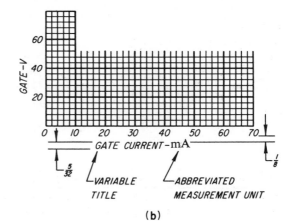

FIG. 17-8. Graph Lettering.   **(b)**

of the graph, the title of the vertical scale is normally read from the right-hand side, Fig. 17-8(b).

The complete title, including the measurement unit designation, should be centered on each axis. Short titles help to make them stand out, to be more specific, and to avoid confusion. As a rule, any abbreviations used in titles should be confined to the abbreviations listed in Chapter 3 or to such transistor abbreviations as V_{CE}, I_E, I_C, etc. Compare such titles to the abbreviated versions on right:

COLLECTOR CURRENT IN MILLIAMPERES	I_C—mA
CURRENT IN AMPERES	I—A
TIME IN MILLISECONDS	TIME—mS
AUDIO-FREQUENCY—HERTZ	A-F—Hz
RADIO-FREQUENCY—KILOHERTZ	R-F—kHz
RESISTANCE IN OHMS	RES—OHMS or Ω
CAPACITANCE IN PICOFARADS	C—pF

The application of the graph will determine whether a full or an abbreviated title should be used. If the graph is to be published in a technical magazine or manual it may be advantageous to give the full title. On the other hand, an abbreviated title should serve the purpose for internal engineering analysis.

Brevity is also desirable in unit designation. In Fig. 17-9(a), the ohm values are given in full, while in Fig. 17-9(b) the same values have been replaced by their equivalents in kilohms (K). A note in the title, such as "K" equals 1000" will further identify the values. Similariy, the use of *MEG* (*MEGOHM*) values for resistance, Fig. 17-9(c), reduces the unit designations from 2,000,000, 4,000,000, etc. to 2, 4, etc. Other methods of simplifying unit designations were given in the first section of Chapter 10. Such brevity is practiced on engineering graphs.

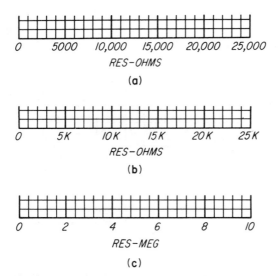

FIG. 17-9. Scale Designation Practices.

17.4 GRAPH LAYOUT

A careful examination should be made of the data to be plotted and the quantities involved before the graph is started.

Most engineering graphs are drawn using the system of rectangular coordinates on uniformly spaced grid paper. Others are drawn on the semilogarithmic or logarithmic patterns previously mentioned.

Quadrant Subdivision

In the system of rectangular coordinates, the two axes are perpendicular to each other and intersect at a point known as the origin or zero point (Fig. 17-10). These axes are known as the X-X or horizontal axis, and Y-Y or vertical axis. They form four quadrants, which are numbered one through four in counterclockwise direction, with quardant one in the upper right-hand corner.

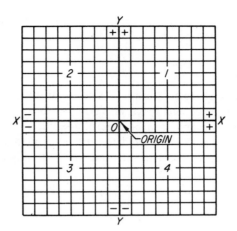

FIG. 17-10. Quadrant Subdivision.

The location of each point on the curve being plotted is determined by its distance from the horizontal and vertical axes. The indicated point location may be positive, negative, or both, depending on the quadrant in which the point falls. As illustrated in Fig. 17-10, distances along the X-X axis are positive to the right of the origin and negative to the left, Distances on the Y-Y axis are positive above the X-X axis and negative below. The point locations on the graphs considered in this chapter are assumed to be in the first quadrant, therefore, all point values are positive.

Point Location

The X-X axis is known as the abscissa and the Y-Y axis as the ordinate. They are also known as base lines and are subdivided into equal spaces on square grid paper. The grid lines extend from these subdivisions to locate the points on the curve numerically.

Points are located on the curve on the basis of their coordinate X and Y values. For example, in Fig. 17-11, the location of point A is identified as X equals 2, Y equals 3. Similarly, for point B, X equals 7 and Y equals 6. The

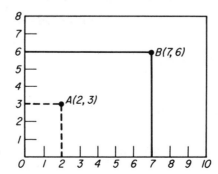

FIG. 17-11. Locating Coordinate Points.

intersection of the grid lines projected from the X-X and Y-Y axes at these values locates each point.

Graph Variables

Test results where a variation in one condition produces a change in another are frequently presented in graph form because a tabulation of these results, transferred to a graph, can be carefully studied and analyzed. Generally, the X-X axis or abscissa is selected for the controlled variable, such as temperature, time, voltage, and current, and the Y-Y axis or ordinate for the dependent variable factor.

Point Symbols

A variety of point symbols are used to show the graph curve. These may be small circles, triangles, squares, or crosses, about $\frac{1}{16}$ inch or larger in size, and either open or filled, Fig. 17-12(a). To give a uniform appearance the filled symbols should be slightly smaller than the open type. A geometrical pattern template can be used to keep symbols of the same type all the same size.

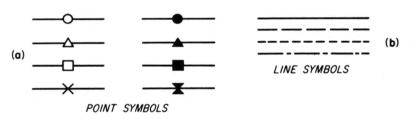

POINT SYMBOLS

FIG. 17-12. Plotting Points and Line Symbols.

Line Symbols

Various line symbols, Fig. 17-12(b), are advisable to clarify the different curves on multiple-curve graphs where curves run close together or cross each other.

Further differentiation can be obtained by drawing lines of various weights, although this method may result in errors unless the line widths are made three or four to one. This is especially true of graphs that are reduced photographically for slides.

Multiple-Curve Identification

The individual curves on multiple-curve graphs (Fig. 17-13) are identified either by legends, Fig. 17-13(a) and (b), or directly on the graph, Fig. 17-13(c). Crossing curves, Fig. 17-13(b), are best distinguished by various line symbols.

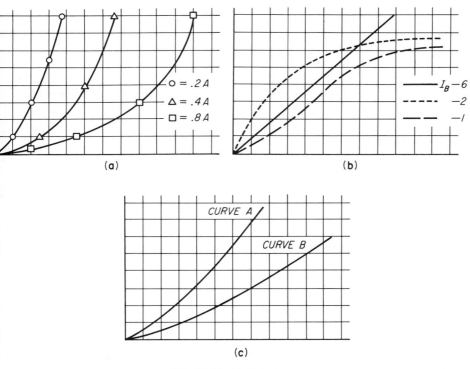

$$O = .2\,A$$
$$\triangle = .4\,A$$
$$\square = .8\,A$$

(a)

$$I_B - 6$$
$$-2$$
$$-1$$

(b)

CURVE A

CURVE B

(c)

FIG. 17-13. Multiple Curves.

Drawing Curves

The position of each point on the curve being drawn is first indicated by a pencil dot or a small circle, and then, after the point location has been checked, by one of the point symbols. When individual points are joined, they may produce a straight line or a curve. However, because of plotting inaccuracies, reading errors when observing instruments during the tabulation, and other factors, the curve may not be smooth. To correct this, a French curve may be used to draw a trial curve, passing through as many successive points as possible. If this trial fitting still fails to establish a uniform curve, three or more plotted points should be joined along a section of the curve, repeating this procedure until all points have been connected in a series of disjointed curves. This establishes the general shape of the curve, and it can be drawn by maintaining approximately equal spacing on each side of these original short trial curves.

In addition to these curves, which may be considered as continuous or extending in the same basic direction, there are other curves that consist of a series of interconnected lines that do not necessarily follow a uniform trend. In these curves a number of points, which indicate a number of independent quantities may be connected even though they do not establish a definite tendency. Curves of this nature usually involve time periods or data tabulations of an irregular nature.

Graph Construction

The graph paper should be selected according to the data and range of values to be plotted. As with schematic diagrams, a rough layout is made on cross-section paper to determine the best arrangement, vertical or horizontal; the space available for the lettering and the title; actual grid space needed for the graph, and so forth.

Some graph construction details are shown in Fig. 17-14. The two axes are drawn on the heavy grid subdivision lines one inch from the edge of the grid pattern and extend beyond the anticipated curve limit. The two remaining sides are filled in to form a rectangular graph border. It is advisable to keep a ratio of 3:4 between the two axes so that the graph will be suitable for possible future use as a 3- by 4-inch projection slide.

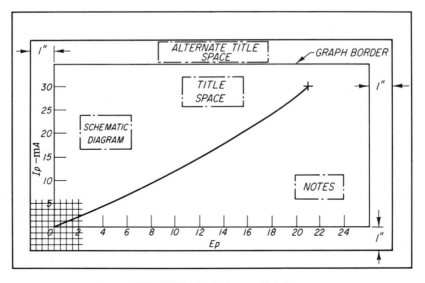

FIG. 17-14. Graph Layout Details.

Two possible locations for graph titles are indicated on the figure. The title may contain such information as descriptive name, job number, drawn by, checked by, and dates, or this information may be given in a preprinted title block located in a corner of the graph sheet.

Such additional information as notes, equations, simplified schematic diagram, and point symbol and line identification may also appear on the graph.

However, these details should be kept to a minimum to avoid a cluttered appearance.

17.5 RECTILINEAR GRAPHS

The square grid graph paper is also known as arithmetic or rectilinear graph paper. It is used for many engineering graphs where quantities increase arithmetically.

Before the actual graph is started, the accumulated data should be arranged in a tabular form, from the smallest number to the largest, to establish the high and low limits for both variables.

Once the independent variable has been determined, a suitable rectilinear paper can be selected and the direction of the abscissa can be established along the long or short side on the graph paper.

A typical rectilinear graph is shown in Fig. 17-15. This is a derating curve for a Variac®,* a manually adjustable autotransformer that delivers an output voltage that is adjustable from 0 to 117 percent of the input voltage. The normal operating temperature of the unit is 50°C. Its output capacity is derated for

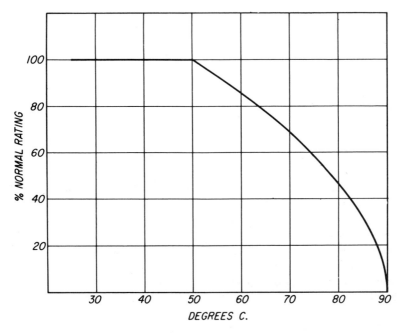

FIG. 17-15. Variac® Derating Graph. (GenRad.)

*® Registered trademark of GenRad.

higher temperatures in accordance with the graph shown in Fig. 17-15. Any intermediate point desired can be determined from such a curve. For example, the power capacity at 70°C is reduced to 67 percent of normal rating, and to 46 percent at 80°C.

17.6 LOGARITHMIC GRAPH LAYOUT

In addition to rectilinear graphs, there are the logarithmic graphs already mentioned, which require semilogarithmic or logarithmic graph paper.

Semilogarithmic Graphs

These graphs are used to represent data in which the successive values of one variable increase in an arithmetical progression, while the other variable increases in an geometric progression or remains at a constant rate. Semilog graph paper is used for such graphs because it can project an extensive range of values of one variable.

For example, copper wire resistance rises with the increases in the gage number of the wire at a constant rate of 26 percent. Selecting a group of wires from #10 through #24, their resistance per 1000 feet is as follows:

AWG No.	Resistance per 1000 feet, (ohms)
10	1.02
12	1.62
14	2.58
16	4.09
18	6.51
20	10.35
22	16.46
24	26.17

The computed resistance of #11 wire would be 1.02 ohms plus 26 percent of 1.02 or 1.285 ohms. In like manner, the resistance of #12 wire would be 26 percent greater than #11 wire or: $1.285 + (1.285 \times .26) = 1.619$ ohms. Thus, a geometric series is established: 1.02, 1.02×1.26, 1.02×1.26^2, and so forth.

When these same data are plotted on rectilinear graph paper, Fig. 17-16(a), it is difficult to read the ends of the curve because of the insignificant change in its slope. However, when they are presented on semilogarithmic paper, Fig. 17-16(b), the curve becomes a straight line to indicate a constant rate of change.

Logarithmic Graphs

There are occasions when a graph requires logarithmic scales on both axes. For example, they may be necessary because large numerical quantities are involved for both the dependent and independent variables or because the re-

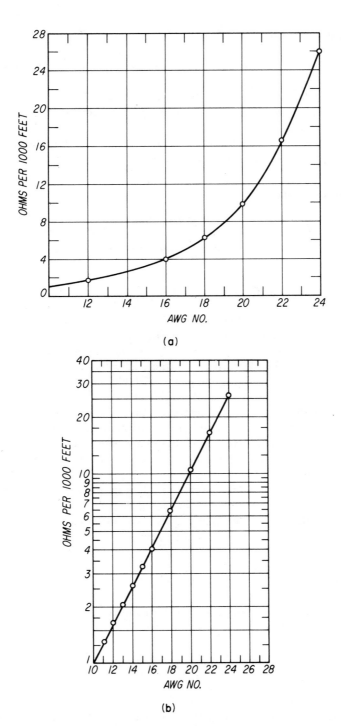

FIG. 17-16. Graph Example. (a) On Rectlinear Paper. (b) On Semilogarithmic Paper.

lationship between them would otherwise result in a straight or an almost straight curve.

A typical logarithmic graph is shown in Fig. 17-17. The points plotted were taken at various frequency values of RF voltage across an *IN1206* diode. A 3- by 5-cycle log graph paper was needed to cover the range of bias voltage and resistance values involved. It is very important to plot points accurately in logarithmic type graphs, especially within the 6–10 portion of the scale.

Logarithmic Charts

Logarithmic or reference charts were developed to determine various electrical and component values and for other similar applications. Their most common use is in catalogs and technical magazines.

17.7 CHARACTERISTIC CURVES

The characteristics of individual electronic components, such as transistors, are determined experimentally by varying two or more variables and noting the resultant change in collector current, base current, and so on. Such changes are shown on graphs, known as characteristic curves of these components, that are prepared by keeping one of the variables constant and then observing and noting the changes in the other variables.

Figure 17-18 shows the characteristic curves for a transistor. In this application, the current to the transistor base was held constant at various selected values from −1 milliampere to −250 milliamperes. A series of curves were drawn to show the change in collector current with the changes in collector-to-emitter voltage.

This figure illustrates the preparation of a graph with three variables, one of which is held constant. The student should become familiar with this method because it can be applied to other graphs involving three variables. The format shown has been adopted by the manufacturer for manuals produced by the firm.

17.8 POLAR GRAPHS

These are used to show the radiation directional characteristics of antennas, heaters, lamps, microphones, or speakers involving one variable of an angular value.

Polar Coordinates

These coordinates are shown in Fig. 17-5 and 17-19. A series of equally spaced radial lines, marked from 0 to 360 degrees, serve as the directional scale, while a series of concentric circles for plotting the linear values serve as the

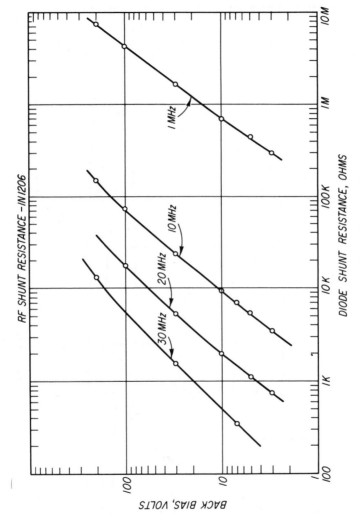

FIG. 17-17. Typical Logarithmic Graph.

507

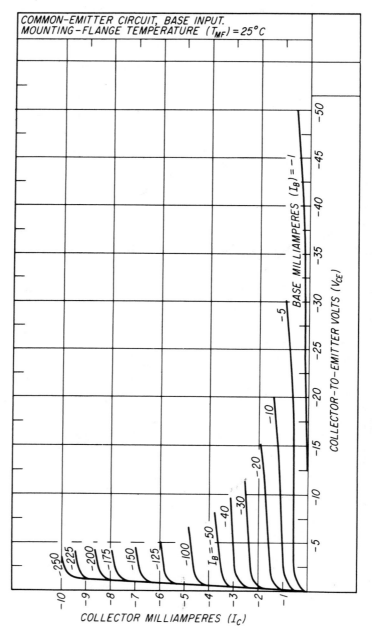

FIG. 17-18. Set of Transistor Characteristic Curves.

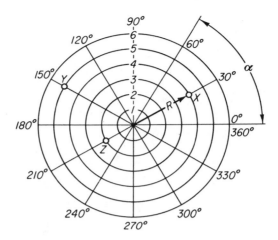

FIG. 17-19. Polar Coordinates.

other scale. A datum line, or *polar axis*, extends through the center from 0° to 180°. Values of one variable are represented by the length of the rotating radius R. Angle α represents the direction of the rotating radius R relative to the polar axis. The source of radiation, such as an antenna or a lamp, is assumed to be located at the *pole* or center of the graph.

Polar coordinates are specified by the length of the radius and reference angle, i.e., point X in Fig. 17-19 is identified as being 4, 30°; point Y as being 5, 150°; and point Z as being 2, 210°.

Polar Scales

The angular scale in a polar graph is the equivalent of the abscissa or X-X axis on which the independent variable is plotted in rectangular graphs. The distance (radius R) from the pole is the equivalent of the ordinate or Y-Y axis on which the dependent variable is plotted. This calibration may be response in decibels, voltage gain, or some other criterion.

Polar Graphs

A typical polar graph is shown in Fig. 17-20. This is a compilation of laboratory measurements of a quarter-wave whip antenna and a twin whip observed aboard a ship in both vertical and horizontal positions. The radiation from the quarter-wave whip was omni-directional, as evidenced by the full circle, while the twin whip antenna radiation varied according to the relative position and angular direction of the antenna. The scale was calibrated in relative gain or loss in decibels. Line symbols were used to differentiate the three curve plottings instead of plotting symbols, and a legend was included to define the test conditions.

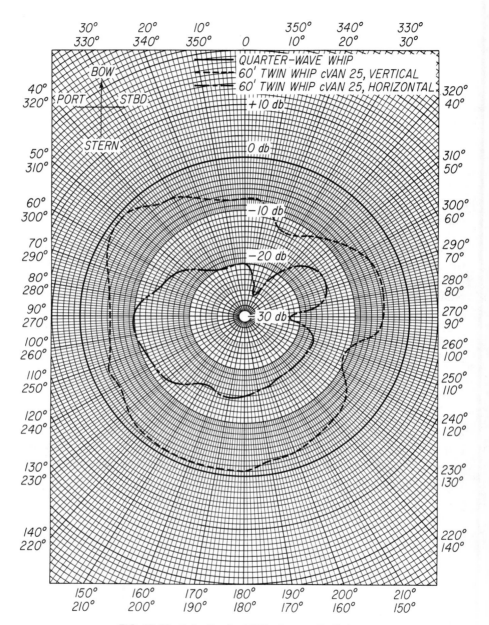

FIG. 17-20. Polar Graph of Whip Antenna Radiation.

17.9 GRAPHS FOR PROJECTION SLIDES

Certain precautions should be observed in preparing graphs that are to be used for projection slides or for publication.

Slide Size

The overall size of projection slides has been standardized at $3\frac{1}{4}$ by 4 inches, with a maximum subject area of $2\frac{1}{4}$ by $3\frac{1}{4}$ inches by using a suitable mask (Fig. 17-21). Other masks are available with desired opening size.

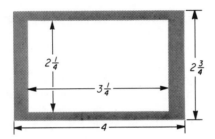

FIG. 17-21. Standard Slide Opening.

Graph Details

Graphs and block or schematic diagrams that are to be made into slides should be drawn several times larger than the subject area of the slide and reduced to the correct size photographically.

The following pointers should prove helpful in preparing such graphs:

1. Graphs should be prepared with mechanical-type instruments and India ink.

2. Titles should be kept short and direct.

3. Only one set of curves should be presented on each graph, and the number kept to a minimum.

4. Grid lines should be kept to a minimum, and they should not intersect lettering or data points.

5. Curves should be identified by individual titles placed close to the curve instead of by a keyed list. The use of arrows should be kept to a minimum.

6. Graphs should be drawn three times their actual subject matter size, or $6\frac{3}{4}$ by $9\frac{3}{4}$ inch dimensions that are convenient for an $8\frac{1}{2}$- by 11-inch *A*-size graph sheet.

7. The minimum lettering height for the graph size in item 6 should be $\frac{7}{32}$ inch. Lettering line widths should be somewhat finer than normal.

8. Curve-line widths should be about $\frac{3}{64}$ inch, axes and reference lines about $\frac{1}{32}$ inch, and grids and arrows about .025 inch.

9. Sketches and diagrams should be drawn with line widths of $\frac{3}{64}$ inch.

17.10 NOMOGRAPHS

A nomograph, or alignment chart, is a special type of chart used to derive solutions or answers to a specific problem. It is prepared by using three or more vertical or inclined lines, suitably calibrated, and arranged in proper geometrical relation.

A simple alignment chart consists of three graduated vertical scales, each graphically representing a variable. When two of the variables are known, the third is found by aligning a straightedge on the known value on each of the respective scales and reading the third variable scale at that point, to obtain the unknown value.

An example of a typical alignment chart is shown in Fig. 17-22. The chart was laid out on an *A*-size sheet, and the log scales were transferred from log graph paper. The length of the vertical scale is selected by choosing 1-, 2-, or 3-cycle paper to fit the required graph accuracy. For ease of reading, the subdivisions on the scales should not be closer than $\frac{1}{8}$ inch. When a special vertical scale length, subdivided logarithmically, is desired, the subdivisions may be made by slipping the logarithmic graph paper under the graph drawing sheet and tilting it to secure the desired scale length. Scale calibrations are projected to the new scale by parallel lines.

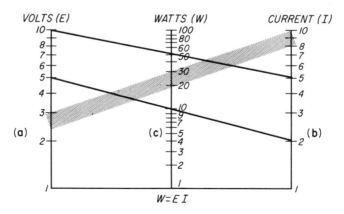

FIG. 17-22. Basic Multiplication and Division Nomograph.

The nomograph shown in Fig. 17-22 has been constructed for the equation:

$$E \times I = W \qquad \frac{W}{I} = E \qquad \frac{W}{E} = I$$

The variable scale (a) represents volts, E, and the variable scale (b) represents current in amperes, I. The resultant power in watts is read from the middle scale (c).

Accuracy is of prime importance in scale layout to obtain accurate readings on the center scale. The proper selection of scale length also contributes to accuracy.

The center scale (c) is calibrated by selecting various values on scale (a), multiplying them by scale (b) values, and noting the result on the middle or (c) scale with the aid of a straightedge.

If one end of a straightedge is placed at 3 volts on scale (a) and the other end at 10 amperes on scale (b), the resultant power on scale (c) is 30 watts. Placing the end of the straightedge at the 50-watt mark on (c) and tilting the straightedge until it touches the 10-volt mark on (a), or dividing power by voltage, will give a reading of 5 amperes on scale (b).

The basic formula and unit designations should be included on the nomograph together with the title and any other necessary data.

SUMMARY

As part of his work, the electronic draftsman may be required to present technical data by charts, graphs, or diagrammatically. Graph representation may be on square grid, semilogarithmic, or logarithmic paper, depending upon the type of data to be presented. There are certain accepted practices that must be followed in graph layout, especially in graphs for projection slide use.

QUESTIONS

Sketches should be included with answers when necessary.

17.1 What are the methods of showing engineering data graphically? Briefly describe each form.

17.2 What is square grid-type paper? What is its primary application?

17.3 Describe semilogarithmic and logarithmic papers and sketch a comparison of uniform and logarithmic scales.

17.4 What factors determine the choice of scales?

17.5 Describe the practices to be followed in graph lettering.

17.6 Describe the following parts of a basic rectilinear graph: (a) origin; (b) quadrants; (c) abscissa; (d) ordinate; (e) coordinate axes.

17.7 What is a graph variable?

17.8 What is the advantage of using blue-line graph paper?

17.9 Describe uses of the following, and sketch two examples of each: (a) line symbols; (b) plotting points; (c) legends.

17.10 Describe the process of fitting a curve.

17.11 When might a discontinuous curve be required?

17.12 What kind of data would require the following graph paper types: (a) semilog; (b) logarithmic; (c) rectilinear?

17.13 List some of the additional data that may be required on a graph.

17.14 What is a characteristic curve?

17.15 Describe polar coordinates.

17.16 What is the standard size of a projection slide?

17.17 Describe some of the precautions that should be taken in preparing graphs for use as projection slides.

17.18 Describe a typical nomograph.

17.19 When would it be desirable to construct a nomograph?

EXERCISES

17.1 Make a sketch to illustrate the coordinate values of a point.

17.2 Make a sketch of four quadrants. Identify each quadrant, and indicate the positive and negative relationships.

17.3 Make a sketch of a simple graph to show the following: (a) scale designations on abscissa and ordinate; (b) scale divisions; (c) margins; (d) general title; (e) a typical curve; (f) border.

17.4 Draw a sketch of the following: (a) polar scales; (b) polar axis; (c) pole.

17.5 Plot the following coordinate values, allocating two units per inch: (a) $X9$, $Y4$; (b) $X - 6.2$, $Y4$; (c) $X6$, $Y - 6$; and (d) $X - 9$, $Y - 5$. The horizontal and vertical coordinates should be 6 inches long and drawn on an A-size sheet graph paper with 10 divisions per inch.

17.6 From the following data, draw a multiple-curve graph of the typical characteristics of a transistor collector:

Curve A (base microamperes -5):

X	-1	-2	-4	-6	-8	-10
Y	-0.96	-1.00	-1.10	-1.20	-1.30	-1.40

Curve B (base microamperes -10):

X	-1	-2	-4	-6	-8	-10
Y	-1.48	-1.56	-1.76	-1.90	-2.00	-2.2

Curve C (base microamperes -15):

X	-1	-2	-4	-6	-8	-10
Y	-2.00	-2.25	-2.35	-2.55	-2.75	-2.90

Use graph paper with 10 divisions per inch and select an appropriate scale to fill the graph space. Designate the curves by collector base-current values.

17.7 Plot a graph of voltage versus current where a fixed 10K resistor is connected across a variable transformer adjustable in 5-volt steps from zero to 70 volts.

17.8 The following points represent a typical video-amplifier voltage gain versus frequency:

Frequency	1Hz	10Hz	100Hz	1kHz	10kHz	100kHz	1mHz	5mHz
Voltage Gain	0	100	1000	990	1000	1000	1200	3000

Plot these values on: (a) rectilinear paper; (b) semilogarithmic paper; (c) logarithmic paper.

17.9 Plot the cathode gate trigger voltage for a silicon controlled rectifier with variations in ambient temperature. The values are as follows:

Trigger Voltage, V_{gtc}	.75	.65	.54	.35	.28
Ambient Temperature (°C)	−40	0	+40	+90	+130

Draw the curve on rectilinear paper with ten divisions per inch, and plot these points with circular symbols. Select a scale that will fill an *A*-size sheet.

17.10 Draw a polar graph showing the response pattern of a triangular dipole UHF antenna, using the following information:

Angle	90°	75°	60°	45°	30°	15°	0°
Response (%)	0	10	22	44	69	86	100

Locate zero degrees at the top of the vertical axis. The same data, repeated in the other three quadrants, will provide a complete antenna pattern.

17.11 Draw a 90-degree type nomograph (*X* and *Y* axes at 90 degrees) to determine the value of two resistors that are connected in parallel or $R_x = R_a R_b / R_a + R_b$. The *R* scale ranges from 0 to 600 ohms. Use paper with 10 divisions per inch and make the scales 100 ohms per inch. Draw the 45-degree diagonal bisecting line for the R_x scale and subdivide it so each division equals 1.41 of the length of the vertical and horizontal scale divisions. Include the equation on the chart. Draw a sample line for 300 and 450 ohms in parallel.

17.12 Draw a multiple-curve graph from the following data:

Curve A:

X	.25	.50	1.00	1.50	2.00	2.20	
Y	.44	.81	1.41	1.69	1.87	1.96	

Curve B:

X	.25	.50	.80	1.25	1.75	2.00	2.20
Y	.30	.55	.75	1.02	1.31	1.38	1.45

Curve C:

X	.40	.70	1.00	1.40	1.75	2.00	2.20
Y	.30	.30	.71	.95	1.09	1.19	1.23

Use a grid with ten divisions per inch and scale spacings of one unit for every four inches. Identify the curves by: (a) curve labels or (b) line symbols and include a suitable legend.

17.13 Plot the maximum power output in milliwatts of a class A single-ended amplifier using transistors *2N1414* and *2N1449*. The values are as follows:

2N1414 Transistor

Power Output—Milliwatts	.5	1.0	2.0	4.0	10.0	25	60	100
Power Gain—Decibels	35	34.6	34.3	34	33.2	32	30	27.9

2N1449 Transistor

Power Output—Milliwatts	.5	1.0	2.0	4.0	10.0	25	60	100
Power Gain—Decibels	38.6	38.5	38.4	38.1	37.0	36	34.3	33.1

Draw the curves on semilogarithmic paper, and plot the points with circular symbols. Select a scale to fill an *A*-size sheet.

18

BLOCK DIAGRAMS

A diagram that uses graphic symbols to represent the mechanical and electronic components is known as an elementary or schematic diagram. A block diagram, on the other hand, uses blocks or rectangles to represent the components, groups of components, or units of equipment. The diagram is still referred to as a block diagram even though it may combine both graphic and block symbols.

This type of diagram is frequently used to present complex equipment in a simple format because the complex circuits in each stage or unit of the equipment can be reduced to a simple block form. This simple format makes it possible to see at a glance the relationships between various parts of the electronic or other equipment, their interconnections, and other details.

18.1 SYMBOLS

A square or rectangular block is the basic symbol on block diagrams, although in some instances it is supplemented by such graphic symbols as the circle, triangle, resistor, capacitor, or other symbols used on schematic diagrams.

Blocks

To a large extent, the lettering that has to be enclosed within the block determines the size of the block because the lettering should not be crowded or break through the block outline.

Square or rectangular blocks, drawn horizontally, are usually used. Ideally, all blocks should be the same size and of dimensions selected in multiples of .100-inch grid spacing. However, an additional size may be used if extensive lettering or the multiplicity of interconnections warrant its inclusion.

Graphic Symbols

Occasionally, conventional graphic symbols and circuit portions are used with the block symbols to clarify the diagram. These may include transistors, switches, resistors, or capacitors.

An example of such a diagram is shown in Fig. 18-1. In it, amplifiers are represented by the triangle symbol, the meter by a circle, etc.

Symbols on the block diagram follow a sequence similar to those on a schematic diagram. The input is at the left, near the center, or in the upper left-hand corner. The output is on the right-hand side, near the center, or below.

As in schematic diagrams, such auxiliary circuits as power supplies are placed below the main body of the diagram.

Special Symbols

The logic diagram, shown in Fig. 11-15, might be considered a special type of block diagram. In it, specially shaped blocks represent flip-flops, shift registers, and single-shot functions. Other symbols are special graphic symbols devised for logic diagrams.

18.2 BLOCKS

Because block diagrams are frequently reproduced in technical magazines, the size of the blocks and lettering should allow for the required reduction. This should be determined before starting the layout.

Layout

Although the circuit path in block diagrams should follow from left to right, such an arrangement is not always possible. It may sometimes be necessary to have more than one row of blocks to keep the diagram within a reasonable length, or paper area.

Some typical layouts are shown in Fig. 18-2. The conventional left-to-right layout is shown in Fig. 18-2(a), while Fig. 18-2(b) illustrates a layout arranged in two rows to shorten the overall length. Examples of auxiliary blocks feeding into the main line of blocks are given in Fig. 18-2(c) and (d). The complex diagram shown in Fig. 18-2(e) uses a heavy line to indicate the signal and a dotted or light line to indicate the auxiliary flow.

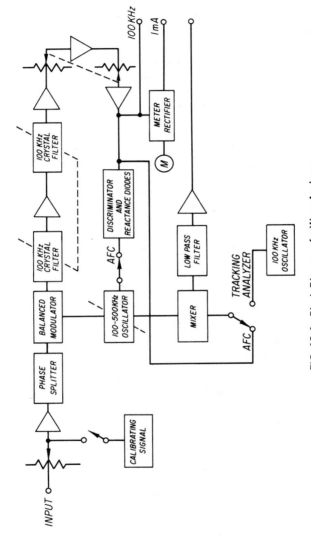

FIG. 18-1. Block Diagram of a Wave Analyzer.

519

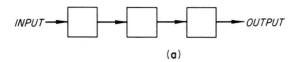

(a)

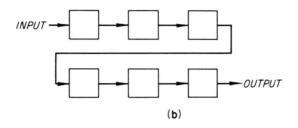

(b)

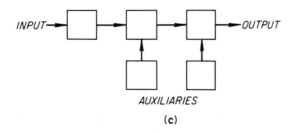

(c)

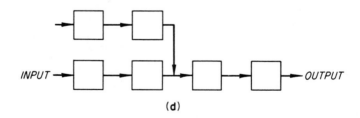

(d)

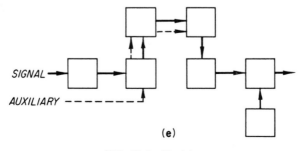

(e)

FIG. 18-2. Block Layouts.

To eliminate any misunderstanding, it is customary on block diagrams to indicate the signal or other flow by arrowheads even though it may be assumed they are from left to right. Confusion could arise in complex diagrams if the direction of flow is not indicated.

Block Proportions

Rectangular shaped blocks with a $2:1$ or $2\frac{1}{2}:1$ ratio may be used on the diagram when square blocks are not suitable. In special cases, the ratio may be considerably higher. For instance, if the block has several input or output lines that have been widely separated to fit the diagram layout, bends in these lines can be eliminated by extending the block length.

Block Spacing

Sometimes several blocks will have approximately the same amount of lettering to go inside or next to the block. If a number of these blocks of equal weight are to be distributed on the diagram, they may be positioned with spaces equal to block size, Fig. 18-3, or one and one-half block size.

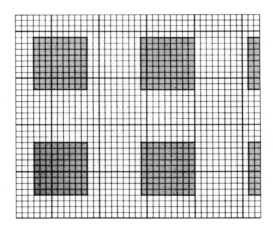

FIG. 18-3. Block Spacing.

For emphasis, the size of one or more blocks may be increased, while the spacing around them remains the same, as in Fig. 18-4(a). Increasing the block size may also help eliminate line bends when several connection lines emerge from the block, Fig. 18-4(b).

Blocks may be arranged in vertical columns or horizontal rows to maintain a symmetrical appearance. If they all have about the same amount of lettering, they should be evenly spaced. In special cases, the diagram may be subdivided into two or more sections with a vertical or horizontal open space between to emphasize the major divisions.

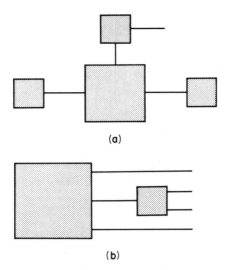

(a)

(b)   **FIG. 18-4.** Block Sizes.

18.3 LINES

A medium-weight line is suitable for block outlines, connection lines, and lettering.

Flow Direction

As already mentioned, the input to each block is considered to be on the left-hand side and the output on the right. The signal flow proceeds horizontally through each block at about the center. The power and control lines are generally shown entering or leaving on the top or bottom of each block.

Open or closed arrowheads, placed on or above the connection lines, are used to indicate the flow direction. When there are multiple connection lines at the block (Fig. 18-5) the arrowheads should be staggered to assist reading the flow.

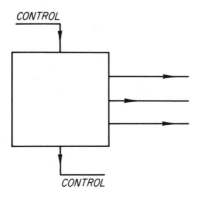

FIG. 18-5. Arrowheads on Multiple Connection Lines.

Connection Lines

Customarily, all connection lines on block diagrams are drawn vertically or horizontally and have 90-degree corners. When space limitations require joining several lines together into a single trunk line, sloping lines are used to indicate the direction of the lines entering or emerging from the trunk (Fig. 18-6). This practice is also followed on connection diagrams.

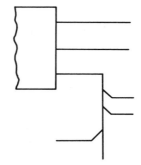

FIG. 18-6. Connection Line Grouping into a Trunk Line.

Line Spacing

Multiple connection lines extending from individual blocks should be separated on the basis of multiples of grid spacing of .100 inch. Because it is hard to follow a number of closely drawn lines on a diagram, it is advisable to separate these lines into groups of threes. The space between each group should be equal to two line spaces, Fig. 18-7(a). Crossovers and bends should also be kept to a minimum.

When an appreciable number of long connection lines present a problem on a complex block diagram, they may be terminated in short, identified lines that point to their termination points at the other end, Fig. 18-7(b). This practice also applies to connection diagrams.

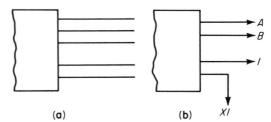

FIG. 18-7. (a) Line Grouping and (b) Line Terminations.   **(a)**   **(b)**   *XI*

Line Details

A heavy connection line is used to distinguish a major signal circuit, a common or ground line, or a power line (Fig. 18-8). A lighter solid line is used to identify control circuits, and broken lines are used for auxiliary circuits.

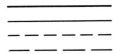

FIG. 18-8. Line Variations.

Colors, such as red, green, or blue, may be used to emphasize or to simplify tracing the connections on a complex block diagram, as long as these colors are not used with too many connection lines, which reduces the desired contrast.

Whenever possible, blocks should be located to avoid crossovers of connection lines, thus eliminating possible misinterpretation of the circuit.

18.4 LETTERING

The size of the blocks is largely determined by the lettering to be included. The length of the block is established by the longest word. If the size of the diagram precludes enclosing long words within the block or if the symbol will not take a word enclosure, the lettering may be located adjacent to the symbol, Fig. 18-9(a).

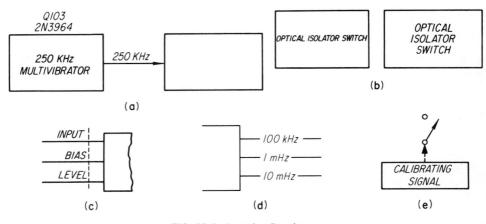

FIG. 18-9. Lettering Practices.

Lettering should be centered within each block, with enough margin to prevent breaking into the block outline. When possible, long words should be abbreviated according to the standards outlined in Chapter 3. The length of rectangular blocks is not to be reduced for short words. All block titles should either be spelled out or abbreviated to maintain a consistent pattern throughout the diagram.

Dividing extensive lettering into several lines avoids jamming the words or reducing the size of letters and achieves a better balance, Fig. 18-9(b).

Block diagrams made for reproduction in technical journals should be lettered in ink with a lettering aid or appliques. The letter size selected should allow for reduction.

Any lettering next to connection lines should be at least $\frac{1}{32}$ inch away from the identified line, Fig. 18-9(c), and stop at a uniform distance from the block outline. Another method is to break the connection lines and insert the lettering, Fig. 18-9(d). To secure better balance, the breaks should be staggered.

Lettering should read from either the bottom of the drawing or from the right-hand side. If possible, special lettering identifying functions or switch positions should be enclosed and have an arrow extended to the component to distinguish it from the block symbols, Fig. 18-9(e).

As the amplifier symbol and many logic symbols are not suited to lettering, any identifying lettering is generally placed adjacent to these symbols.

18.5 TYPICAL BLOCK LAYOUT

The project engineer generally furnishes the electronic draftsman with a free-hand sketch of the block diagram, a practice also followed in schematic diagram work. The draftsman should examine this sketch carefully and obtain full information about the size requirements of the finished diagram, the type of lettering, the possibilities of reduction for publication purposes, the need for complete titles or abbreviations, etc.

It may be necessary to make one or more rough layouts to fit the diagram on a specified drawing sheet size, to simplify the inter-block connections, or to attain a balanced layout.

A typical block diagram, representing a modular receiver separated into basic circuit assemblies, is shown in Fig. 18-10. In this diagram, trial lettering of such words as *DEMODULATOR* and *PRESELECTOR* established the block length. Two-line titles were selected to give a balanced appearance. The block size that was finally chosen was $1\frac{1}{4}$ by 2 inches, which fitted the blocks within the drawing area without crowding. Since the diagram was to be reproduced in a technical publication, all lettering was done with a mechanical aid.

The final layout was made on a drawing sheet with a grid-spaced pattern sheet underneath. The antenna and speaker symbols were the only graphic symbols used in addition to the rectangular blocks.

18.6 BLOCK DIAGRAMS IN CIRCUIT LAYOUT

Drawing a block diagram of a complex circuit is very helpful in developing a neat and compact arrangement for a schematic diagram. Cardboard blocks, labelled for each circuit subdivision, can be arranged and rearranged to develop the best circuit and space allocation. Since circuit subdivisions vary in complexity and thus in space required, the individual cardboard blocks should have the approximate proportions of the expected circuit space requirements.

A block diagram of a television receiver is shown in Fig. 18-11. The circuit

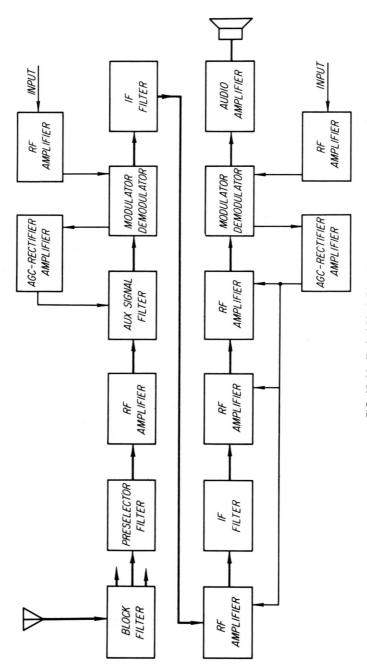

FIG. 18-10. Typical Block Diagram.

526

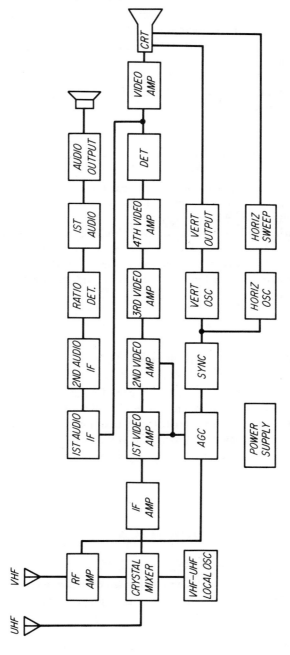

FIG. 18-11. Television Receiver Block Diagram.

has been subdivided into four major horizontal lines of circuits, beginning with the tuning system and the audio portion of the receiver. The crystal mixer, the video amplifiers, and the picture or *CRT* tube follow in the second line. The third line has the automatic gain control (*AGC*), synchronization, and vertical and local oscillators. The power supply and the horizontal sweep circuits are given in the last line. This, or a similar arrangement, prevents the schematic diagram from being elongated horizontally, while the available vertical space goes unused.

18.7   MISCELLANEOUS PRACTICES

A partly pictorial approach is employed occasionally, by introducing depth, or a third dimension, to the diagram. This aids in the visualization of some of the components on the block diagram.

Thus, cylindrical objects are shown as cylinders, instead of by rectangular symbols. An isometric layout of the block gives depth to the appearance of the conventional block symbol. This technique helps to define the location of such items as cables or terminal boards on individual blocks or to establish various dimensions.

SUMMARY

A block diagram is made up of a number of square or rectangular blocks that represent a series of components or equipments and may include graphic symbols. Following the natural left-to-right reading sequence, the blocks are arranged to have the circuitry follow in the same manner. The block with the largest amount of lettering determines the block size to be used throughout the diagram.

QUESTIONS

18.1 Describe a block diagram.

18.2 What are some of the symbols commonly used on block diagrams?

18.3 Select four of the following and describe each briefly: (a) symbol sequence; (b) symbol size; (c) block proportions; (d) flow direction; (e) line spacing; (f) crossovers.

18.4 What factors contribute to a well-balanced block layout?

18.5 What factors determine the block size?

18.6 Describe line-work details on block diagrams.

18.7 List typical uses for the following: (a) a heavy line; (b) a broken line; (c) a medium line.

18.8 In what respect is a logic diagram similar to a block diagram?

18.9 Illustrate a partially pictorial diagram approach, showing the following: power-supply system, a power transformer, a rectifier unit, a voltage divider, and an output terminal board.

EXERCISES

Use cross-section paper for the following exercises and develop layouts freehand. Block titles should use uppercase lettering at least $\frac{1}{8}$ inch high. Block title words may be abbreviated when required to fit within given block space, using abbreviations given in Chapter 3. Blocks are to be rectangular with a $1:1\frac{1}{2}$ ratio. Graphic symbols should be used as indicated. Exercises should be drawn in pencil or ink as required, using mechanical lettering aids when specified by the instructor. Connection lines should be solid and of medium weight. Lines for feedback and other purposes should be in the form of dash lines.

18.1 Sketch the use of arrowheads on a block diagram.

18.2 Make a sketch of a simple block diagram showing the following: (a) input; (b) auxiliaries; (c) signal flow; (d) output; (e) block lettering; (f) line lettering; (g) three graphic symbols. Identify each item by corresponding letter.

18.3 Sketch the method used to handle multiple connection lines in a crowded area.

18.4 Letter the following items in Fig. 18-12(a): (1) line gain; (2) line AF amplifier; (3) line AF output; (4) line meter; (5) line audio.

18.5 Letter the following items in Fig. 18-12(b): (1) 455-kHz IF input; (2) 1st IF amplifier; (3) 2nd IF amplifier; (4) 3rd IF amplifier; (5) 4th IF amplifier; (6) To detector.

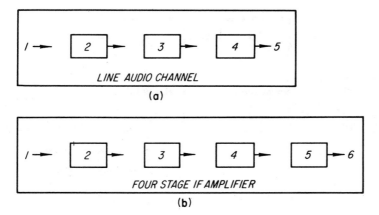

(a)

(b)

FIG. 18-12. Block Diagram Exercises.

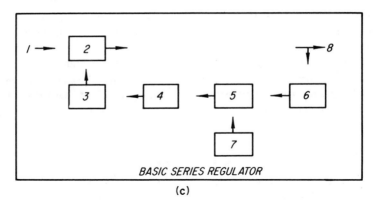

BASIC SERIES REGULATOR

(c)

FIG. 18-12. (Continued)

18.6   Letter the following items in Fig. 18-12(c): (1) dc input; (2) series control element; (3) emitter followers; (4) voltage amplifier circuit; (5) comparison circuit; (6) sampling circuit; (7) reference element; (8) regulated output.

18.7   Draw a block diagram of the receiver circuit shown in Fig. 11-1. Show sections *A* through *H* as rectangular blocks and add interconnections of 14.4 V to them. Add ground connections to individual blocks and label each block and terminal connection.

18.8   Item 1 in Fig. 18-13(a) is an antenna symbol. Item 8 is a graphic symbol for a

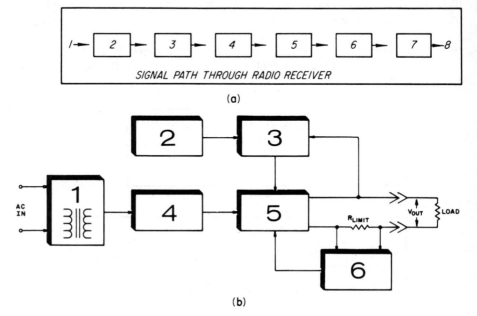

(a)

(b)

FIG. 18-13. Block Diagram Exercises.

loudspeaker. Letter the following blocks: (2) preselector; (3) mixer; (4) filter; (5) amplifier; (6) detector; (7) audio amplifier.

18.9 Letter the following items in Fig. 18-13(b): (1) isolation transformer; (2) voltage reference; (3) error detector and amplifier; (4) rectifier and filter; (5) series pass regulator; (6) overload protector.

18.10 Draw a block diagram of a transistorized power supply system with 25 V dc input and 7000 V dc output. Indicate the following by symbols or blocks: (1) input terminals; (2) and (3) two GE-4 transistors connected between input terminals; (4) power transformer primary terminals; (5) secondary output terminals on power transformer; (6) rectifier and capacitor assembly; (7) output resistor network; (8) output terminals from resistor network identified as follows: +7KV at the top, +400 V dc in the middle; and ground symbol at the bottom. Identify the block diagram as *Transistorized Power Supply*, 25 V *DC* to 7 KV.

18.11 Letter the following items from Fig. 18-14: (1) dipole antenna symbol; (2) RF amplifier; (3) mixer; (4) local oscillator; (5) sound IF; (6) limiter; (7) discriminator; (8) audio amplifier; (9) loudspeaker symbol; (10) video-sound separator; (11) video IF; (12) video detector; (13) video amplifier; (14) picture-tube graphic symbol with inputs from items (13), (18), and (19); (15) sync separator; (16) vertical sweep oscillator; (17) horizontal sweep oscillator; (18) vertical sweep amplifier; (19) horizontal sweep amplifier. ..

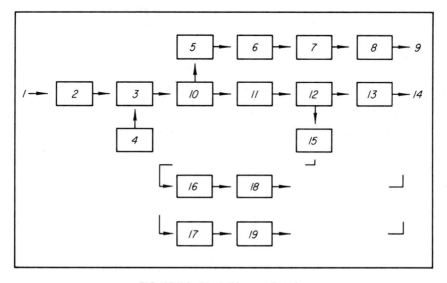

FIG. 18-14. Block Diagram Exercise.

18.12 Letter the following items in Fig. 18-15: (1) voltage reference; (2) error detector and amplifier; (3) isolation transformer; (4) rectifier and filter; (5) series pass regulator; (6) overload protector; (7) voltage reference; (8) error detector and amplifiers; (9) rectifiers and filters; (10) series pass regulators; (11) overload protector.

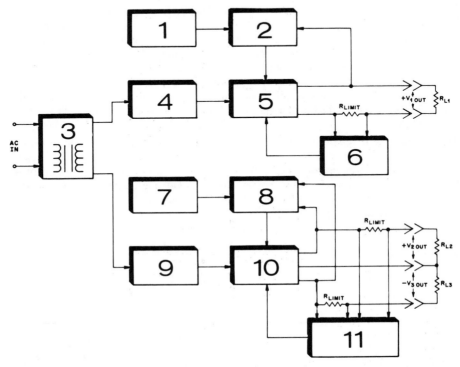

FIG. 18-15. Block Diagram Exercise.

18.13 Draw a block diagram of a transistorized 27-mHz transmitter. Indicate the following by symbols or blocks: (1) RF oscillator; (2) RF driver; (3) RF power amplifier; (4) audio input amplifier; (5) audio driver; (6) modulator; (7) output terminals. Show +15 V connections to items 1, 4, 5, and 6. Identify the block diagram as 27-*mHz, 5-watt Citizens-Band Transmitter.*

SELECTED BIBLIOGRAPHY

ELECTRICAL AND ELECTRONICS DRAWING

Basic Electronic and Electrical Drafting. J. D. Bethune; Prentice-Hall, Inc., 1980, Englewood Cliffs, N.J.

Electrical and Electronics Drafting. H. W. Richter; John Wiley & Sons, Inc., 1977, New York.

Electrical and Electronics Drawing. C. J. Baer; McGraw-Hill Book Company, 1980, New York.

Electronic Drafting. G. Shiers; Prentice-Hall, Inc., 1962, Englewood Cliffs, N.J.

ENGINEERING DRAWING

Fundamentals of Engineering Drawing and Graphic Technology. T. E. French and C. J. Vierck; McGraw-Hill Book Company, 1972, New York.

Fundamentals of Engineering Drawing for Design, Product Development and Numerical Control. W. J. Luzzader; Prentice-Hall, Inc., 1977, Englewood Cliffs, N.J.

Technical Drafting: Metric Design and Communication. W. Spence and M. R. Atkins; Chas. A. Bennett Company, Inc., 1980, Peoria, Ill.

Technical Drawing. F. E. Giesecke, A. Mitchell, and H. C. Spencer; The Macmillan Publishing Company, Inc., 1974, New York.

Technical Illustration. T. A. Thomas; McGraw-Hill Book Company, 1968, New York.

ENVIRONMENTAL

Human Factors Engineering and Design. E. J. McCormick; McGraw-Hill Book Company, 1975, New York.

GRAPHICS

Graphic Science and Design. T. E. French and C. J. Vierck; McGraw-Hill Book Company, 1970, New York.
Innovative Design with an Introduction to Design Graphics. W. J. Luzzader; Prentice-Hall, Inc., 1975, Englewood Cliffs, N.J.

HANDBOOKS

Handbook of Electronic Packaging. C. A. Harper, ed.; McGraw-Hill Book Company, 1969, New York.
Maintainability Engineering Handbook, NAVORD OD 39223. Department of the Navy, 1969, Washington, D.C.
Printed Circuit Handbook. C. F. Coombs, Jr.; McGraw-Hill Book Company, 1978, New York.
Radio Amateur's Handbook. American Radio Relay League (annually), Hartford, Conn.
Radio Engineering Handbook. K. Henney; McGraw-Hill Book Company, 1959, New York.
SEM Program Applications Handbook, MIL-HDBK-239. Naval Publications and Forms Center, Philadelphia, Pa.
Shock and Vibration Control Handbook. C. M. Harris and C. E. Crede; McGraw-Hill Book Company, 1976, New York.
Standard Handbook for Electronic Engineers. D. G. Fink; McGraw-Hill Book Company, 1978, New York.

INTEGRATED CIRCUITS

Design Guide for Integrated Circuits, NAVSHIPS 0900–004–4000. Department of the Navy, Washington, D.C.
Integrated Circuits—A Basic Course for Engineers and Technicians. R. G. Hibberd; McGraw-Hill Book Company, 1969, New York.
Semiconductor Electronics. J. F. Gibbons; McGraw-Hill Book Company, 1966, New York.
Thin Film Technology. R. W. Berry, P. M. Hall, and M. T. Harris; Van Nostrand Reinhold Company, 1968, New York.

COMPONENT DATA MANUALS

Electronic Engineers Master (EEM). United Technical Publications (annually), Garden City, N.Y.
Electronics Buyers' Guide. McGraw-Hill Book Company (annually), New York.
SCR Manual. General Electric Company, Chicago.
Solid-State Devices Manual. RCA/Solid State Division, Somerville, N.J.

ELECTRONIC AND PRODUCT DESIGN MAGAZINES

Circuits Manufacturing. Benwill Publishing Corp., 167 Corey Road, Brookline, Mass. 02146.
Electronic Component News. Radnor Park, Pa. 19089.
Electronic Design. Hayden Publishing Company, Inc., 50 Essex Street, Rochelle Park, N.J. 07622.
Electronic News. 7E. 12th Street, New York, N.Y. 10003.
Electronic Packaging and Production. Kiver Publications, Inc., 222 West Adams Street, Chicago, Ill. 60606.
Electronics. McGraw-Hill Book Company, Inc., 1221 Avenue of the Americas, New York, N.Y. 10020.
Machine Design. 1111 Chester Avenue, Cleveland, Ohio 44119.
Reprographics. United Business Publications, Inc., 730 Third Avenue, New York, N.Y. 10017.

APPENDIX

Dimensional Outlines

TO-5 Style 8-Lead Package

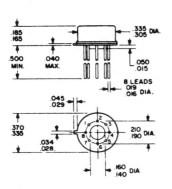

TO-5 Style 10-Lead Package

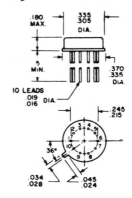

TO-5 Style 12-Lead Package   TO-5 Style 10 Formed-Lead Package

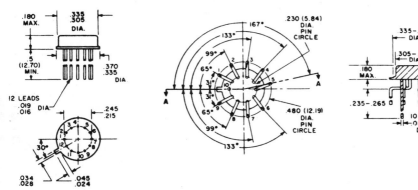

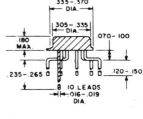

14-Lead Ceramic-to-Metal Flat Package

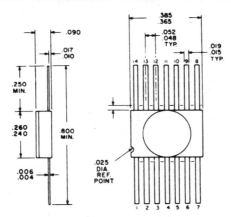

FIG. A-1. Integrated Circuits—Dimensional Outlines (RCA)

14-Lead Dual-In-Line Ceramic Package
JEDEC-TO-116

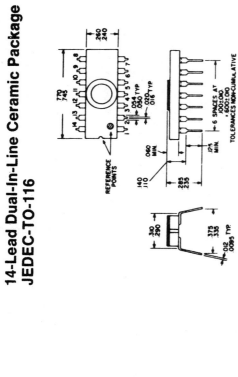

16-Lead Dual-In-Line Ceramic Package

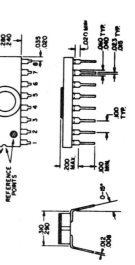

14-Lead Dual-In-Line Plastic Package

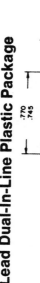

16-Lead Dual-In-Line Plastic Package

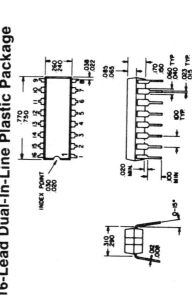

FIG. A-1. (Continued)

538

Table A-1 Head Dimensions (in.) of Slotted Round and Flat Head Machine Screws (ANSI B18.6.3-1972)

Round Head

Nominal Size	Max Screw Diam	Head Diam		Height of Head		Width of Slot		Depth of Slot	
		Max	Min	Max	Min	Max	Min	Max	Min
0	.0600	.113	.099	.053	.043	.023	.016	.039	.029
1	.0730	.138	.122	.061	.051	.026	.019	.044	.033
2	.0860	.162	.146	.069	.059	.031	.023	.048	.037
3	.0990	.187	.169	.078	.067	.035	.027	.053	.040
4	.1120	.211	.193	.086	.075	.039	.031	.058	.044
5	.1250	.236	.217	.095	.083	.043	.035	.063	.047
6	.1380	.260	.240	.103	.091	.048	.039	.068	.051
8	.1640	.309	.287	.120	.107	.054	.045	.077	.058
10	.1900	.359	.334	.137	.123	.060	.050	.087	.065
12	.2160	.408	.382	.153	.139	.067	.056	.096	.073
$\frac{1}{4}$	.2500	.472	.443	.175	.160	.075	.064	.109	.082
$\frac{5}{16}$	.3125	.590	.557	.216	.198	.084	.072	.132	.099
$\frac{3}{8}$	.3750	.708	.670	.256	.237	.094	.081	.155	.117
$\frac{7}{16}$	.4375	.750	.707	.328	.307	.094	.081	.196	.148
$\frac{1}{2}$	.5000	.813	.766	.355	.332	.106	.091	.211	.159

Flat Head

Nominal Size	Max Screw Diam	Head Diam		Height of Head		Width of Slot		Depth of Slot	
		Max Sharp	Min Sharp	Abs. Min	Ref	Max	Min	Max	Min
0	.0600	.119	.105	.099	.035	.023	.016	.015	.010
1	.0730	.146	.130	.123	.043	.026	.019	.019	.012
2	.0860	.172	.156	.147	.051	.031	.023	.023	.015
3	.0990	.199	.181	.171	.059	.035	.027	.027	.017
4	.1120	.225	.207	.195	.067	.039	.031	.030	.020
5	.1250	.252	.232	.220	.075	.043	.035	.034	.022
6	.1380	.279	.257	.244	.083	.048	.039	.038	.024
8	.1640	.332	.308	.292	.100	.054	.045	.045	.029
10	.1900	.385	.359	.340	.116	.060	.050	.053	.034
12	.2160	.438	.410	.389	.132	.067	.056	.060	.039
$\frac{1}{4}$	.2500	.507	.477	.452	.153	.075	.064	.070	.046
$\frac{5}{16}$	.3125	.635	.600	.568	.191	.084	.072	.088	.058
$\frac{3}{8}$	.3750	.762	.722	.685	.230	.094	.081	.106	.070
$\frac{7}{16}$	.4375	.812	.767	.723	.223	.094	.081	.103	.066
$\frac{1}{2}$	.5000	.875	.831	.775	.223	.106	.091	.103	.065

Table A-2 Head Dimensions (in.) of Slotted Pan Head Machine
Screws (ANSI B18.6.3-1972)

Nominal Size	Max Screw Diam	Head Diam		Height of Head		Width of Slot		Depth of Slot		Radius
		Max	Min	Max	Min	Max	Min	Max	Min	Nominal
2	.086	.167	.155	.053	.045	.031	.023	.031	.022	.035
4	.112	.219	.205	.068	.058	.039	.031	.040	.030	.042
6	.138	.270	.256	.082	.072	.048	.039	.050	.037	.046
8	.164	.322	.306	.096	.085	.054	.045	.058	.045	.052
10	.190	.373	.357	.110	.099	.060	.050	.068	.053	.061
12	.216	.425	.407	.125	.112	.067	.056	.077	.061	.078
$\frac{1}{4}$	.250	.492	.473	.144	.130	.075	.064	.087	.070	.087

Table A-3   Decimal Equivalents of Wire and Sheet Gages* (Pheoll Mfg. Co.)

Number Gage	American or Brown & Sharpe Gage	Music Wire Gage American Steel & Wire Co.	United States Standard Gage	Machine and Wood Screw Gage
7/0	—			—
6/0	.5800			—
5/0	.5165			—
4/0	.4600	.006	.4063	—
3/0	.4096	.007	.3750	—
2/0	.3648	.008	.3438	—
0	.3249	.009	.3125	.060
1	.2893	.010	.2813	.073
2	.2576	.011	.2656	.086
3	.2294	.012	.2500	.099
4	.2043	.013	.2344	.112
5	.1819	.014	.2188	.125
6	.1620	.016	.2031	.138
7	.1443	.018	.1875	.151
8	.1285	.020	.1719	.164
9	.1144	.022	.1563	.177
10	.1019	.024	.1406	.190
11	.0907	.026	.1250	.203
12	.0808	.029	.1094	.216
13	.0720	.031	.0938	—
14	.0641	.033	.0781	.242
15	.0571	.035	.0703	—
16	.0508	.037	.0625	.268
17	.0453	.039	.0563	—
18	.0403	.041	.0500	.294
19	.0359	.043	.0438	—
20	.0320	.045	.0375	.320
21	.0285	.047	.0344	—
22	.0253	.049	.0313	—
23	.0226	.051	.0281	—
24	.0201	.055	.0250	.372
25	.0179	.059	.0219	—
26	.0159	.063	.0188	—
27	.0142	.067	.0172	—
28	.0126	.071	.0156	—
29	.0113	.075	.0141	—
30	.0100	.080	.0125	.450
31	.0089	.085	.0109	—
32	.0080	.090	.0102	—
33	.0071	.095	.0094	
34	.0063	.100	.0086	
35	.0056	.106	.0078	
36	.0050	.112	.0070	

*Use of Gages:
American or Brown & Sharpe Gage: copper wire, brass, copper alloys and nickel silver wire and sheet, also aluminum sheet, rod, and wire.
Music Wire Gage, American Steel & Wire Co.: music spring wire.
United States Standard Gage: steel, nickel and Monel metal sheets.
The use of decimals of an inch for dimensions specifying sheet and wire is recommended to avoid confusion.

Table A-4   Decimal Equivalents of Twist Drills (in.)

Size	Drill Diam	Size	Drill Diam	Size	Drill Diam	Size	Drill Diam	Size	Drill Diam
1	.2280	17	.1730	33	.1130	49	.0730	65	.0350
2	.2210	18	.1695	34	.1110	50	.0700	66	.0330
3	.2130	19	.1660	35	.1100	51	.0670	67	.0320
4	.2090	20	.1610	36	.1065	52	.0635	68	.0310
5	.2055	21	.1590	37	.1040	53	.0595	69	.0292
6	.2040	22	.1570	38	.1015	54	.0550	70	.0280
7	.2010	23	.1540	39	.0995	55	.0520	71	.0260
8	.1990	24	.1520	40	.0980	56	.0465	72	.0250
9	.1960	25	.1495	41	.0960	57	.0430	73	.0240
10	.1935	26	.1470	42	.0935	58	.0420	74	.0225
11	.1910	27	.1440	43	.0890	59	.0410	75	.0210
12	.1890	28	.1405	44	.0860	60	.0400	76	.0200
13	.1850	29	.1360	45	.0820	61	.0390	77	.0180
14	.1820	30	.1285	46	.0810	62	.0380	78	.0160
15	.1800	31	.1200	47	.0785	63	.0370	79	.0145
16	.1770	32	.1160	48	.0760	64	.0360	80	.0135

Letter Sizes

A	.234	G	.261	L	.290	Q	.332	V	.377
B	.238	H	.266	M	.295	R	.339	W	.386
C	.242	I	.272	N	.302	S	.348	X	.397
D	.246	J	.277	O	.316	T	.358	Y	.404
E	.250	K	.281	P	.323	U	.363	Z	.413
F	.257								

Table A-5   Decimal Equivalents

1/64	.015625	17/64	.265625	33/64	.515625	49/64	.765625
1/32	.031250	9/32	.281250	17/32	.531250	25/32	.781250
3/64	.046875	19/64	.296875	35/64	.546875	51/64	.796875
1/16	.062500	5/16	.312500	9/16	.562500	13/16	.812500
5/64	.078125	21/64	.328125	37/64	.587125	53/64	.828125
3/32	.093750	11/32	.343750	19/32	.593750	27/32	.843750
7/64	.109375	23/64	.359375	39/64	.609375	55/64	.859375
1/8	.125000	3/8	.375000	5/8	.625000	7/8	.875000
9/64	.140625	25/64	.390625	41/64	.640625	57/64	.890625
5/32	.156250	13/32	.406250	21/32	.656250	29/32	.906250
11/64	.171875	27/64	.421875	43/64	.671875	59/64	.921875
3/16	.187500	7/16	.437500	11/16	.687500	15/16	.937500
13/64	.203125	29/64	.453125	45/64	.703125	61/64	.953125
7/32	.218750	15/32	.468750	23/32	.718750	31/32	.968750
15/64	.234375	31/64	.484375	47/64	.734375	63/64	.984375
1/4	.250000	1/2	.500000	3/4	.750000	1	1.000000

Table A-6   Metric Small Drills

Metric Drill Size (mm)	Dia. (in.)	Metric Drill Size (mm)	Dia. (in.)	Metric Drill Size (mm)	Dia. (in.)
.35	.0138	3.25	.1280	7.60	.2992
.40	.0158	3.30	.1299	7.70	.3031
.45	.0177	3.40	.1339	7.75	.3051
.50	.0197	3.50	.1378	7.80	.3071
.55	.0217	3.60	.1417	7.90	.3110
.60	.0236	3.70	.1457	8.00	.3150
.65	.0256	3.75	.1476	8.10	.3189
.70	.0276	3.80	.1496	8.20	.3228
.75	.0295	3.90	.1535	8.25	.3248
.80	.0315	4.00	.1575	8.30	.3268
.85	.0335	4.10	.1614	8.40	.3307
.90	.0354	4.20	.1654	8.50	.3346
.95	.0374	4.25	.1673	8.60	.3386
1.00	.0394	4.30	.1693	8.70	.3425
1.05	.0413	4.40	.1732	8.75	.3445
1.10	.0433	4.50	.1772	8.80	.3465
1.15	.0453	4.60	.1811	8.90	.3504
1.20	.0472	4.70	.1850	9.00	.3543
1.25	.0492	4.75	.1870	9.10	.3583
1.30	.0512	4.80	.1890	9.20	.3622
1.35	.0531	4.90	.1929	9.25	.3642
1.40	.0551	5.00	.1968	9.30	.3661
1.45	.0571	5.10	.2008	9.40	.3701
1.50	.0591	5.20	.2047	9.50	.3740
1.55	.0610	5.25	.2067	9.60	.3780
1.60	.0630	5.30	.2087	9.70	.3819
1.65	.0650	5.40	.2126	9.75	.3839
1.70	.0669	5.50	.2165	9.80	.3858
1.75	.0689	5.60	.2205	9.90	.3898
1.80	.0709	5.70	.2244	10.00	.3937
1.85	.0728	5.75	.2264	10.50	.4134
1.90	.0748	5.80	.2283	11.00	.4331
1.95	.0768	5.90	.2323	11.50	.4528
2.00	.0787	6.00	.2362	12.00	.4724
2.05	.0807	6.10	.2402	12.50	.4921
2.10	.0827	6.20	.2441	13.00	.5118
2.15	.0846	6.25	.2461	13.50	.5315
2.20	.0866	6.30	.2480	14.00	.5512
2.25	.0886	6.40	.2520	14.50	.5709
2.30	.0906	6.50	.2559	15.00	.5905
2.35	.0925	6.60	.2598	15.50	.6102
2.40	.0945	6.70	.2638	16.00	.6299
2.45	.0965	6.75	.2657	16.50	.6496
2.50	.0984	6.80	.2677	17.00	.6693
2.60	.1024	6.90	.2717	17.50	.6890
2.70	.1063	7.00	.2756	18.00	.7087
2.75	.1083	7.10	.2795	18.50	.7283
2.80	.1102	7.20	.2835	19.00	.7480
2.90	.1142	7.25	.2854	19.50	.7677
3.00	.1181	7.30	.2874	20.00	.7874
3.10	.1220	7.40	.2913		
3.20	.1260	7.50	.2953		

Table A-7 Metric Screw Threads and Tap Drill Sizes

Coarse				Fine			
Metric Screw Size		Tap Drill Size		Metric Screw Size		Tap Drill Size	
Dia.	Pitch	mm	in.	Dia.	Pitch	mm	in.
M1.6 × .35		1.25	.049				
M1.8 × .35		1.45	.057				
M2. × .40		1.60	.063				
M2.2 × .45		1.75	.069				
M2.5 × .45		2.05	.081				
M3 × .50		2.5	.098				
M3.5 × .60		2.9	.114				
M4.0 × .70		3.3	.130				
M4.5 × .75		3.75	.148				
M5. × .80		4.2	.166				
M6 × 1.00		5.0	.199				
M7 × 1.00		6.0	.236				
M8 × 1.25		6.75	.266	M8 × 1.00		7.0	.276
M10 × 1.5		8.5	.335	M10 × 1.25		8.75	.344
M12 × 1.75		10.25	.403	M12 × 1.25		10.50	.413
M14 × 2		12.0	.472	M14 × 1.5		12.50	.492
M16 × 2		14.0	.551				
M16 × 1.5		14.5	.571				
M18 × 2.5		15.5	.610	M18 × 1.5		16.50	.650
M20 × 2.5		17.5	.689	M20 × 1.5		18.50	.728
M20 × 1.5		18.5	.728				
M22 × 2.5		19.5	.768	M22 × 1.5		20.5	.807
M24 × 3		21.0	.827	M24 × 2		22.0	.866
M27 × 3		24.0	.942	M27 × 2		25.0	.984

Table A-8 Metric Conversion: Inches to Millimeters

in.	mm	in.	mm	in.	mm
$\frac{1}{64}$	.3969	$\frac{1}{2}$	12.7000	5	127.000
$\frac{1}{32}$	.7937	$\frac{17}{32}$	13.4937	6	152.400
$\frac{3}{64}$	1.1906	$\frac{9}{16}$	14.2875	8	203.200
$\frac{1}{16}$	1.5875	$\frac{19}{32}$	15.0812	10	254.001
$\frac{5}{64}$	1.9844	$\frac{5}{8}$	15.8750	12	304.801
$\frac{3}{32}$	2.3812	$\frac{21}{32}$	16.6687		
$\frac{7}{64}$	2.7881	$\frac{11}{16}$	17.4625		
$\frac{1}{8}$	3.1750	$\frac{23}{32}$	18.2562		
$\frac{9}{64}$	3.5719	$\frac{3}{4}$	19.0500		
$\frac{5}{32}$	3.9687	$\frac{25}{32}$	19.8437		
$\frac{3}{16}$	4.7625	$\frac{13}{16}$	20.6375		
$\frac{7}{32}$	5.5562	$\frac{27}{32}$	21.4312		
$\frac{1}{4}$	6.350	$\frac{7}{8}$	22.2250		
$\frac{9}{32}$	7.1437	$\frac{29}{32}$	23.0187		
$\frac{5}{16}$	7.9375	$\frac{15}{16}$	23.8125		
$\frac{11}{32}$	8.7312	$\frac{31}{32}$	24.6062		
$\frac{3}{8}$	9.5250	1	25.4001		
$\frac{13}{32}$	10.3187	2	50.8001		
$\frac{7}{16}$	11.1125	3	76.2002		
$\frac{15}{32}$	11.9062	4	101.600		

Table A-9 Metric Conversion: Decimals (Inches) to Millimeters

in.	mm	in.	mm	in.	mm
.001	.025	.140	3.56	.550	13.97
.002	.051	.150	3.81	.600	15.24
.003	.076	.160	4.06	.650	16.51
.004	.102	.170	4.32	.700	17.78
.005	.127	.180	4.57	.750	19.05
.006	.152	.190	4.83	.800	20.32
.007	.178	.200	5.08	.850	21.59
.008	.203	.210	5.33	.900	22.86
.009	.229	.220	5.59	.950	24.13
.010	.254	.230	5.84	1.000	25.40
.020	.508	.240	6.10	2.000	50.80
.025	.635	.250	6.35	3.000	76.20
.030	.762	.260	6.60	4.000	101.60
.035	.889	.280	7.11	5.000	127.00
.040	1.016	.300	7.62		
.045	1.143	.320	8.13		
.050	1.270	.340	8.64		
.060	1.524	.360	9.14		
.070	1.778	.380	9.65		
.080	2.032	.400	10.16		
.090	2.286	.420	10.67		
.100	2.540	.440	11.18		
.110	2.794	.460	11.68		
.120	3.048	.480	12.19		
.130	3.302	.500	12.70		

Table A-10 Metric Conversion: Millimeters to Inches

mm	in.	mm	in.
1	.0394	21	.8268
2	.0787	22	.8661
3	.1181	23	.9055
4	.1575	24	.9449
5	.1968	25	.9842
6	.2362	26	1.0236
7	.2756	27	1.0630
8	.3150	28	1.1024
9	.3543	29	1.1417
10	.3937	30	1.1811
11	.4351	35	1.3779
12	.4724	40	1.5748
13	.5118	45	1.7716
14	.5512	50	1.9685
15	.5905	75	2.9527
16	.6299	100	3.9370
17	.6693		
18	.7087		
19	.7480		
20	.7874		

INDEX